D0722172

Molecular Evolutionary Genetics

MOLECULAR EVOLUTIONARY GENETICS

MASATOSHI NEI

COLUMBIA UNIVERSITY PRESS
New York

Library of Congress Cataloging-in-Publication Data

Nei, Masatoshi.
Molecular evolutionary genetics.

Bibliography: p.
Includes index.
1. Molecular genetics.   2. Evolution.   I. Title.
QH430.N45   1987   574.87'328   86-17599
ISBN 0-231-06320-2

Columbia University Press
New York   Guildford, Surrey
Copyright © 1987 Columbia University Press
All rights reserved

Printed in the United States of America

c 10 9 8 7 6 5 4 3 2

To my family,
Nobuko, Keitaro, *and* Maromi

Contents

Preface

During the last ten years, spectacular progress has occurred in the study of molecular evolution and variation mainly because of the introduction of new biochemical techniques such as gene cloning, DNA sequencing, and restriction enzyme methods. Studies at the DNA level have led to many intriguing discoveries about the evolutionary change of genes and populations. These discoveries have in turn generated several new evolutionary theories. Furthermore, the molecular approach is now being used for studying the evolution of morphological, physiological, and behavioral characters.

The purpose of this book is to summarize and review recent developments in this area of study. Previously, molecular evolution and population genetics were studied as separate scientific disciplines. In this book, an attempt will be made to unify these two disciplines into one which may be called *molecular evolutionary genetics*. While emphasis is placed on the theoretical framework, experimental data will also be discussed to present a comprehensive view of the subject. There are highly developed mathematical theories related to the study of molecular evolution and variation. To make the book accessible to a wide audience, however, only those theories that are useful for interpretation and analysis of data are presented. When a sophisticated theory is needed, the meaning of the theory is discussed without going into detail. On the other hand, some detailed explanations will be given of statistical methods that are useful for data analysis.

Molecular evolutionary genetics is an interdisciplinary science dependent upon knowledge from many different areas of biology. Particularly important are the evolutionary history of life and the basic structure of genes and their mutations. Therefore, two chapters are devoted to a brief discussion of these subjects. The discussion is based on recent studies, and I hope it is useful even for professional workers. Although the purpose of this book is to discuss the recent development of molecular evolutionary genetics, it is important to know its implications for the general theory of evolution. The final chapter is therefore devoted to a discussion of this problem. The subjects chosen and the views presented in this chapter are quite personal, but they will give some general idea about the relationship between the current study of molecular evolution and the classical study of morphological evolution.

I am deeply indebted to my colleagues who have collaborated with me during the last ten years. Several of them helped me in developing statistical methods which are included in this book, whereas others conducted extensive data analysis. I am particularly grateful to Wen-Hsiung Li, Ranajit Chakraborty, Paul Fuerst, Takeo Maruyama, Yoshio Tateno, Fumio Tajima, Dan Graur, Takashi Gojobori, Clay Stephens, Naoyuki Takahata, and Naruya Saitou. Some of them kindly read and commented on drafts of this book. I am also grateful to many authors who sent me unpublished manuscripts so that I could include the newest information in the book. James Crow, David Jameson, Motoo Kimura, Pekka Pamilo, Robert Selander, and Peter Smouse read the entire text and offered valuable suggestions. Draft versions of certain chapters were read by Arthur Cain (1, 2, 14), Joseph Felsenstein (11), Walter Fitch (11), Stephen Gould (14), William Provine (1, 14), and Robert Sokal (11). All of them offered numerous suggestions for improving the book. Needless to say, however, none of them is responsible for errors which will undoubtedly be found in the book, particularly because their advice was not always heeded. I owe special thanks to Sandra Starnader who patiently typed many versions of the manuscript. My thanks also go to Bett Stap who drew most of the illustrations in this book.

Masatoshi Nei

□ CHAPTER ONE □
INTRODUCTION

In the study of evolution, there are two major problems. One is to clarify the evolutionary histories of various organisms, and the other is to understand the mechanism of evolution. Until the mid-1960s, the first problem was studied mainly by paleontologists, embryologists, and systematists, and the second by population geneticists. In the study of the first problem, it was customary to consider a species (sometimes even a genus, family, or order) as the unit of evolution and to ignore the genetic variation within species. The main task was to reconstruct evolutionary trees of organisms as accurately as possible. The ideal approach to this problem was to examine fossil records, but since there are not enough fossils for most groups of organisms, morphological and physiological characters were studied. Using this approach, classical evolutionists were able to infer the major aspects of evolution. However, the evolutionary change of morphological and physiological characters is usually so complex that this approach does not produce a clear-cut picture of evolutionary history, and the details of the evolutionary trees reconstructed were almost always controversial.

The mechanism of evolution was speculated by a number of authors, notably by J. B. Lamarck, in the early nineteenth century, but it was Charles Darwin (1859) who started a serious work on this problem. Without knowing the source of genetic variation, he proposed that evolution occurs by natural selection in the presence of variation. Later, when genetic variation was shown to be generated by spontaneous mutation, Darwin's theory was transformed into neo-Darwinism or the synthetic theory of evolution (see Mayr and Provine 1980). According to this theory, mutation is the primary source of variation, but the major role of creating new organisms is played by natural selection. The theoretical basis of neo-Darwinism was the mathematical theory of population genetics developed by Fisher (1930), Wright (1931), and Haldane (1932). From the 1930s to the 1950s, great efforts were made to provide an empirical basis for neo-Darwinism (Dobzhansky 1937, 1951; Huxley 1942; Mayr 1942, 1963; Simpson 1944, 1953; Stebbins 1950; Ford

1964). However, it was often difficult to obtain experimental verification of population genetics theory because investigators' lifetimes are too short to observe a substantial genetic change of populations except under special circumstances. Interspecific hybridization was occasionally used to study the long-term genetic change of populations, but this was possible only for very closely related species.

The situation suddenly changed in the mid-1960s when molecular techniques were introduced in the study of evolution. Since the chemical substance of genes was now shown to be deoxyribonucleic acid (DNA) [ribonucleic acid (RNA) in some viruses] and all developmental information was shown to be stored in DNA, one could study the evolution of organisms by examining the nucleotide sequences of DNAs from various organisms. Molecular techniques removed the species boundary in population genetics studies and allowed investigators to study the evolutionary change of genes within and between species quantitatively by using the same statistical measure. Of course, sequencing of nucleotides was not easy until around 1977, and many investigators initially studied the evolutionary change of genes by examining amino acid sequences of proteins. This is because all proteins are direct products of genes and amino acid sequences are determined by nucleotide sequences of DNA.

As soon as the amino acid sequences of proteins from diverse organisms were determined, it became clear that for a given protein the number of amino acid substitutions between a pair of species increases approximately linearly with the time since divergence between the species studied (Zuckerkandl and Pauling 1962; Margoliash 1963). This finding of the *molecular clock* has had an important implication for the study of evolution; it can be used not only for obtaining rough estimates of evolutionary times of various groups of organisms but also for constructing evolutionary trees. Indeed, immediately after the discovery of the molecular clock, amino acid sequencing was used extensively for the study of long-term evolution of organisms (e.g., Fitch and Margoliash 1967a; Dayhoff 1969).

One problem in using amino acid sequencing for evolution is that it is time-consuming and expensive. For this reason, various other methods were also developed. One of them was to use the relationship between the extent of immunological reaction and the number of amino acid substitutions (Goodman 1962; Sarich and Wilson 1966), and another was to use the DNA hybridization method (Kohne 1970). All these methods are still useful for finding phylogenetic relationships of organisms.

The molecular approach also introduced a revolutionary change in the study of genetic polymorphism within populations in the mid-1960s. In the study of polymorphisms, we must examine many individuals from a population, and thus amino acid sequencing is too costly. For this reason, a simpler method of studying protein variation, i.e., electrophoresis, was used. This method detects only a fraction of amino acid changes in proteins, yet it showed that most natural populations have a high degree of genetic variation at the protein level (Harris 1966; Lewontin and Hubby 1966). This discovery resulted in a great controversy over the mechanism of maintenance of genetic variability in natural populations (see Kimura and Ohta 1971a; Lewontin 1974; Nei 1975; Ayala 1976). Particularly heated was the controversy over Kimura's (1968a) neutral theory, which proclaimed that most nucleotide substitutions in evolution occur by mutation and random genetic drift and that a large proportion of molecular variation within populations is neutral or nearly neutral. This controversy has not yet been completely resolved.

In recent years, there has been another technical breakthrough in molecular biology and in the study of evolution. The techniques introduced this time are gene cloning, rapid DNA sequencing, and restriction enzyme methods. These techniques have generated a revolution in molecular biology and uncovered many unexpected properties of the structure and organization of genes (e.g., exons, introns, flanking regions, repetitive DNA, pseudogenes, gene families, and transposons). It is now clear that most genes in higher organisms do not exist as a single copy in the genome but rather in clusters and that the number of genes in a cluster varies extensively from cluster to cluster. Comparison of nucleotide sequences from diverse organisms indicates that the rate of sequence change in evolution varies considerably with the DNA region examined and that the more important the function of the DNA region, the lower the rate of sequence change. Furthermore, the extent of genetic variation undetectable by protein electrophoresis is enormous. Evolutionists now face a new challenge to explain all these observations coherently.

The boundary between the two areas of evolutionary study, i.e., the evolutionary history of life and the mechanism of evolution, was theoretically removed when the techniques of amino acid sequencing and electrophoresis were introduced. In practice, however, most evolutionists were concerned with only one of the two problems even after the mid-1960s. Thus, biochemical evolutionists were mainly interested in constructing evolutionary trees for distantly related organisms, whereas traditional population geneticists were engaged in measuring the extents of protein polymorphism within and between populations. The real erosion of the boundary between the two areas of study started to occur only after the techniques of DNA sequencing and restriction enzyme methods were introduced. Biochemical evolutionists now realize that the extent of DNA polymorphism within species is enormous and cannot be neglected in the study of evolution of higher-order taxa such as genera or families, whereas population geneticists have come to know that polymorphic alleles (DNA sequences) are often older than the species itself. It should also be noted that while long-term evolution is essentially an accumulation of consecutive short-term evolutions, the pattern of evolutionary change of organisms is often seen more clearly when long-term change is examined. In the near future, the boundary between the two areas of study is expected to disappear completely.

The study of evolution at the DNA level has just begun, and the patterns of nucleotide substitution and DNA polymorphism have been examined only for a limited number of genes from a small group of organisms. Although these examinations have revealed some interesting features of nucleotide substitution and polymorphism (Kimura 1983a; Nei and Koehn 1983), we must study many more genes to learn the general patterns. As mentioned earlier, many genes exist as multiple copies in the genome, and they seem to be subject to frequent unequal crossover or gene conversion. This makes it difficult to identify homologous genes between different species and creates a problem in measuring the rate of nucleotide substitution, unless all multiple copies are studied. The mechanism of maintenance of DNA polymorphism has scarcely been studied. Although natural selection is generally considered to operate for eliminating deleterious mutations at the DNA level, the pattern of polymorphism in some genes (e.g., immunoglobulin genes) does not seem to be compatible with this hypothesis. Clearly, a more detailed study is necessary to understand the mechanism of evolution and maintenance of genetic polymorphism.

It has often been stated that the study of amino acid substitution and protein polymorphism has not contributed to the understanding of morphological or physiological evolution. This statement is incorrect, since there are many examples in which the change in function or activity of a protein can be related to a particular amino acid substitution. Nevertheless, it seems true that a majority of amino acid substitutions do not change protein function appreciably. This led Wilson (1975) and King and Wilson (1975) to propose the hypothesis that morphological evolution is caused mainly by the change of regulatory genes rather than of structural genes. They presented several examples of bacterial adaptation caused by regulatory gene mutations. They could not produce direct evidence for their hypothesis in higher organisms, however. Techniques are now available to study this problem at the molecular level. Indeed, many molecular biologists are currently investigating the regulatory mechanism of gene function. Once this mechanism is elucidated, evolutionists will be able to study the molecular basis of morphological evolution.

As the study of evolution has become molecular, it has been realized that quantitative approaches are necessary. Since the basic process of molecular evolution is the change in genome size and DNA sequence, we need mathematical and statistical methods to quantify the evolutionary change of DNA. Mathematical methods are also necessary to understand the process of evolution, because this process can only be inferred from information on extant organisms. Mathematical and statistical methods have therefore become an essential part of the study of molecular evolution. A number of authors (e.g., Kimura and Ohta 1971a; Nei 1975; Ewens 1979; Kimura 1983a) have realized the importance of integrating the mathematical theory of DNA or protein evolution with the classical theory of population genetics.

As mentioned earlier, the mathematical theory of population genetics played an important role in formulating neo-Darwinism. This is because evolution is affected by many factors in a complicated way and it is difficult to see the final outcome of the action of these factors intuitively. Initially, mathematical theory was used mainly to understand the possible effects of mutation, selection, and random genetic drift on the frequencies of alleles or chromosome types in populations. By the mid-1960s, many elaborate mathematical theories on the population dynamics of genes had been developed. As mentioned earlier, however, most of these theories were rarely used to interpret observed or experimental data on evolution except under special circumstances.

The situation changed abruptly when molecular data on the evolutionary change of genes became available. Such theories as those of the probability of fixation of mutant genes and of heterozygosities suddenly became useful for computing the expected rate of amino acid substitution, expected heterozygosity, etc., under various conditions. Thus, the theories could be used for testing alternative hypotheses on the mechanism of evolution. This interaction between theory and data stimulated further works on mathematical theories of molecular evolution and population genetics that can be used for hypothesis testing. Particularly important was the development of theories for testing the "null hypothesis" of neutral mutations.

As these mathematical theories were being developed, statistical tests of various hypotheses of evolution were also conducted by many biologists as well as by statisticians. In these tests, new statistical methods often had to be introduced. Various new statistical methods were also developed for measuring and testing the extent of protein and DNA polymorphism within and between populations. Using these methods, one can now compare the extent of genetic polymorphism between any pair of species.

Another important statistical development in the last fifteen years was the theory of estimation of the number of amino acid or nucleotide substitutions from observed sequence data or restriction site maps. This theory has proved to be useful for studying long-term evolution. A related problem is the quantification and estimation of genetic distance between populations. The concept of genetic distance was developed as early as 1953 by Sanghvi in a study of the genetic differentiation of human populations and was later refined by Cavalli-Sforza and Edwards (1967), Steinberg et al. (1967), and others. However, a distance measure that is appropriate for studying protein evolution was developed only after electrophoretic studies became popular. In recent years, statistical properties of various distance measures have been studied.

As mentioned earlier, many different kinds of molecular data can be used for constructing phylogenetic trees. To construct a phylogenetic tree, however, some statistical methods are required. Before molecular evolutionists started tree-making, numerical taxonomists had already developed various methods for constructing trees from morphological characters. Some molecular evolutionists are using these methods directly for molecular data, whereas others have invented new methods that are more appropriate for these data. Nevertheless, there are many unsolved problems in this area, and intensive study is currently under way.

As is clear from the above brief survey, the study of evolution has become increasingly analytical since molecular techniques were introduced, and for a proper analysis of molecular data various mathematical and statistical methods are necessary. Furthermore, a substantial part of the theory of molecular evolution can now be written in unambiguous mathematical terms. The mathematical and statistical methods used are, however, quite diversified, and some of the theories, particularly the stochastic theory of population genetics, require a high level of mathematics. For many biologists, this has been an obstacle to appreciating the importance and usefulness of the mathematical theories of molecular evolution. In practice, the essential conclusions obtained from mathematical studies are relatively simple when properly stated, and the mathematical formulas developed can easily be used for data analysis.

It should be mentioned, however, that while mathematical formulation is important for developing a scientific theory of evolution, it depends on a number of simplifying assumptions. If these assumptions are not satisfied in reality, mathematical formulation may lead to an erroneous conclusion. It is necessary, therefore, to check the validity of the assumptions by examining empirical data. Fortunately, empirical data are increasing rapidly, and they are now being used not only for checking the validity of the assumptions but also for examining the predictability of a theory. Only through this process can we make progress in understanding the mechanism of evolution. It should also be mentioned that there are still many evolutionary events that are not amenable to mathematical treatment either because they are not well characterized or because their occurrence is irregular. These events are currently described in a qualitative manner.

☐ CHAPTER TWO ☐

EVOLUTIONARY HISTORY OF LIFE

In recent years, substantial progress has been made in the study of the evolutionary history of life through the efforts of paleontologists, geologists, molecular biologists, and systematists. It is now possible to present a reasonable account of the major aspects of the evolutionary history of life from the origin of prokaryotes to the development of higher organisms. Yet the account is full of conjecture, and the details are highly controversial. In this chapter, I discuss only the major aspects of evolution that are useful for understanding subsequent chapters.

Evidence from Paleontology and Comparative Morphology

It is believed that the earth was formed about 4.5 billion (10^9) years ago. It is not known exactly when the first life or self-replicating substance was formed. Since Barghoorn and Schopf (1966) reported the discovery of "probable" fossilized bacteria, numerous claims of microfossils from the Precambrian period (more than 570 million years ago) have been reported. Although most of them have not been sustained by careful reexamination (Schopf and Walter 1983; Hoffmann and Schopf 1983), the bacteria-like microfossils reported by Awramik et al. (1983) seem to be authentic; they have been dated 3.5 billion years old. Walsh and Lowe (1985) have also reported 3.5 billion year old bacteria-like fossils. Considering these microfossils and other fossilized organic matters, Schopf et al. (1983) suggest that life probably arose around 3.8 billion years ago (figure 2.1). By 3.5 billion years ago, both anaerobic and photosynthetic bacteria seem to have originated. The next two billion years were the age of prokaryotes. According to Schopf et al.'s (1983) "best guess scenario" for the early history of life, unicellular mitotic eukaryotes originated around 1.5 billion years ago, and the divergence between animals and plants occurred somewhere between 600 million years (MY) and 1 billion years ago, probably close to the latter time.

There are rather extensive fossil records from the Phanerozoic ("visible life"), and the major evolutionary events in this era can be reconstructed from these fossils (figure 2.2). The fossils in the early Cambrian era show

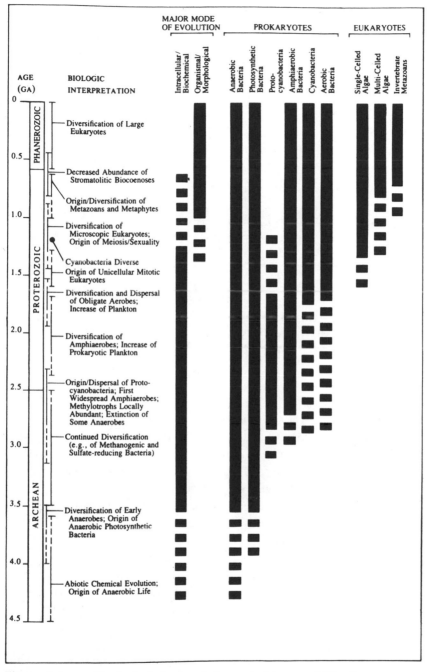

Figure 2.1.   Geological time and the early history of life. From Schopf et al. (1983). Reprinted by permission of Princeton University Press.

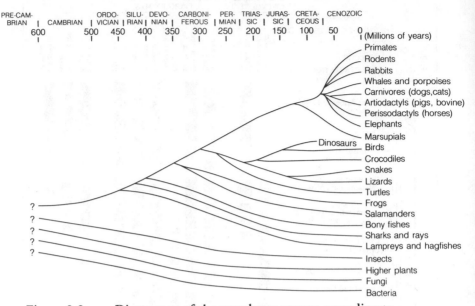

Figure 2.2.   Divergence of the vertebrate groups according to geological and morphological evidence. Modified from McLaughlin and Dayhoff (1972).

that most living phyla of animals and plants were present at that time. This indicates that they were differentiated before the Cambrian era. The phylogeny of representative vertebrates that can be constructed from paleontological data is given in figure 2.2. Lampreys and mammals diverged about 450 MY ago, whereas teleosts diverged from mammals about 400 MY ago. The divergence of amphibia and reptiles from the mammalian line seems to have occurred about 350 and 300 MY ago, respectively. Among mammals, marsupials branched off from eutherians about 125 MY ago, and the eutherian radiation, i.e., the divergence of various eutherian orders, seems to have occurred between 60 and 80 MY ago, just before or after dinosaurs became extinct. Birds are considered to have evolved from a dinosaur line about 150 MY ago. Estimates of divergence times for some other groups of organisms are presented in table 2.1.

It should be noted that the details of the evolutionary tree in figure 2.2 are not known with as much confidence as the sharp lines might suggest. Furthermore, the evolutionary trees of families, genera, and

Table 2.1   Times of divergence for various groups of organisms which have been used for the study of molecular evolution.

Organisms involved	Time (MY)	Authors
Animal/plant	1,000	Dayhoff (1978)
Mammal/arthropod	700	"
Sea urchin: Echinidae/		
Strongylocentrotidae	65	Busslinger et al. (1982)
Horse/cow, pig, sheep	54	Romero-Herrera et al. (1973)
New world/old world monkeys	50	"
Apes/old world monkeys	30	"
Man/orangutan	13–16	Sibley and Ahlquist (1984)
Cow/goat	18–20	Romero-Herrera et al. (1973)
Goat/sheep	5–7	Novacek (1982)
Mouse/rat	10–25	Britten (1986)
Baboon/macaque	5	Romero-Herrera et al. (1973)
Horse/donkey	2	Langley and Fitch (1974)
Mono-/dicotyledons	100–200	Shinozaki et al. (1983)
Corn/barley	50	Zurawski et al. (1984)

species are usually much more difficult to construct from fossil records than those of classes and orders. Therefore, the trees for them are usually made from morphological data. However, since the evolutionary changes of morphological characters are complicated, this method usually does not give very reliable trees; it almost never gives estimates of evolutionary times. For this reason, details of the evolutionary relationships of most present-day organisms remain unclarified.

Evidence from Molecular Biology

As soon as the molecular basis of genes was elucidated, it became obvious that the evolutionary relationships of organisms can be studied by comparing nucleotide sequences in DNA or amino acid sequences in proteins (Crick 1958). Zuckerkandl and Pauling (1962, 1965) and Margoliash and Smith (1965) later showed that the rate of amino acid substitution in proteins is approximately constant when time is measured in years. This finding has provided a new method of constructing phylogenetic trees. Furthermore, the principle of constant rate of gene substitution was quickly extended to RNAs and DNAs, and many authors—notably Dayhoff (1969, 1972) and her associates—have used this method to clarify phylogenetic relationships of many different groups of organisms. Although the evolutionary trees constructed by this method are subject to large sampling errors as well as to systematic biases, the results obtained are usually quite reasonable. Recent studies indicate that the molecular clock does not run as regularly as was previously thought (chapters 4 and 5), but this does not seriously affect the utility of molecular data for constructing evolutionary trees (chapter 11).

One advantage of molecular methods is that the rate of nucleotide or amino acid substitution varies considerably among different genes and thus one can study both short-term and long-term evolution by choosing appropriate genes. This is similar to radioactive isotope dating, where various radioactive elements with different decay rates are used. The genes for transfer RNAs (tRNAs), 5S RNA, and 16S ribosomal RNAs (rRNAs) in nuclear genomes evolve very slowly and have been used extensively for clarifying the early stages of biotic radiation (McLaughlin and Dayhoff 1970; Kimura and Ohta 1973a; Fox et al. 1977; Hori and Osawa 1979). For example, Hori and Osawa (1979) studied the nucleotide differences of 5S RNAs from various eukaryotic and prokaryotic species and constructed the phylogenetic tree given in figure 2.3. The time scale given in this figure was derived under the assumption of a constant rate of nucleotide substitution. According to this tree, prokaryotes and eukaryotes diverged about 1.5 billion years ago, whereas plants and animals separated about 1.0 billion years ago. Although these esti-

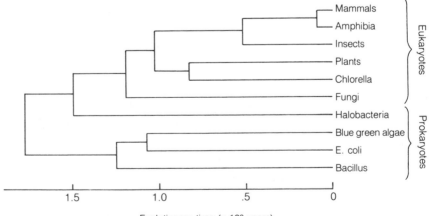

Evolutionary time (x 10⁹ years)

Figure 2.3.   Phylogenetic tree of various eukaryote and prokaryote species inferred from nucleotide sequence differences of 5S RNAs. Adapted from Hori and Osawa (1979).

mates are dependent on a number of assumptions, they are not unreasonable in view of the available fossil record. It is also interesting to note that various groups of prokaryotes diverged a long time ago and that some of them diverged before the prokaryote-eukaryote divergence. According to Woese and his colleagues (Fox et al. 1980; Woese 1981), present-day bacteria can be divided into two groups, i.e., archebacteria and eubacteria, the former being more ancient than the latter, though this view has not been universally accepted (Hori and Osawa 1980).

At this point, it is interesting to note that until around 1950 it was customary to divide life forms into two kingdoms: animals and plants. This classification was later found to be inadequate, and five kingdoms, i.e., animals, plants, fungi, algae, and bacteria, were proposed (Whittaker 1969). Recent molecular data indicate that even this classification is not entirely satisfactory (Lake 1985).

For higher organisms, numerous evolutionary trees have been constructed by using various proteins and genes (e.g., Fitch and Margoliash 1967a; Dayhoff 1969, 1972; Goodman et al. 1974, 1982, 1983; Sibley and Ahlquist 1981, 1984; Sarich and Wilson 1967; Maxson and Wilson, 1975; Lakovaara et al. 1972; Ferris et al. 1981; Brown et al. 1982; Avise et al. 1983). These trees are subject to large sampling errors, and none of them seems to be definitive. Yet they have already provided

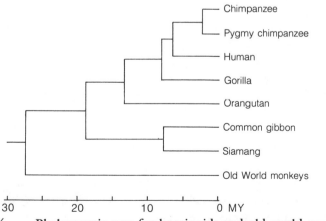

Figure 2.4.   Phylogenetic tree for hominoids and old world monkeys. The evolutionary time scale was obtained by assuming that the orangutan diverged from the human lineage 13 MY ago [see Note added in proof in Sibley and Ahlquist (1984)]. Adapted from Sibley and Ahlquist (1984).

some new insight into the evolutionary history of many organisms. Figure 2.4 gives one such example in which the evolutionary tree of humans and apes constructed from DNA-DNA hybridization data (chapter 5) is presented. Before molecular data were available, humans and chimpanzees were considered to have diverged about 30 MY ago. Therefore, when Sarich and Wilson (1967) showed, on the basis of immunological study, that the divergence time could be as recent as 5 MY ago, most anthropologists criticized the work. However, a more careful paleontological study of the time of divergence between humans and great apes has led to the conclusion that molecular data and fossil records are not contradictory and the divergence time between humans and chimpanzees could indeed be as small as 5 MY (Pilbeam 1984). The results presented in figure 2.4 support this view. The evolutionary tree in this figure is also in rough agreement with the trees obtained from data on protein sequences (Goodman et al. 1982), chromosome banding pattern (Yunis and Prakash 1982), and mitochondrial DNA (chapter 11).

Major Geological Events and Evolution

Although fossils are the only direct evidence of the evolution of organisms aside from various observations of short-term genetic changes of populations, information on geological events such as continental drift, island formation, and glaciation is also useful for the study of evolution. It is now generally accepted among geologists that continents on the earth continually drifted in the past, splitting and merging several times (see Calder 1983). At the time of the Cambrian era, about 550 MY ago, there were four continents, but by the time of the Jurassic, about 200 MY ago, they had merged to form a single supercontinent called Pangaea. About 20 MY later, however, Pangaea began to break into two supercontinents: Laurasia and Gondwanaland (figure 2.5). Laurasia consisted of the present North America, Greenland, and Eurasia north of the Alps and the Himalayas, whereas Gondwanaland included the present South America, Africa, India, Australia, and Antarctica. The supercontinents later split into smaller land masses, but the rifts between them did not become effective barriers to the movement of land animals until well into the Cretaceous. By about 75 MY ago, however, North America was completely separated from Asia, and Gondwanaland split into South America, Australia, Antarctica, etc. Several important aspects of the present geographical distribution of animals and plants can be explained by this continental drift. Continental drift continued to

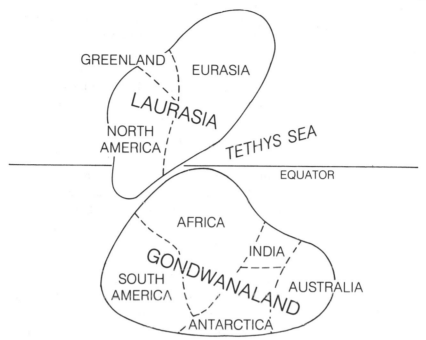

Figure 2.5. Distribution of the world's land mass during the Mesozoic.

operate, and even in the Cenozoic era, about 60 MY ago, the land forms were not the same as those of the present time. The present form of land distribution was attained only about 40 MY ago.

Information on continental drift has been used by a number of authors to date the times of divergence between different groups of organisms. Maxson and Wilson (1975) constructed the phylogeny of tree frogs from North America, South America, Australia, and Asia by using the albumin clock for mammalian organisms (see chapter 4). They showed that South American and Australian species were separated about 75 MY ago, in agreement with the time estimated from information on the separation of the two continents. Sibley and Ahlquist (1981, 1984) used the time of separation, about 80 MY ago, between the ostrich in Africa and the common rhea in South America to calibrate their molecular clock for DNA hybridization data. Similarly, Vawter et al. (1980) used the time of formation of a land bridge between North and South America for calibrating the electrophoretic clock (chapter 9).

Another type of geological event that is important for evolutionary studies is glaciation. Glaciation or deglaciation causes drastic changes of planetary temperature, which in turn affect the geographical distributions of animals and plants. Glaciation also changes the sea level. During an ice age, sea level is lowered so that two islands or continents which were previously isolated may be connected. This allows land animals and plants to move across previous barriers. During an interglacial period, migration between two geographical areas is often terminated, which may lead to formation of new species.

In the last 2 MY (Pleistocene), there were many cycles of glaciation, five of which occurred during the last 500,000 years (figure 2.6). In this period, many species became extinct, while many new species emerged. Particularly in the last glaciation (Wisconsin-Würm period), which ended about 10,000 years ago, more than 50 percent of mammalian genera in North America became extinct. At the times of maximum glaciation, sea level was lowered by about 100 meters. This caused many islands to connect with each other or with their neighboring continents. Thus, in the last glaciation New Guinea and Tasmania were connected to Australia, Cuba to Florida, the Lesser Antilles to South America, Sumatra, Java, and Borneo to Eurasia, and North America to Eurasia (see Calder 1983). It is, therefore, important to know how these glaciations affected the distribution of organisms and gene migration in the study of speciation and gene differentiation. In table 2.2, I have listed several examples of geological events that have been used for evolutionary studies. Although dating of geological events is not necessarily very accurate, it is useful for calibrating molecular clocks.

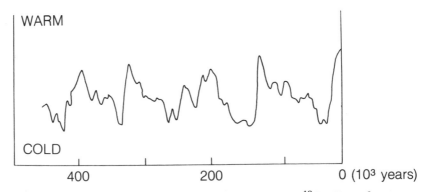

Figure 2.6.   Ice ages recorded by heavy oxygen (^{18}O). Data from Hays et al. (1976).

Table 2.2   Times of various geological events that have been used for calibrating molecular clocks.

Continents or islands involved	Time (MY ago)	Authors
I. Separation		
Africa/South America	80	Sibley and Ahlquist (1984)
New Zealand/Australia	80	"
South America/Australia	75	Maxson and Wilson (1975)
Pacific/Gulf of Mexico	2–5	Vawter et al. (1980)
II. Island formation		
Hawaii	0.8	Hunt et al. (1981)
Oahu	4	"
Kauai	6	"
Lesser Antilles	3–5	Kim et al. (1976)
Galapagos	0.5–4	Yang and Patton (1981)

Geological data indicate that after the Cambrian era there have been five great biological crises in which many groups of organisms died out. The most recent is the mass extinction that occurred at the end of the Cretaceous about 65 MY ago. At that time, a large proportion of marine plants and animals disappeared. Many species of land animals, particularly dinosaurs, also became extinct. The reasons for this "mass extinction" have been debated for many decades, but no agreement has been reached among geologists. Alvarez et al. (1980) have recently proposed the asteroid impact theory. According to this theory, the earth was hit by an asteroid with a diameter of about 10 km at this time, and a large crater with a diameter of 100 km was produced. The impact of this asteroid injected about 60 times the object's mass into the atmosphere as pulverized rock, and a fraction of this dust stayed in the stratosphere for several years and was distributed worldwide. The resulting darkness suppressed photosynthesis and disturbed the food chain of organisms, leading to mass extinction of organisms. This theory is supported by a marked increase of the extraterrestrial element iridium in the strata of soil dating to the end of the Cretaceous. Alvarez et al. also proposed that the four earlier great extinctions were probably caused by similar asteroid impacts.

This theory is still controversial (e.g., Stanley 1984), but if it is correct, it will have profound implications for evolutionary studies. For example, Wyles et al. (1983) suggest that because of the asteroid (or comet) impact discussed above most species of birds also became extinct about 65 MY ago and that the majority of present-day orders and families in birds were formed during the last 65 MY. If this is the case, it would change the current calibration of molecular clocks in birds (see chapter 4).

Recently, examining the temporal distribution of major extinctions of marine animals over the past 250 MY, Raup and Sepkoski (1984) suggested that extinction occurred periodically with a mean interval of about 26 MY and that this might be caused by extraterrestrial forces. This suggestion is again controversial, and we need more data to confirm it (Benton 1985). However, if mass extinction of organisms occurs periodically and it wipes out many families and orders of animals and plants, irrespective of their Darwinian fitness, it will have a profound effect on the current theory of evolution. Particularly, if mass extinction is caused by extraterrestrial forces, extinction or survival of organisms will be unpredictable. Examining the evolutionary patterns of marine mollusks during the Cretaceous, Jablonski (1986) has concluded that the survivorships during normal times and extinction times are quite different and that perfectly adapted organisms during normal times may be eliminated during extinction times.

□ CHAPTER THREE □
GENES AND MUTATION

Evolution is "descent with modification" (Darwin 1859). Modification occurs when genetic variability exists, and genetic variability is produced by mutation. It is, therefore, important to understand the nature of mutation and how it occurs. In the early days of genetic studies, any genetic change in phenotypic characters was called a mutation, with no knowledge of the cause of the change. At present, we know that various factors are involved in causing genetic changes of phenotypes. They can be studied at three different levels, i.e., molecular, chromosomal, and genomic levels. In this chapter, I shall briefly discuss the nature and the rate of occurrence of mutations at the molecular level. Before the discussion of mutation, however, I would like to review the molecular structure and function of genes. The reader who is interested in more details of this subject may refer to Lewin's (1985) book, *Genes*.

Structure and Function of Genes

Our concept of genes or genomes has changed drastically in recent years mainly because the molecular structure of genes has been elucidated at the DNA level. Elucidation of the gene structure has led to new insight into gene function, though our knowledge of the regulation of gene function and its relation to morphogenesis is still severely limited.

Types of Genes

In terms of their function, genes can be classified into two groups: protein-coding (structural) genes and RNA-coding genes. Protein-coding genes produce messenger RNAs (mRNA) of which the genetic information is translated into the amino acid sequence of proteins. RNA-coding genes are those which produce transfer RNAs (tRNA), ribosomal RNAs (rRNA), small nuclear RNAs (snRNA), etc. These nonmessenger RNAs are the final products of the genes that code for them. Ribosomal

RNAs are components of ribosomes that are a part of the machinery of protein synthesis, whereas tRNAs are essential in transferring the genetic information of mRNAs into amino acid sequences of proteins. The function of snRNAs is not well understood, but they are apparently important for mRNA transcription and processing (formation of mature mRNAs from precursor mRNAs).

Basic Structure of Genes

The basic structure of a protein-coding gene in eukaryotes is given in figure 3.1. It is a linear arrangement of four nucleotides, adenine (A), thymine (T), cytosine (C), and guanine (G), and consists of a transcribable part of DNA and the 5' and 3' nontranscribed flanking regions. The flanking regions seem to be necessary for controlling transcription, i.e., production of precursor messenger RNAs (mRNAs). The CAAT box and the TATA box (often called the Hogness box) in the 5' flanking region exist in virtually all eukaryotic genes and are considered to be recognition sites for RNA polymerase. Transcription of precursor mRNAs starts from the cap site and ends at the poly(A) addition site.

A precursor mRNA consists of coding regions and noncoding regions. Coding regions contain information for encoding amino acids in the polypeptide produced by the gene, whereas noncoding regions contain information necessary for regulation of polypeptide production. Some segments of noncoding regions are spliced out in the process of produc-

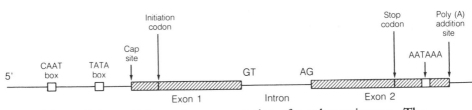

Figure 3.1.   Schematic representation of a eukaryotic gene. The CAAT box (a short nucleotide sequence involving CAAT or its minor variation) and the TATA box (another short nucleotide sequence involving TATA or its minor variation) in the 5' flanking region are believed to control the initiation of transcription. The conserved sequence AATAAA (or its minor variation) near the poly (A) addition site might be important for poly (A) addition or RNA processing. Transcription starts from the cap site.

cccctgtggagccacaccctagggttgg[ccaat]ctactcccaggagcagggagggcaggagccagggctggg[cataaaa]

gtcagggcagagccatctattgcttACATTTGCTTCTGACACAACTGTGTTCACTAGCAACCTCAAACAGACACC[ATG]

ValHisLeuThrProGluGluLysSerAlaValThrAlaLeuTrpGlyLysValAsnValAspGluValGlyGlyGlu
GTGCACCTGACTCCTGAGGAGAAGTCTGCCGTTACTGCCCTGTGGGGCAAGGTGAACGTGGATGAAGTTGGTGGTGAG

AlaLeuGlyArg
GCCCTGGGCAGGTTGGTATCAAGGTTACAAGACAGGTTTAAGGAGACCAATAGAAACTGGGCATGTGGAGACAGAGAAG

                                                      LeuLeuValValTyr
ACTCTTGGGTTTCTGATAGGCACTGACTCTCTCTGCCTATTGGTCTATTTTCCCACCCTTAGGCTGCTGGTGGTCTAC

ProTrpThrGlnArgPhePheGluSerPheGlyAspLeuSerThrProAspAlaValMetGlyAsnProLysValLys
CCTTGGACCCAGAGGTTCTTTGAGTCCTTTGGGGATCTGTCCACTCCTGATGCTGTTATGGGCAACCCTAAGGTGAAG

AlaHisGlyLysLysValLeuGlyAlaPheSerAspGlyLeuAlaHisLeuAspAsnLeuLysGlyThrPheAlaThr
GCTCATGGCAAGAAAGTGCTCGGTGCCTTTAGTGATGGCCTGGCTCACCTGGACAACCTCAAGGGCACCTTTGCCACA

LeuSerGluLeuHisCysAspLysLeuHisValAspProGluAsnPheArg
CTGAGTGAGCTGCACTGTGACAAGCTGCACGTGGATCCTGAGAACTTCAGGGTGAGTCTATGGGACCCTTGATGTTTT

CTTTCCCCTTCTTTTCTATGGTTAAGTTCATGTCATAGGAAGGGGAGAAGTAACAGGGTACAGTTTAGAATGGGAAAC

AGACGAATGATTGCATCAGTGTGGAAGTCTCAGGATCGTTTTAGTTTCTTTTATTTGCTGTTCATAACAATTGTTTTC

TTTTGTTTAATTCTTGCTTTCTTTTTTTTTCTTCTCCGCAATTTTTACTATTATACTTAATGCCTTAACATTGTGTAT

AACAAAAGGAAATATCTCTGAGATACATTAAGTAACTTAAAAAAAAAACTTTACACAGTCTGCCTAGTACATTACTATT

TGGAATATATGTGTGCTTATTTGCATATTCATAATCTCCCTACTTTATTTTCTTTTATTTTTAATTGATACATAATCA

TTATACATATTTATGGGTTAAAGTGTAATGTTTTAATATGTGTACACATATTGACCAAATCAGGGTAATTTTGCATT

TGTAATTTTAAAAAATGCTTTCTTCTTTTAATATACTTTTTTGTTTATCTTATTTCTAATACTTTCCCTAATCTCTTT

CTTTCAGGGCAATAATGATACAATGTATCATGCCTCTTTGCACCATTCTAAAGAATAACAGTGATAATTTCTGGGTTA

AGGCAATAGCAATATTTCTGCATATAAATATTTCTGCATATAAATTGTAACTGATGTAAGAGGTTTCATATTGCTAA

TAGCAGCTACAATCCAGCTACCATTCTGCTTTTATTTTATGGTTGGGATAAGGCTGGATTATTCTGAGTCCAAGCTAG
                                                      LeuLeuGlyAsnValLeuValCysValLeuAla
GCCCTTTTGCTAATCATGTTCATACCTCTTATCTTCCTCCCACAGCTCCTGGGCAACGTGCTGGTCTGTGTGCTGGCC

HisHisPheGlyLysGluPheThrProProValGlnAlaAlaTyrGlnLysValValAlaGlyValAlaAsnAlaLeu
CATCACTTTGGCAAAGAATTCACCCCACCAGTGCAGGCTGCCTATCAGAAAGTGGTGGCTGGTGTGGCTAATGCCCTG

AlaHisLysTyrHis
GCCCACAAGTATCAC[TAA]GCTCGCTTTCTTGCTGTCCAATTTCTATTAAAGGTTCCTTTGTTCCCTAAGTCCAACTAC

TAAACTGGGGGATATTATGAAGGGCCTTGAGCATCTGGATTCTGCCTAATAAAAAACATTTATTTTCATTGCaatgat

gtatttaaattatttctgaatattttactaaaaagggaatgtgggaggtcagtgcatttaaaacataaagaaatgatg

agctgttcaaaccttgggaaaatacactatatcttaaactccatgaaagaaggtgaggctgcaaccagctaatgcaca

ttggcaacagcccctgatgcctatgccttattcatccctcagaaaaggattcttgtagaggcttgatttgcaggttaa

agtttgctatgctgtattttacattacttattgttttagctgtcctcatgaatgtcttttcactacccatttgctta

tcctgcatctctctcagccttgact

Figure 3.2. Nucleotide sequence of the human β-globin gene. The nucleotide sequence of the mRNA strand (noncoding strand) of the gene is shown from the 5′ to the 3′ direction. Large and small capital letters represent sequences corresponding to exons (mature mRNA) and introns, respectively. Flanking sequences are in lower-case letters. Two boxed sequences ccaat and cataaaa in the 5′ flanking region are the CAAT and TATA boxes, respectively. The ATG initiator and TAA terminator are also boxed. The amino acid sequence of β globin is shown on a line above the coding sequence. Nucleotide redundancies of intron-exon junctions are underlined, and the GT and AG dinucleotides at the beginning and end of the introns are overlined. From Lawn et al. (1980).

tion of a mature mRNA. These segments are called *introns* or *intervening sequences* (IVS). The remaining segments are called *exons*. For example, the β globin gene in mammals has three exons and two introns (figure 3.2). The number of exons in a gene varies from gene to gene. The largest number of exons so far observed is more than fifty in the $\alpha2$ type I collagen gene from chicken (Wozney et al. 1981). The role of introns is not known; it is possible that they are historical remnants and do not have any important functions. It should be noted that an intron always starts from a dinucleotide GT and ends with AG. These dinucleotides are considered to be recognition sites of an intron-removing enzyme (RNA splicing enzyme).

The region between the cap site and the translation initiation site in mRNA is called the leader region. This region is considered to be important for controlling RNA processing and translation. It contains a segment which is highly homologous to a part of 18S rRNA and snRNAs. The 3' side noncoding region of mRNA is also considered to be important for RNA processing and translation as well as for transcription.

The structure of protein-coding genes in prokaryotes is different from that in eukaryotes in one important aspect; prokaryotic genes have no introns. Because of the absence of introns, prokaryotes apparently do not possess the RNA-splicing enzyme. Therefore, a eukaryotic gene introduced into bacteria by genetic engineering techniques does not produce a functional protein unless the introns have been spliced out before insertion.

The structure of RNA-producing genes is simpler than that of protein-coding genes and is essentially the same for eukaryotes and prokaryotes. The RNA-coding region is usually a stretch of DNA; tRNAs, rRNAs, snRNAs, etc., are directly produced by the action of RNA polymerase. In some primitive organisms, such as ciliates and slime molds, RNA-producing genes have introns that must be removed before making an active gene product. In some genes (e.g., 5S RNA and tRNA genes in vertebrates), the production of RNAs is controlled by a sequence of nucleotides inside the gene (chapter 6).

Protein Synthesis

As mentioned earlier, the genetic information carried by the nucleotide sequence of a gene is first transferred to mRNA by a simple process of one-to-one transcription of the nucleotides. The genetic information transferred to mRNA determines the amino acid sequence of the protein produced. Nucleotides of mRNA are read sequentially, three at a time. Each such triplet or *codon* is translated into one particular amino acid in the growing protein chain through the genetic code. Synthesis of proteins occurs at the sites where ribosomes are attached to the mRNA. This is a rather complicated process, and each amino acid is added to the growing polypeptide sequence with the aid of tRNAs.

The genetic code for nuclear genes seems to be "universal" for both prokaryotes and eukaryotes with a few exceptions which will be mentioned later. The same genetic code is used for chloroplast genes, but mitochondrial genes use slightly different genetic codes. The nuclear genetic code is given in table 3.1. In this table, amino acids are represented by three-letter codes (see table 3.2). There are $4^3 = 64$ possible codons for four different nucleotides, uracil (U), cytosine (C), adenine (A), and guanine (G). Three of them (UAA, UAG, UGA) are, however, *amino acid terminating* (or *nonsense*) *codons* and do not code for any amino acid. Each of the remaining 61 codons *(sense codons)* codes for a particular

Table 3.1 The genetic code for nuclear genes.

Codon	Amino Acid	Codon	Amino Acid	Codon	Amino Acid	Codon	Amino Acid
UUU	Phe	UCU	Ser	UAU	Tyr	UGU	Cys
UUC	Phe	UCC	Ser	UAC	Tyr	UGC	Cys
UUA	Leu	UCA	Ser	UAA	Ter	UGA	Ter
UUG	Leu	UCG	Ser	UAG	Ter	UGG	Trp
CUU	Leu	CCU	Pro	CAU	His	CGU	Arg
CUC	Leu	CCC	Pro	CAC	His	CGC	Arg
CUA	Leu	CCA	Pro	CAA	Gln	CGA	Arg
CUG	Leu	CCG	Pro	CAG	Gln	CGG	Arg
AUU	Ile	ACU	Thr	AAU	Asn	AGU	Ser
AUC	Ile	ACC	Thr	AAC	Asn	AGC	Ser
AUA	Ile	ACA	Thr	AAA	Lys	AGA	Arg
AUG	Met	ACG	Thr	AAG	Lys	AGG	Arg
GUU	Val	GCU	Ala	GAU	Asp	GGU	Gly
GUC	Val	GCC	Ala	GAC	Asp	GGC	Gly
GUA	Val	GCA	Ala	GAA	Glu	GGA	Gly
GUG	Val	GCG	Ala	GAG	Glu	GGG	Gly

Table 3.2   Twenty amino acids that compose proteins and their three- and one-letter codes.

Name	Code Three-letter	Code One-letter	Name	Code Three-letter	Code One-letter
1. Alanine	Ala	A	11. Leucine	Leu	L
2. Arginine	Arg	R	12. Lysine	Lys	K
3. Asparagine	Asn	N	13. Methionine	Met	M
4. Aspartic acid	Asp	D	14. Phenylalanine	Phe	F
5. Cysteine	Cys	C	15. Proline	Pro	P
6. Glutamine	Gln	Q	16. Serine	Ser	S
7. Glutamic acid	Glu	E	17. Threonine	Thr	T
8. Glycine	Gly	G	18. Tryptophan	Trp	W
9. Histidine	His	H	19. Tyrosine	Tyr	Y
10. Isoleucine	Ile	I	20. Valine	Val	V

amino acid, but since there are only twenty amino acids used for making proteins, there are many codons which code for the same amino acid. Codons coding for the same amino acid are called *synonymous* codons. In the genetic code table, codon AUG codes for methionine, but this codon is also used as the initiation codon. Therefore, every mRNA has this codon at the beginning of its coding region. The methionine encoded by the initiation codon is in a modified form and is later removed from the polypeptide.

Table 3.3 shows the genetic code for mammalian mitochondrial genes. There are a few differences between this genetic code and the nuclear genetic code. In the mitochondrial genetic code, codon UGA is not a termination codon but codes for tryptophan. On the other hand, codons AGA and AGG are terminating codons instead of coding for arginine. AUA, which codes for isoleucine in the nuclear code, is used for methionine. The genetic code of yeast mitochondrial genes is slightly different from that of the mammalian mitochondrial code (table 3.3). In yeast, CUU, CUC, CUA, and CUG code for threonine instead of leucine, and AUA codes for isoleucine instead of methionine. The genetic code of *Drosophila* mitochondrial genes is identical with that of mammalian mitochondrial genes, except that AGA codes for serine and the quadruplet ATAA is used in initiation of translation (de Bruijn 1983).

In recent years, a further loss in generality of the "universal" genetic code has occurred. In ciliated protozoans such as *Tetrahymena* and *Para-*

Table 3.3 The genetic code for mammalian mitochondrial genes. The codons for yeast are identical with those for mammals except the ones in parentheses.

Codon	Amino Acid	Codon	Amino Acid	Codon	Amino Acid	Codon	Amino Acid
UUU	Phe	UCU	Ser	UAU	Tyr	UGU	Cys
UUC	Phe	UCC	Ser	UAC	Tyr	UGC	Cys
UUA	Leu	UCA	Ser	UAA	Ter	UGA	Trp
UUG	Leu	UCG	Ser	UAG	Ter	UGG	Trp
CUU	Leu(Thr)	CCU	Pro	CAU	His	CGU	Arg
CUC	Leu(Thr)	CCC	Pro	CAC	His	CGC	Arg
CUA	Leu(Thr)	CCA	Pro	CAA	Gln	CGA	Arg
CUG	Leu(Thr)	CCG	Pro	CAG	Gln	CGG	Arg
AUU	Ile	ACU	Thr	AAU	Asn	AGU	Ser
AUC	Ile	ACC	Thr	AAC	Asn	AGC	Ser
AUA	Met(Ile)	ACA	Thr	AAA	Lys	AGA	Ter(Arg)
AUG	Met	ACG	Thr	AAG	Lys	AGG	Ter(Arg)
GUU	Val	GCU	Ala	GAU	Asp	GGU	Gly
GUC	Val	GCC	Ala	GAC	Asp	GGC	Gly
GUA	Val	GCA	Ala	GAA	Glu	GGA	Gly
GUG	Val	GCG	Ala	GAG	Glu	GGG	Gly

mecium, UAA and UAG do not appear to be termination codons even in nuclear genes but code for glutamine. Furthermore, in the prokaryotic organism *Mycoplasma capricolum* the usual termination codon UGA is apparently used for encoding tryptophan (see Fox 1985 for a review).

Jukes (1983) has presented a theory of evolution of the genetic code. According to his theory, the primodial genetic code consisted of 16 $(=4^2)$ codons, the third base in a codon having no effect on the choice of amino acid. As additional amino acids were used for protein synthesis at a later evolutionary time, the so-called universal genetic code evolved. Once this "universal" genetic code evolved, it was frozen because mutation changing codon assignments would cause errors in many protein sequences (Crick 1968). In mitochondrial DNA, however, the number of proteins encoded is so small (about ten) that such a mutation may be accepted. In Jukes' view, the evolutionary change of codons in mitochondrial DNA occurred in the direction of the primordial code. In protozoans and mycoplasma, the change in genetic code occurred in nuclear genes, but it was probably harmless because previous termination codons were changed to amino acid-coding codons.

Types of Mutations at the DNA Level

Since all morphological and physiological characters of organisms are
controlled by the genetic information carried by DNA, any mutational
changes in these characters are due to some change in DNA molecules.
There are four basic types of changes in DNA. They are *substitution* of a
nucleotide for another nucleotide (figure 3.3b), *deletion* of nucleotides
(figure 3.3c), *insertion* of nucleotides (figure 3.3d), and *inversion* of nucleotides (figure 3.3e). Insertion, deletion, and inversion may occur with
one or more nucleotides as a unit. Insertion and deletion may shift the
reading frames of the nucleotide sequence, if they occur in the coding
region. These insertions and deletions are called *frameshift mutations*. Nucleotide substitutions can be divided into two different classes, i.e.,
transition and transversion. *Transition* is the substitution of a purine
(adenine or guanine) for another purine or the substitution of a pyrimidine (thymine or cytosine) for another pyrimidine. Other types of nucleotide substitutions are called *transversions* (see figure 3.4).

Recent data indicate that insertion and deletion occur quite often,
particularly in the noncoding region of DNA. The number of nucleotides involved in an insertion or deletion varies from a few nucleotides
to several thousand nucleotides. Short insertions or deletions are apparently caused by errors in DNA replication. Long insertions or deletions

(a) Wild type	ACC	TAT	TTG	CTG	
	Thr	Tyr	Leu	Leu	
		↓			
(b) Substitution	ACC	TCT	TTG	CTG	
	Thr	Ser	Leu	Leu	
		↓			
(c) Deletion	ACC	TTT	TGC	TG-	
	Thr	Phe	Cys	---	
		↓			
(d) Insertion	ACC	TAC	TTT	GCT	G--
	Thr	Tyr	Phe	Ala	
		↓	↓		
(e) Inversion	ACC	TTT	ATG	CTG	
	Thr	Phe	Met	Leu	

Figure 3.3.   Four basic types of mutations at the nucleotide level.
The nucleotide sequence is represented in units of codons or nucleotide
triplets in order to show how the amino acids coded for are changed by
the nucleotide changes.

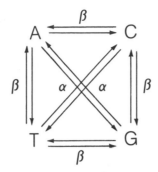

Figure 3.4.   Transitional (A ⇄ G and T ⇄ C) and transversional (all others) nucleotide substitutions.

seem to be mainly due to unequal crossover or DNA transposition. DNA transposition—the movement of a DNA sequence from one chromosomal position to another—occurs through transposons or transposable elements (see chapter 6). Another possible mechanism of gene insertion is *horizontal gene transfer,* which will be discussed in chapter 6.

The possible role of unequal crossover in increasing the number of genes in the genome has been known for many years (Bridges 1936). Only recently, however, has it been realized that it plays an important role in evolution in increasing or decreasing DNA content. Particularly in multigene families such as immunoglobulin genes and ribosomal RNA genes, unequal crossover seems to play a predominant role (Hood et al. 1975). A phenomenon related to unequal crossover is *gene conversion.* This is believed to occur by the repair of mismatched bases in heteroduplex DNA (Radding 1982), and the total number of nucleotides is not altered. Slightom et al. (1980) noted that part of the second intron of the human $^A\gamma$ globin gene is very similar to that of the nonhomologous gene $^G\gamma$ from the same individual. They hypothesized that the part of the $^A\gamma$ gene had been "converted" by an intergenic exchange to become more like the $^G\gamma$ gene. Although the molecular mechanism of gene conversion is not well understood, the partial similarity between nonhomologous genes, which has subsequently been observed in many other genes, is considered to be due to gene conversion.

Pattern of Nucleotide Substitution

As mentioned above, there are many different types of DNA changes in eukaryotic genes. However, if we consider only coding regions of

DNA, the majority of the changes are nucleotide substitutions. It is, therefore, interesting to know the pattern of nucleotide substitution in coding regions.

If mutation occurs at random among the four nucleotides A, T, C, and G, we would expect that there are two times more transversional changes than transitional changes (see figure 3.4). In practice, this is not the case, and transitional changes occur more often than expected (Fitch 1967; Vogel 1972). Recently, Gojobori et al. (1982b) and Li et al. (1984) studied the relative mutational changes among the four nucleotides by using pseudogenes. *Pseudogenes* are nonfunctional (dead) genes and are believed to have been produced by silencing of a duplicate gene (chapter 6). In pseudogenes, mutations accumulate without being subject to any purifying selection. Indeed, the rate of nucleotide substitution in pseudogenes is known to be extremely high compared with that in functional genes (chapter 6). Thus, the nucleotide differences between a pseudogene and its functional counterpart are considered to be mainly due to the mutations that occurred in the pseudogene unless the time since divergence between them is very long. Therefore, one can determine the approximate relative mutation rates among the four nucleotides by comparing the nucleotide sequences of the two genes. Note that when there is no selection, the rate of nucleotide substitution in evolution is equal to the mutation rate (chapter 13).

The relative mutation rates obtained in this way are presented in table 3.4. It is clear that the frequency (59.30%) of transitional changes (T$\rightleftharpoons$C and A$\rightleftharpoons$G) is much higher than the expected (33.3%). However, the four types of transitional changes do not occur with equal frequency, the frequency of C$\rightarrow$T being higher than that of the other changes. The

Table 3.4   Relative mutation rates among the four nucleotides A, T, C, and G in pseudogenes. These values are based on substitution data from 16 pseudogenes in mammals. From Li et al. (1984).

Mutant nucleotide	Original nucleotide			
	A	T	C	G
A		4.4 ± 1.1	6.5 ± 1.1	20.7 ± 2.2
T	4.7 ± 1.3		21.0 ± 2.1	7.2 ± 1.1
C	5.0 ± 0.7	8.2 ± 1.3		5.3 ± 1.0
G	9.4 ± 1.3	3.3 ± 1.2	4.2 ± 0.5	

high frequency of C→T change seems to be partially related to methylation of cytosine in the CG dinucleotides in the genome, but other factors are also involved. It is interesting to note that the eight types of transversional changes occur with more or less the same frequency, though the G→T and C→A changes tend to be higher. The reader who is interested in the biochemical reasons for the unequal mutation rates among the four nucleotides should consult Topal and Fresco (1976) and Gojobori et al. (1982b). For our purpose, it is sufficient to keep in mind that the mutation rates among the four nucleotides are not equal.

Mutations and Amino Acid Substitutions

Since the amino acid sequence of a polypeptide is determined by the nucleotide sequence of a protein-coding gene, any change in amino acid sequences is caused by a mutation occurring in DNA. On the other hand, a mutational change in DNA does not necessarily result in a change in amino acid sequence because of the degeneracy of the genetic code (synonymy of codons). Mutations that result in synonymous codons are called *synonymous* or *silent mutations,* whereas others are called *nonsynonymous* or *amino acid altering mutations.*

When nucleotide changes occur at random, the proportion of nonsynonymous mutations can be computed by using the (nuclear) genetic code. If we assume that all codons are equally frequent in the genome and the probability of substitution is the same for all pairs of nucleotides, the proportion of nonsynonymous mutations in nuclear genes becomes 71 percent, excluding nonsense mutations (Nei 1975). That is, about 29 percent of the mutations occurring at the nucleotide level cannot be detected by examining amino acid changes. This figure does not change very much, even if we make more realistic assumptions about the codon frequencies and the probabilities of nucleotide changes.

The genetic code in table 3.1 indicates that synonymous codons occur mainly because of the redundancy of the third nucleotide position. Indeed, a computation similar to the one mentioned above shows that about 72 percent of nucleotide changes at the third position are synonymous. Synonymous mutations also occur at the first nucleotide position, but the proportion of synonymous changes is about 5 percent. At the second nucleotide position, there is no synonymous substitution.

In population genetics and molecular taxonomy, protein differences are often studied by electrophoresis. Although this technique does not detect all amino acid differences, it facilitates a large-scale study of genetic variation within and between populations. The electrophoretic mobility of a protein is largely determined by the net charge of the protein. At the ordinary pH value at which electrophoresis is conducted, lysine and arginine are positively charged, whereas aspartic acid and glutamic acid are negatively charged. All other amino acids are neutral. Therefore, it is possible to compute the proportion of amino acid changes that result in charge change. Considering the genetic code and actual data on amino acid substitution, Nei and Chakraborty (1973) estimated that the proportion is 25 to 30 percent. A similar estimate has also been obtained by Marshall and Brown (1975) under various biochemical conditions.

It should be noted, however, that electrophoretic detectability of protein differences is greatly enhanced by heat treatment (Bernstein et al. 1973), application of various electrophoretic conditions (Singh et al. 1976; Ramshaw et al. 1979; Coyne 1982; McLellan 1984), etc. However, these techniques are too cumbersome to be used for large-scale population survey. A method that can be used for a population survey to detect more protein variation than standard electrophoresis is isoelectric focusing electrophoresis. This method, particularly immobilized gradient isoelectric focusing, seems to detect a large fraction of neutral amino acid changes (Whitney et al. 1985). Another method that is occasionally used is two-dimensional electrophoresis (O'Farrell 1975). In this method, many different proteins can be examined by the same electrophoretic gel, so that it is suitable for a large-scale survey of genetic variability. Goldman et al. (1987) were able to study 383 polypeptides in humans and apes. However, the proportion of polymorphic loci detectable by this method is lower than that detectable by ordinary electrophoresis (e.g., Leigh Brown and Langley 1979; Aquadro and Avise 1981; Ohnishi et al. 1983; McLellan et al. 1983). The reason for this seems to be that the proteins studied by this technique are mainly structural proteins such as actin, tubulin, etc., and these proteins are less variable because of stronger functional constraints (Kimura 1983a).

Mutation Rate

Direct Methods

Determination of the rate of spontaneous mutation is fundamental for the study of evolution, yet this crucial parameter has not been measured with sufficient accuracy at the molecular level. This is because the mutation rate is generally very low, and to get an accurate estimate a large number of genes must be studied. Classical geneticists have determined that the rate of mutations that alter phenotypic characters or induce lethal effects is of the order of 10^{-5} per locus per generation in higher organisms such as man, *Drosophila,* and corn. However, little is known about the nature of these mutations at the DNA level. Furthermore, most of these mutations are deleterious and contribute little to the evolutionary change of organisms.

In recent years, there have been a number of attempts to measure the mutation rate at the molecular level. One of the most extensive studies was carried out by Mukai and his coworkers (Mukai and Cockerham 1977; Voelker et al. 1980). They accumulated spontaneous mutations that affected electrophoretic mobility in *Drosophila melanogaster,* maintaining 500 lines for 211 to 224 generations. Using 7 enzyme loci and examining 3,111,598 gene generations (the number of genes examined × the number of generations), Voelker et al. (1980) estimated the rates of electrophoretic mobility changes and null mutations to be 1.28×10^{-6} and 3.86×10^{-6} per locus per generation, respectively. Unfortunately, the *Drosophila* lines they used were later found to carry mutator genes that increased the rate of lethal mutations about ten times. Because of this special property of the lines used, it is not clear whether these estimates are representative of the mutation rate for enzyme loci. Nevertheless, their estimates show that the mutation rate is somewhat lower than the rate for morphological characters or lethal genes.

A similar study of mutation rate was done by Kahler et al. (1984) with barley. Examining the electrophoretic mobility at 5 enzyme loci for 841,260 gene generations, they found no mutant allele. On the basis of this observation, they concluded that the mutation rate must be lower than 3.6×10^{-6} per locus per generation at the 95 percent probability level. They also studied the rate of nondeleterious mutations in morphological characters, examining 3,386,850 gene generations. They again found no mutant and estimated that the mutation rate cannot be higher than 8.6×10^{-7} per locus per generation at the 95 percent probability level.

Extensive electrophoretic studies on mutation rate were also conducted in human populations (Harris et al. 1974; Neel et al. 1980). Although a total of 522,119 genes were examined, no mutation on electrophoretic mobility was found. From these data, Neel et al. (1980)

concluded that the mutation rate must be lower than 6×10^{-6} per locus per generation with a probability of 95 percent.

Indirect Methods

A number of authors (e.g., Neel 1973; Nei 1977a; Neel and Rothman 1978; Bhatia et al. 1979) have attempted to estimate the mutation rate for electrophoretic alleles indirectly, considering the balance among mutation, selection, and genetic drift. For example, Nei (1977a) obtained an estimate of 2.3×10^{-6} per locus per generation, using data on 29 protein loci from Japanese macaques. Similar estimates were obtained for various human populations. These estimates again include deleterious mutations. Extending Nei's work, Kimura (1983c) estimated the proportion of deleterious mutation among the total mutations. His results suggest that 80 to 96 percent of new mutations in electrophoretic loci are deleterious. In the case of Japanese macaques, he estimated that the rate of neutral mutation is 1.65×10^{-7} per locus per generation. Although these estimates are quite reasonable, they are dependent on the assumption of population equilibrium. Since there is no guarantee for the validity of this assumption, these estimates should be regarded as tentative ones.

Estimates from the Rate of Gene Substitution

If we assume that most gene substitutions occur by random genetic drift of neutral mutations, the rate of mutations can be estimated from data on gene substitution. This is because the rate of gene substitution (α) for neutral mutations is equal to the mutation rate (v) (chapter 13). In practice, a majority of fresh mutations seem to be deleterious, but they are quickly eliminated from the population by means of purifying selection. Thus, if we estimate the mutation rate under the assumption of $v = \alpha$, it would give an underestimate of the total mutation rate, unless the gene studied is subject to no selection as in the case of pseudogenes. However, this estimate can be used for testing the "null hypothesis" of neutral mutations, because in the neutral theory of molecular evolution only those mutations that contribute to gene substitution or polymorphism are considered (Kimura 1983a,b).

The genomes of higher organisms contain many pseudogenes. Since pseudogenes apparently do not have any biological function, the total mutation rate can be estimated from the rate of nucleotide substitution, if the time since nonfunctionalization is known (chapter 6). Some estimates of the rate of nucleotide substitution for pseudogenes (b) are given in table 6.4 together with the rates for functional genes. In that table, a_1, a_2, and a_3 refer to the rates of nucleotide substitutions for the first, second, and third positions of codons in functional genes. It is clear that pseudogenes have a considerably higher rate of nucleotide substitution than functional genes, and under the null hypothesis of neutral mutation the total mutation rate is estimated to be 4.7×10^{-9} per nucleotide site per year. The mammalian α globin genes consist of 423 nucleotides (141 codons), so that the total mutation rate per gene (coding regions only) is estimated to be 2.0×10^{-6} per year. During the last one million years, the average generation time for the human lineage was probably about 20 years. Therefore, the total mutation rate for the coding region of the human α globin gene is estimated to be 4.0×10^{-5} per generation. This is quite a high rate.

In functional genes, however, the majority of these mutations are apparently eliminated by purifying selection. Indeed, the rate of substitution for functional globin genes is much lower than the pseudogene rate in all three nucleotide positions (table 6.4). Note also that in functional genes the substitution rate for the first and second nucleotide positions of codons is considerably lower than that for the third position. This apparently reflects the difference in intensity of purifying selection between these two groups of nucleotide positions, because the proportion of synonymous substitutions is virtually 0 for the former but is about 72 percent for the latter.

The above result suggests that synonymous nucleotide substitutions occur more frequently than nonsynonymous substitutions, possibly at the same rate as the pseudogene rate. The rate of synonymous substitution can be estimated by the methods presented in chapter 5. Using these methods, Miyata et al. (1980) and Li et al. (1985b) have shown that the rate of synonymous substitution is quite similar for many genes (α and β globins, growth hormones, etc.) and that the average rate for mammalian groups is 4.6×10^{-9} per nucleotide site per year (table 5.6). This rate is virtually identical with the rate for pseudogenes. This suggests that synonymous substitutions are subject to virtually no purifying selection, though this matter is still controversial (chapter 5). If this is the case, the total mutation rate under the null hypothesis of neutral mutation can also be estimated from synonymous substitutions.

So far, we have considered only nuclear genes. Recent studies suggest that the mutation rate in mitochondrial genes in mammals is much higher than that in nuclear genes (Brown et al. 1979, 1982). The rate of silent nucleotide substitution has been estimated to be 4.7×10^{-8} per site per year (Brown et al. 1982). This is about ten times higher than that of nuclear pseudogenes. Extremely high rates of mutation at the nucleotide level have also been observed in RNA viruses (Air 1981; Holland et al. 1982; Nei 1983; Gojobori and Yokoyama 1985). Studies of the rate of nucleotide substitution in the influenza A virus suggest that the mutation rate is of the order of 0.01 per nucleotide site per year. This is about 2 million times higher than the rate for nuclear genes in higher organisms. It is believed that these high rates of mutation in nonnuclear genes are caused by the lack of the repair mechanism of mutational damage of DNA or RNA (Holland et al. 1982).

In the study of the mechanism of evolution, it is important to know the rate of mutations that contribute to amino acid substitutions in evolution or protein polymorphisms in populations. Under the assumption

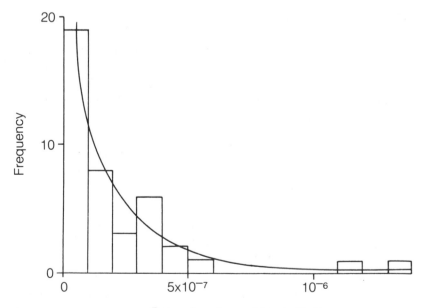

Figure 3.5.   Distribution of the rate of amino acid substitution per polypeptide per year. The number of polypeptides used is 41.

of neutral mutations, this mutation rate can be estimated from the rate of amino acid substitution given in table 4.4. That is, the mutation rate for a given polypeptide is given by the substitution rate per amino acid site multiplied by the number of amino acids in the polypeptide. This rate varies greatly with polypeptide, and the distribution of the rate among the 41 polypeptides used in Chakraborty et al.'s (1978) study is given in figure 3.5. This distribution roughly follows the gamma distribution with mean 2.47×10^{-7} and standard deviation 2.51×10^{-7}. Variation in the substitution rate per polypeptide is due to the difference either in the number of amino acids or in the substitution rate per site. [Note that the mean substitution (mutation) rate is approximately equal to the product of the average substitution rate (1×10^{-9}) in table 4.4

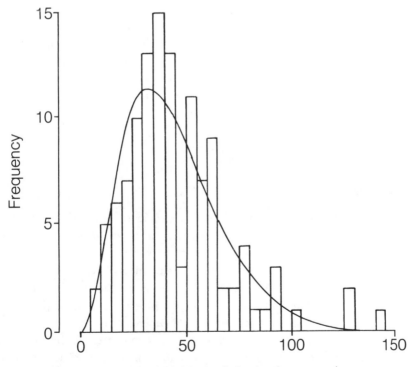

Molecular weight in thousands

Figure 3.6.   Distribution of molecular weights of protein subunits in mammalian species. The total number of subunits used is 119. The gamma distribution fits the data very well ($\chi^2_{(9)} = 6.43$; $P > 0.65$).

and the average number of amino acids per polypeptide (240).] The difference in the substitution rate per site is considered to reflect the variation in the intensity of purifying selection (chapter 4).

It should be noted that the proteins of which the substitution rate has been studied are generally of small size, the average number of amino acids per polypeptide being 240. Most proteins which are used for electrophoretic studies are considerably larger than these proteins. Figure 3.6 shows the distribution of molecular weight for 119 protein subunits in mammalian species (Darnell and Klotz 1975). The mean and standard deviation of this distribution are 45,102 and 24,531, respectively. Since the average molecular weight of an amino acid is about 110, the average polypeptide seems to have about 410 amino acids. The distribution is considerably less leptokurtic than that in figure 3.5. This is partly because the rate of amino acid substitution per site varies from protein to protein and this factor has not been considered here. At any rate, if we assume that the rate of amino acid substitution per site is the same for both groups of polypeptides in figures 3.5 and 3.6, the mean rate of amino acid substitution for average polypeptide is estimated to be 4×10^{-7} per locus per year. Previously, we noted that electrophoresis detects only about one-quarter of all amino acid changes. Therefore, the rate of amino acid substitutions detectable by electrophoresis is estimated to be about 10^{-7} per locus per year. Under the "null hypothesis" of neutral mutations, this is equal to the mutation rate.

Mutation Rate per Year

As mentioned earlier, the rate of amino acid or nucleotide substitution seems to be constant per year rather than per generation. If the mutation rate at the DNA level is constant per *year,* this is easily understandable in the framework of the neutral theory (Kimura and Ohta 1971a; Nei 1975). However, classical genetics has established that the mutation rate is generally constant per *generation* rather than per year. How can we then explain the constancy of molecular evolution per year?

One possible explanation is that the mutations studied by classical geneticists are different from those contributing to amino acid or nucleotide substitution. In classical genetics, the mutation rate was measured mostly by using deleterious mutations. It is possible that a majority of these mutations are due to deletion, insertion, or frameshift at the

molecular level. In fact, Magni (1969) showed in yeast that most deleterious mutations are frameshifts that occur at the time of meiosis. Muller
(1959) also showed that the majority of lethal mutations in *Drosophila*
occur at the meiotic stage. Thus, we would expect that the rate of deleterious mutations is constant per generation rather than per year. In
contrast, the evolutionarily acceptable mutations appear to be a small
fraction of the total mutations and may occur at any time. In classical
genetics, these mutations were almost never measured.

In microorganisms, there is evidence that mutation rates depend largely
on chronological time rather than generation time. In Drake's (1966)
experiment with bacteriophage T4 stored in extracellular media, where
neither DNA replication nor repair occurs, the frequency of spontaneous
mutation increased linearly with time. In a chemostat experiment of
Escherichia coli, Novick and Szilard (1950) also showed that the frequency
of mutation from the wild-type to the phage-resistant type is proportional to chronological time, though in this case the number of genomes
per cell seems to have increased proportionally with time (Kubitschek
1970).

In general, however, the mutation frequency depends on DNA replication as well as on absolute time. According to Kubitschek (1970),
replication-dependent mutations include those caused by caffeine and
some other chemicals, whereas time-dependent mutations include those
induced by ultraviolet rays. In the latter case, what accumulates with
time would not be real mutations but premutational DNA lesions, and
only when DNA replication occurs do they seem to be converted into
real mutations.

If mutation occurs mainly at the time of DNA replication, we would
not expect a strict linear relationship between mutation frequency and
chronological time. It would be more reasonable to assume that the
mutation frequency is proportional to the number of DNA replications.
However, the number of DNA replications for a given period of time
does not seem to vary with organism as much as the number of generations does. For example, the average generation times in man and rodents are about 30 and 0.5 years, respectively, at the present time, but
the number of DNA replications for the rodent lineage seems to be only
about two times as large as that for the human lineage if we consider
long-term evolution (Wilson et al. 1977a). Therefore, if the mutation
frequency is proportional to the number of DNA replications, the num-

ber of nucleotide substitutions may be correlated more closely with chronological time than with the number of generations (see chapter 5 for further discussion).

Nevertheless, the approximate constancy of nucleotide or amino acid substitution for both prokaryotes and eukaryotes, as observed with 5S RNA and cytochrome c, suggests that there are mutations that occur in proportion to chronological time. They could be due to ultraviolet radiation or to background radiation. Unfortunately, we know very little about the real mechanism of mutation in nature. It should also be noted that the production of mutations is an extremely complicated process involving DNA lesion and repair and it is difficult to make a priori predictions about the mutation frequency from the molecular mechanism alone (Kondo 1977).

Finally, it should be noted that even if the mutation rate is generation dependent, it is possible to explain the linear relationship between molecular evolution and chronological time by using the theory of slightly deleterious mutations (Kimura 1979), though the explanation is somewhat contrived. At any rate, the mechanism for the constant rate of molecular evolution remains an unresolved problem at the present time. Furthermore, the constant rate is an approximate property and there are many exceptions, as will be discussed later.

□ CHAPTER FOUR □

EVOLUTIONARY CHANGE
OF AMINO ACID SEQUENCES

Although the basic genetic change occurs in DNA, it is important to know the evolutionary change of proteins, since proteins are molecules essential for building morphological characters and carrying out physiological functions. It should also be noted that the rate of amino acid substitution in proteins is approximately constant, so that information on amino acid substitution is useful for elucidating the evolutionary relationship of organisms. The mathematical formulation of the evolutionary change of amino acid sequence is also simpler than that of nucleotide substitution, since backward and parallel mutations occur with a lower probability for amino acid sites than for nucleotide sites. Technically, nucleotide sequencing is less time-consuming than amino acid sequencing. However, once the coding regions of DNA are sequenced, the corresponding amino acid sequence can easily be inferred.

Proportion of Different Amino Acids and the Number of Amino Acid Substitutions

The study of the evolutionary change of proteins or polypeptides starts from the comparison of two or more amino acid sequences of a given polypeptide from different organisms. A simple quantity to measure the evolutionary divergence between a pair of amino acid sequences is the proportion (p) of different amino acids between the two sequences. This proportion is estimated by

$$\hat{p} = n_d/n, \qquad (4.1)$$

where n is the total number of amino acids compared and n_d is the number of different amino acids. If all amino acid sites are subject to substitution with an equal probability, n_d follows the binomial distribution. Therefore, the variance of $\hat{p}$ is given by

$$V(\hat{p}) = p(1-p)/n. \tag{4.2}$$

When p is small, it is approximately equal to the number of amino acid substitutions per site. When p is large, however, it is no longer a good measure of this number, because there might have been two or more amino acid substitutions at sites where amino acids are different between the two sequences. To estimate the number of amino acid substitutions from p, we need a mathematical model.

Poisson Process

A simple mathematical model that can be used for relating p to the expected number of amino acid substitutions per site is the Poisson process in probability theory. Let λ be the rate of amino acid substitution per year at a particular amino acid site and assume for simplicity that it is the same for all sites. This assumption does not necessarily hold in reality, but, as will be seen later, the error introduced by this assumption is small unless a very long evolutionary time is considered. The mean number of amino acid substitutions per site during a period of t years is then λt, and the probability of occurrence of r amino acid substitutions at a given site is given by the following Poisson distribution:

$$P_r(t) = e^{-\lambda t}(\lambda t)^r/r!. \tag{4.3}$$

Therefore, the probability that no change has occurred at a given site is $P_0(t) = e^{-\lambda t}$. Thus, if the number of amino acids in a polypeptide is n, the expected number of unchanged amino acids is $ne^{-\lambda t}$.

In reality, we generally do not know the amino acid sequence for an ancestral species, so that (4.3) is not applicable. The number of amino acid substitutions is usually computed by comparing homologous polypeptides from two different organisms that diverged t years ago. Since the probability that no amino acid substitution occurs during t years is $e^{-\lambda t}$, the probability (q) that neither of the homologous sites of the two polypeptides undergoes substitution is

$$q = e^{-2\lambda t}. \tag{4.4}$$

This probability can be estimated by $\hat{q} \equiv 1 - \hat{p} = n_i/n$, where n_i is the number of identical amino acids between the two polypeptides. The equation $q = e^{-2\lambda t}$ is approximate because backward mutations and parallel mutations (the same mutations occurring at the homologous amino acid sites in two different evolutionary lines) are not taken into account. But the effects of these mutations are generally very small unless a long evolutionary time is considered.

If we use (4.4), the total number of amino acid substitutions per site for the two polypeptides ($d = 2\lambda t$) can be estimated by

$$\hat{d} = -\log_e \hat{q}. \tag{4.5}$$

Therefore, if we know t, λ is estimated by $\hat{\lambda} = \hat{d}/(2t)$. On the other hand, if we know λ, t is estimated by $\hat{t} = \hat{d}/(2\lambda)$. The large-sample variance of $\hat{d}$ is

$$V(\hat{d}) = \left[\frac{dd}{dq}\right]^2 V(q)$$

$$= (1-q)/(qn), \tag{4.6}$$

since $V(\hat{q}) = q(1-q)/n$ (see Elandt-Johnson 1970). Obviously, the variances of $\hat{\lambda}$ and $\hat{t}$ are given by $V(\hat{d})/(2t)^2$ and $V(\hat{d})/(2\lambda)^2$, respectively.

It should be noted that if we knew the numbers of amino acid substitutions for all amino acid sites, the variance of the number of amino acid substitutions per site would have been $2\lambda t/n$ under the Poisson process. (The variance of a Poisson variable is equal to the mean.) In practice, it is impossible to know these numbers, so we must estimate d by equation (4.5). Since (4.5) is based on incomplete information on amino acid substitutions, (4.6) gives a variance larger than $2\lambda t/n$.

In the above formulation, we assumed that the rate of amino acid substitution is the same for all amino acid sites. This assumption usually does not hold, since the rate is higher at functionally less important sites than at functionally more important sites (see next section). Indeed, Fitch and Margoliash (1967b) and Uzzell and Corbin (1971) have shown that the distribution of the number of amino acid substitutions has a larger variance than the Poisson variance. However, equation (4.5) is quite robust and approximately holds even if the rate varies considerably from site to site. This can be seen by considering the extreme case, where proportion a of amino acid sites is invariable and proportion $1 - a$ is subject to the Poisson change. The expected proportion of identical amino acids for this case is given by

$$q = a + (1-a)e^{-2\lambda t}.$$

Strictly speaking, therefore, $d \equiv -\log_e q$ is not linear with evolutionary time. However, as can be seen from figure 4.1, d is approximately linear when $2\lambda t \leq 1$. It should be noted that when $2\lambda t \ll 1$, $e^{-2\lambda t} \simeq 1 - 2\lambda t$. Therefore, $d \simeq -\log_e[1 - 2(1-a)\lambda t] \simeq 2(1-a)\lambda t$. Since $(1-a)\lambda$ is the average rate $(\bar{\lambda})$ of amino acid substitution for all amino acid sites, d can be written as $2\bar{\lambda}t$. Namely, when λ varies with amino acid, $d = 2\bar{\lambda}t$ still holds unless $2\bar{\lambda}t$ is large.

Of course, if a is large and $2\lambda t > 1$, the relationship between d and q is no longer linear. In this case, a curve similar to figure 5.3 may be obtained. Recently, Lee et al. (1985) obtained a curvilinear relationship between evolutionary time and d for superoxide dimutase for yeast, fruitfly, horse, cow, and man. It is possible that this is caused by variation in the rate of amino acid substitution among different sites.

In the above formulation, we assumed that the two amino acid se-

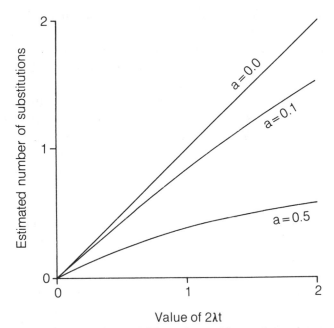

Figure 4.1.   Effects of invariable amino acids on the estimate of amino acid substitutions (d). a = proportion of invariable amino acids in a polypeptide.

quences to be compared have the same number of amino acids and that the divergence between them occurred solely by amino acid substitution. When the two sequences are distantly related, however, insertions and deletions are often involved. In this case, we must first identify the locations of insertions and deletions. When the number of insertions or deletions involved is small, as in the case of the following example, this can be done relatively easily. When the number of insertions and deletions is relatively large, however, the alignment of amino acid sequences is quite troublesome. Since this problem is usually more serious with DNA sequences and the problem is nearly identical for the two types of data, I shall discuss this problem in the next chapter.

EXAMPLE

Figure 4.2 shows the amino acid sequences of hemoglobin α chains from the human, horse, bovine, and carp. Here, amino acids are represented by one-letter codes rather than by usual three-letter codes (see table 3.2). The three mammalian hemoglobins consist of 141 amino acids, whereas the carp hemoglobin has 142 amino acids. Comparison of these sequences suggests that deletions or insertions occurred at three different positions after the divergence between fish and mammals. If we ignore these deletions/insertions, the proportion of different amino acids

```
Human    VLSPADKTNVKAAWGKVGAHAGEYGAEALERMFLSFPTTKTYFPHF-DLSHGSAQVKGHG
Horse    VLSAADKTNVKAAWSKVGGHAGEYGAEALERMFLGFPTTKTYFPHF-DLSHGSAQVKAHG
Bovine   VLSAADKGNVKAAWGKVGGHAAEYGAEALERMFLSFPTTKTYFPHF-DLSHGSAQVKGHG
Carp     SLSDKDKAAVKIAWAKISPKADDIGAEALGRMLTVYPQTKTYFAHWADLSPGSGPVK-HG
```

```
Human    KKVA-DALTNAVAHVDDMPNALSALSDLHAHKLRVDPVNFKLLSHCLLVTLAAHLPAEFT
Horse    KKVA-DGLTLAVGHLDDLPGALSDLSNLHAHKLRVDPVNFKLLSHCLLSTLAVHLPNDFT
Bovine   AKVA-AALTKAVEHLDDLPGALSELSDLHAHKLRVDPVNFKLLSHSLLVTLASHLPSDFT
Carp     KKVIMGAVGDAVSKIDDLVGGLASLSELHASKLRVDPANFKILANHIVVGIMFYLPGDFP
```

```
Human    PAVHASLDKFLASVSTVLTSKYR
Horse    PAVHASLDKFLSSVSTVLTSKYR
Bovine   PAVHASLDKFLANVSTVLTSKYR
Carp     PEVHMSVDKFFQNLALALSEKYR
```

Figure 4.2.   Amino acid sequences in the α chains of hemoglobins from four vertebrate species. Amino acids are expressed in terms of one-letter codes. The hyphens indicate the positions of deletions or insertions.

($\hat{p}$) and the estimate of the number of amino acid substitutions for each pair of organisms becomes as given in table 4.1. For example, in the case of the human and carp $\hat{p} = 68/140 = 0.486$, and $\hat{d} = -\log_e(1 - \hat{p}) = 0.666$. On the other hand, the variance of $\hat{d}$ becomes $V(\hat{d}) = 0.486/(0.514 \times 140) = 0.006754$, the standard error being 0.082.

Table 4.1 indicates that $\hat{d}$ is nearly the same for the three pairs of mammalian species, whereas the $\hat{d}$ values between the carp and the mammalian species are considerably larger. This observation is in agreement with the view that the number of amino acid substitutions is roughly proportional to evolutionary time, since the human, horse, and bovine diverged about 75 MY ago, whereas the carp (bony fish) and mammals diverged about 400 MY ago (figure 2.2). The average $\hat{d}$ for the three pairs of mammalian species is 0.135, whereas the average for the pairs of the carp and the three mammalian species is 0.642, the latter being about five times the former. This ratio is close to the ratio of the corresponding divergence times.

Amino Acid Substitution Matrix

The above Poisson process method for estimating the number of amino acid substitutions gives quite an accurate estimate as long as the evolutionary time considered is relatively short. When the amino acid sequences from distantly related organisms are compared, however, the effects of backward and parallel mutations cannot be neglected, and the Poisson process method is expected to give an underestimate. Unequal

Table 4.1 Numbers of amino acid differences (above the diagonal) between hemoglobin α chains from the human, horse, bovine, and carp. Deletions and insertions were excluded from the computation, the total number of amino acids used being 140. The figures in parentheses are the proportions of different amino acids. The values given below the diagonal are estimates of the average number of amino acid substitutions per site between two species $(\hat{d})$.

	Human	Horse	Bovine	Carp
Human		18(0.129)	17(0.121)	68(0.486)
Horse	0.138 ± 0.032		18(0.129)	66(0.471)
Bovine	0.129 ± 0.031	0.138 ± 0.032		65(0.464)
Carp	0.666 ± 0.082	0.637 ± 0.080	0.624 ± 0.079	

rates of substitution at different amino acid sites would also contribute to the inaccuracy of the estimate obtained. To take care of these problems, Dayhoff et al. (1978) proposed another method. In this method, the amino acid substitution matrix for a relatively short period of time is considered, and the relationship between the proportion of identical amino acids and the number of amino acid substitutions is derived empirically. The amino acid substitution matrix Dayhoff et al. used was derived from empirical data for many proteins such as hemoglobins, cytochrome c, fibrinopeptides, etc. They first constructed an evolutionary tree for closely related amino acid sequences and then inferred the relative frequencies of substitutions among various amino acids. From these data, they constructed an empirical amino acid substitution matrix (**M**) for the twenty amino acids (see figure 82 of Dayhoff et al. 1978).

An element (m_{ij}) of this substitution matrix gives the probability that the amino acid in column i will be replaced by the amino acid in row j during one evolutionary time unit. The time unit used in the matrix is the time during which on the average one amino acid substitution per 100 residues occurs. Dayhoff et al. (1978) measured the number of amino acid substitutions in terms of "PAM," which is an acronym for *accepted point mutations* and represents one amino acid substitution per 100 residues. Therefore, their substitution matrix gives the amino acid substitution probabilities for one PAM.

The amino acid substitution matrix **M** can be used for predicting amino acid changes for any evolutionary time if we know the initial amino acid frequencies. Let g_0 be a column vector of the relative frequencies of 20 amino acids for a polypeptide at time 0. The amino acid frequencies at time t or for t PAMs are then given by

$$g_t = M_t g_0, \tag{4.7}$$

where $M_t = M^t$. Here, we note that element $m_{t(ij)}$ of matrix M_t gives the probability that the amino acid in column i at time 0 will be replaced by the amino acid in row j at time t. In particular, $m_{t(ii)}$ represents the probability that the ith amino acid at time t is the same as the original one. This latter probability can be used for relating the proportion of different amino acids between homologous polypeptides (p) to the number of amino acid substitutions per site (d_D). Namely, p is given by

$$p = 1 - \Sigma_i g_i m_{2t(ii)}, \tag{4.8}$$

where g_i is the initial frequency of the ith amino acid in the polypeptide under investigation. Here, we use $m_{2t(ii)}$ instead of $m_{t(ii)}$ because we are considering a pair of polypeptides which diverged t time units ago. Since $d_D = 0.01 \times t \ (= 0.01$ PAMs) and $m_{2t(ii)}$ can be obtained from $\mathbf{M}_{2t}$, p can be related to d_D.

In practice, $\mathbf{g}_0$ may vary from polypeptide to polypeptide or in a given polypeptide from time to time. To avoid this difficulty, Dayhoff et al. used the amino acid frequencies averaged over many different proteins. This does not take into account the specificity of each polypeptide, but it certainly makes the method applicable for many different proteins. Furthermore, if we note that many different proteins have rather similar amino acid frequencies (Dayhoff et al. 1976), this procedure seems to be acceptable for obtaining a rough estimate of the number of amino acid substitutions.

Using the above method, Dayhoff et al. (1978) derived the relationship between p and d_D. This relationship is given in table 4.2. This table also includes the value of $d = -\log_e(1-p)$ in (4.5). It is clear that when p is smaller than 0.2, d_D is very close to d. However, as p increases, the difference between d_D and d gradually increases. Kimura (1983a) produced an approximate formula for d_D. It is given by

$$d_K = -\log_e(1-p-0.2p^2). \tag{4.9}$$

This formula gives a good approximation as long as $p \leq 0.7$ (table 4.2). Since p rarely exceeds 0.7 in actual data, the above formula is very useful.

Table 4.2   Relationships of the proportion of different amino acids between two homologous polypeptides (p) to $d_D = 0.01$ PAMs, d in equation (4.5), and d_K in equation (4.9).

p	d_D	d_K	d	p	d_D	d_K	d
.01	.01	.01	.01	.45	.67	.67	.60
.05	.05	.05	.05	.50	.80	.80	.69
.10	.11	.11	.11	.55	.94	.94	.80
.15	.17	.17	.16	.60	1.12	1.11	.92
.20	.23	.23	.22	.65	1.33	1.33	1.05
.25	.30	.30	.29	.70	1.59	1.60	1.20
.30	.38	.38	.36	.75	1.95	1.98	1.39
.35	.47	.47	.43	.80	2.46	2.63	1.61
.40	.56	.57	.51				

Pattern of Amino Acid Substitution

The pattern of amino acid substitution in evolution was first studied by Zuckerkandl and Pauling (1962) and Margoliash (1963) in hemoglobin and cytochrome c, respectively. In these studies, the approximate constancy of the rate of amino acid substitution was apparent. Later, many authors have studied detailed aspects of amino acid substitution. In this section, I will discuss a few important aspects of amino acid substitution.

Rates of Amino Acid Substitution

In the past two decades, the relationship between the number of amino acid substitutions and evolutionary time has been studied for a large number of proteins. In most of these proteins, the number of amino acid substitutions has been shown to increase with increasing evolutionary time. Table 4.3 shows the proportions of amino acid differences between cytochrome c sequences obtained from various groups of organisms. It is clear that the cytochromes c from closely related organisms are more similar than those from distantly related species. The similarity is such that the difference between any two organisms depends almost entirely on the time after divergence. For example, the difference between bacterial cytochrome c_2 (this is homologous to cytochrome c in eukaryotes) and cytochrome c of any other higher organism, plant or animal, is virtually the same (62–72%). Similarly, the cytochrome c in the fungus and yeast group is almost equally related to any other higher organism, the amino acid difference being 41 to 50 percent.

Using equation (4.5), Dickerson (1971) studied the relationship between the number of amino acid differences (d) and divergence time (t) for cytochrome c, hemoglobins, and fibrinopeptides. The results obtained are presented in figure 4.3. The d value increases almost linearly with t in all three groups of polypeptides. A similar linear relationship has been obtained in many other proteins, though the linearity is not always as good as in the case of the three polypeptides in figure 4.3. However, there is enormous variation in the rate of substitution among different polypeptides, as is clear from table 4.4. The highest rate of substitution so far observed is that (9×10^{-9} per site per year) for fibrinopeptides, and this is 900 times higher than the lowest rate observed for histone H4. The mean and median of the rate are 1.2×10^{-9} and 0.74×10^{-9}, respectively.

Table 4.3 Amino acid differences (%) in cytochrome c and c_2 between different organisms. The number of positions compared varies with the pair of organisms. All positions are used in computation except those in which both sequences have a gap. Cytochrome c_2 in bacteria is known to be homologous to cytochrome c in eukaryotes. From Dayhoff (1972).

	Human	Pig	Horse	Chicken	Turtle	Bullfrog	Tuna	Carp	Lamprey	Fruit fly	Screw-worm	Silkworm	Sesame	Sunflower	Wheat	C. krusei	Yeast	N. crassa	R. rubrum
Human	0	10	12	13	14	17	20	17	19	27	25	29	35	38	38	46	41	44	65
Pig, bovine, sheep	10	0	3	9	9	11	16	11	13	22	20	25	38	40	40	45	41	43	64
Horse	12	3	0	11	11	13	18	13	15	22	20	27	39	41	41	46	42	43	64
Chicken, turkey	13	9	11	0	8	11	16	14	17	23	21	26	40	41	41	45	41	44	64
Snapping turtle	14	9	11	8	0	10	17	13	18	22	22	26	38	39	41	47	44	45	64
Bullfrog	17	11	13	11	10	0	14	13	20	20	20	27	41	42	43	46	43	45	65
Tuna fish	20	16	18	16	17	14	0	8	18	23	22	30	42	43	44	43	43	45	65
Carp	17	11	13	14	13	13	8	0	12	21	20	25	40	41	42	45	42	43	64
Lamprey	19	13	15	17	18	20	18	12	0	27	26	30	44	44	46	50	45	47	66
Fruit fly	27	22	22	23	22	20	23	21	27	0	2	14	42	41	42	43	42	38	65
Screw-worm fly	25	20	20	21	22	20	22	20	26	2	0	13	41	40	40	43	42	38	64
Silkworm moth	29	25	27	26	26	27	30	25	30	14	13	0	39	40	40	43	44	44	65
Sesame	35	38	39	40	38	41	42	40	44	42	41	39	0	10	13	47	44	48	65
Sunflower	38	40	41	41	39	42	43	41	44	41	40	40	10	0	13	47	43	49	67
Wheat	38	40	41	41	41	43	44	42	46	42	40	40	13	13	0	45	42	48	66
Candida krusei	46	45	46	45	47	46	43	45	50	43	43	43	47	47	45	0	25	39	72
Baker's yeast	41	41	42	41	44	43	43	42	45	42	42	44	44	43	42	25	0	38	69
Neurospora crassa	44	43	43	44	45	45	45	43	47	38	38	44	48	49	48	39	38	0	69
Rhodospirillum rubrum c_2	65	64	64	64	64	65	65	64	66	65	64	65	65	67	66	72	69	69	0

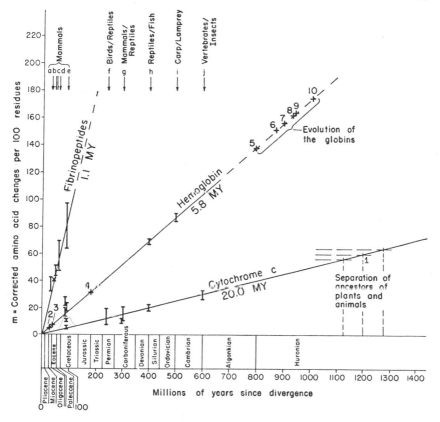

Figure 4.3. Rates of amino acid substitution in fibrinopeptides, hemoglobin, and cytochrome c. Comparisons for which no adequate time coordinate is available are indicated by numbered crosses. Point 1 represents a date of $1,200 \pm 75$ MY (million years) for the separation of plants and animals, based on a linear extrapolation of the cytochrome c curve. Points 2–10 refer to events in the evolution of the globin family. The δ/β separation is at point 3, γ/β is at 4, and α/β is at 500 MY (carp/lamprey). From Dickerson (1971).

The rate of amino acid substitution is usually measured by the number of substitutions per amino acid site per year (λ), as given in table 4.3. Some molecular biologists (e.g., Dickerson 1971), however, have used the *unit evolutionary time* (T_u). This is the average time required for one substitution per 100 amino acid sites. It is related to λ by

$$T_u = 1/(100\lambda). \tag{4.10}$$

Table 4.4   Rates of amino acid substitutions per amino acid site per 10^9 years ($\lambda \times 10^9$) in various proteins. Modified from Dayhoff (1978).

Protein	Rate	Protein	Rate
Fibrinopeptides	9.0	Thyrotropin beta chain	0.74
Growth hormone	3.7	Parathyrin	0.73
Ig kappa chain C region	3.7	Parvalbumin	0.70
Kappa casein	3.3	Protease inhibitors, BP1 type	0.62
Ig gamma chain C region	3.1	Trypsin	0.59
Lutropin beta chain	3.0	Melanotropin beta	0.56
Ig lambda chain C region	2.7	Alpha crystallin A chain	0.50
Complement C3a anaphylatoxin	2.7	Endorphin	0.48
Lactalbumin	2.7	Cytochrome b_5	0.45
Epidermal growth factor	2.6	Insulin (exc. guinea pig and coypu)	0.44
Somatotropin	2.5	Calcitonin	0.43
Pancreatic ribonuclease	2.1	Neurophysin 2	0.36
Lipotropin beta	2.1	Plastocyanin	0.35
Haptoglobin alpha chain	2.0	Lactate dehydrogenase	0.34
Serum albumin	1.9	Adenylate kinase	0.32
Phospholipase A_2	1.9	Triosephosphate isomerase	0.28
Protease inhibitor, PST1 type	1.8	Vasoactive intestinal peptide	0.26
Prolactin	1.7	Corticotropin	0.25
Pancreatic hormone	1.7	Glyceraldehyde 3-PO_4 dehydrogenase	0.22
Carbonic anhydrase C	1.6	Cytochrome c	0.22
Lutropin alpha chain	1.6	Plant ferredoxin	0.19
Hemoglobin alpha chain	1.2	Collagen (exc. nonrepetitive ends)	0.17
Hemoglobin beta chain	1.2	Troponin C, skeletal muscle	0.15
Lipid-binding protein A-II	1.0	Alpha crystallin B chain	0.15
Gastrin	0.98	Glucagon	0.12
Animal lysozyme	0.98	Glutamate dehydrogenase	0.09
Myoglobin	0.89	Histone H2B	0.09
Amyloid AA	0.87	Histone H2A	0.05
Nerve growth factor	0.85	Histone H3	0.014
Acid proteases	0.84	Ubiquitin	0.010
Myelin basic protein	0.74	Histone H4	0.010

Thus, T_u for fibrinopeptides is 1.1×10^6 years, whereas T_u for histone H4 is 10^9 years.

Differences Among Proteins

Why is the rate of amino acid substitution so different among different proteins? The answer to this question seems to be that the functional requirement of each protein determines the rate (Zuckerkandl and Pauling 1965; King and Jukes 1969; Dickerson 1971). For example, fibrinopeptides do not seem to have any particular function after they are cut out of fibrinogen when the latter is converted to fibrin for blood clotting. Thus, virtually any amino acid can be replaced by any other amino acid. That is, almost all mutations occurring in fibrinopeptides seem to be selectively neutral or nearly neutral. The rate of amino acid substitution is, therefore, expected to be close to the mutation rate per locus (chapter 13). The apparently functionless parts of ribonuclease also show a rate of amino acid substitution similar to that of fibrinopeptides (Barnard et al. 1972).

By contrast, cytochrome c seems to require a rather rigid arrangement of amino acids for the protein to function normally (Dickerson 1971). The polypeptide of this protein forms a shell, inside which the heme group is contained with one edge of the heme being exposed outside. Most of the interior amino acids are hydrophobic and apparently cannot be replaced by hydrophilic amino acids. The heme is attached covalently to the protein through cysteines at positions 14 and 17. The amino acids at the surface of this protein are less restrictive but still must form a certain structure to interact with cytochrome oxidase and reductase, both of which are macromolecules much larger than cytochrome c itself. This functional requirement apparently rejects many amino acid changes in this protein, and only at a limited number of amino acid sites are mutational changes freely accepted.

The functional constraint of hemoglobin is intermediate between fibrinopeptides and cytochrome c. This protein also contains the heme group, and its interior amino acids do not easily accept mutational changes. In hemoglobin α chain, there are 19 amino acid sites that are involved in the heme pocket. Replacement of amino acids at these sites is known to cause abnormal function of the hemoglobin molecule in man (Perutz and Lehman 1968). The function of hemoglobin is to bind to O_2 in the lung and interact with CO_2 in the tissue, and the surface of the molecule has no essential function except for holding other important amino acids. Thus, the amino acids at the surface can easily be replaced by other amino acids. Kimura and Ohta (1973b) have shown that the rate of amino acid substitution at the surface is about ten times higher than that at the heme pocket.

As mentioned earlier, histone H4 is a highly conserved protein. There are only two amino acid differences in the sequence of 105 amino acids between calf and pea. If we assume that plants and animals diverged one billion years ago, the rate of amino acid substitution is computed to be

1×10^{-11} per site per year. This is about $1/100$ of the rate of hemoglobin and about $1/40$ of that for cytochrome c. This extremely slow rate of evolution in histone H4 is believed to be due to the important function this protein plays in controlling the expression of genetic information by binding to DNA in the nucleus. An equally slow rate of evolution has been observed for ubiquitin, which also seems to be important in controlling the expression of genetic information (P. M. Sharp and W.-H. Li, personal communication). In addition to these proteins, transfer and ribosomal RNAs are known to evolve very slowly. Since these RNAs play an important role in protein synthesis, many nucleotide substitutions seem to result in deleterious effects.

Some caution, however, should be exercised in the interpretation of the relationship between functional constraint and the rate of amino acid substitution. Because of the above argument, whenever we find a protein with a slow rate of amino acid substitution, we tend to believe that the protein has an important biological function even if its function is not known. Graur (1985) showed that there is a significant negative correlation between the proportion of glycines in proteins and the rate of amino acid substitution, irrespective of the protein function. This correlation is apparently generated by the fact that glycine is smallest among the twenty amino acids and the replacement of this amino acid by another generally impairs the function of the protein.

Nevertheless, a large proportion of the variation in the rate of amino acid substitution can be explained by the differences in functional constraint among proteins. Kimura (1983a) takes this finding as support for his neutral mutation theory. He has hypothesized that almost all amino acid substitutions in fibrinopeptides occur by random fixation of neutral or nearly neutral mutations, and thus the substitution rate per site (λ) is equal to the mutation rate (μ). In most other proteins, however, a certain amount of purifying selection operates because of functional constraints. Therefore, λ may be expressed as

$$\lambda = (1-f)\mu, \tag{4.11}$$

where f is the fraction of deleterious mutations. It would be interesting to test this hypothesis by examining the rates of nucleotide substitution at the three nucleotide positions of codons of the fibrinopeptide gene. If Kimura's hypothesis is correct, we would expect that all three positions show essentially the same rate.

At this point, one might argue that variation in the substitution rate among different proteins is caused by the differences in positive (advantageous) selection rather than negative (purifying) selection. In this argument, however, we must assume that the rate of occurrence of advantageous mutations is higher in rapidly evolving polypeptides such as fibrinopeptides than in slowly evolving polypeptides such as cytochrome c. This assumption does not seem to be reasonable.

Of course, this does not mean that there are no advantageous mutations at the amino acid level. There must be some. Otherwise, adaptive evolution can not occur. However, a relatively small number of advantageous mutations seem to be sufficient for producing an adaptive change of a molecule. For example, crocodilian hemoglobin has lost its old function (the binding of organic phosphate, chloride, and carbamino CO_2) and gained a new function (bicarbonate binding). This functional change represents an adaptive response to the blood acidity that occurs during the prolonged stay of crocodiles under water and can be explained by 5 amino acid substitutions (Perutz et al. 1981). This is a small proportion of the total number of amino acid substitutions (123) between crocodiles and humans, who have not experienced such a change. In general, most amino acid substitutions in hemoglobins do not appear to be related to any significant functional change (Perutz 1983). The functional change of stomach lysozyme of ruminants can also be explained by a small proportion of amino acid changes (Jolles et al. 1984).

In bacteria, there are many examples in which an adaptive change in enzymes (change in substrate specificity) has occurred by one or a few amino acid substitutions (e.g., Clarke 1984). For example, β-lactam antibiotics kill bacteria by inactivating a set of penicillin-binding proteins (PBPs) that are essential for cell division. Some mutants of *E. coli* are resistant to these antibiotics because of the reduction in affinity between antibiotics and PBPs. Hedge and Spratt (1985) have shown that this reduction in affinity is caused by one to four amino acid substitutions in the active center of a PBP and that the majority of other amino acid substitutions do not affect antibiotic susceptibility.

Is the Rate of Amino Acid Substitution Constant in a Given Protein?

Although the pattern of amino acid substitution shows clocklike behavior, the accuracy of the molecular clock has been controversial. Some workers discovered cases in which the clock apparently failed. This subject has been reviewed by Wilson et al. (1977a), Goodman et al. (1982), Dickerson and Geis (1983), and Kimura (1983a), and the reader may refer to them for the details. Here, I present only a brief summary of the subject.

It should first be remembered that the clock is empirical and its theoretical basis is not well founded, though the neutral theory can provide the basis once this is firmly established. Second, the molecular clock is stochastic and has a large variance when the number of amino acids examined is small. Third, there are many exceptions even if the stochastic factor is considered. A notable exception is guinea pig insulin, whose rate (5.3×10^{-9}) of amino acid substitution is more than ten times higher than that (0.33×10^{-9}) of other organisms (King and Jukes 1969). This high rate seems to have been caused by the relaxation of purifying selection, which resulted from a change in the tertiary structure of guinea pig insulin (Kimura and Ohta 1974). Fourth, in the computation of absolute rates of amino acid substitution, it is necessary to have geological dates of divergence times, but these dates are often inaccurate (Wilson et al. 1977a; Joysey 1981). Therefore, great caution is required in testing the constancy of absolute rates of substitution.

A simple method of testing the molecular clock is to estimate the number of amino acid or codon substitutions for each branch of the evolutionary tree for a group of organisms and relate the estimate to the evolutionary time. It is difficult to know the exact number of substitutions for each branch, but the number can be estimated by parsimony methods (e.g., Goodman et al. 1974) or some other technique (chapter 11). Romero-Herrera et al. (1978) used this method to study the rate of codon substitution in myoglobin, using amino acid sequence data from 19 vertebrate species (see also Joysey 1981). From information on fossils and anatomical differences, they first constructed a phylogenetic tree for these organisms. They then estimated the number of codon substitutions (n_c) for all branches and related them to the evolutionary times known from the fossil records. They considered a time span of 80 MY with an interval of 10 MY. There were several independent estimates of codon substitutions at each of the 8 evolutionary times considered. Their results indicate that the average number of codon substitutions increases almost linearly with evolutionary time but that the variation around the linear relationship is quite large. Because of this large variation, the number of codon substitutions does not always give the correct estimate of divergence time when it is used as a clock. For example, at the evolutionary time of 80 MY there were 14 observations of n_c, but the number varied from 11 to 34, the mean being about 22. Therefore, the largest value of n_c gives a time estimate which is about three times greater than the smallest. From this observation, Joysey (1981) concluded that myoglobin cannot be used as a molecular clock.

It should be noted, however, that myoglobin is a relatively small polypeptide (about 150 amino acids long) and that stochastic errors alone produce a large variance relative to the mean when the number of substitutions is small. Under the assumption of the Poisson process, the expected variance of n_c is equal to the mean. Therefore, the expected standard error for a mean of 22 is $\sqrt{(22)} = 4.69$. Thus, the expected range (confidence interval) of n_c with a probability of 95 percent is 12.62 to 31.38 (mean $\pm$ 2 standard errors). This covers all observations except the two extreme ones (11 and 34). Thus, only two of the 14 observations are significantly different from the expected number. Indeed, similar computations suggest that a large portion of the variance of n_c observed by Romero-Herrera et al. is due to stochastic errors. It should also be noted that Romero-Herrera et al.'s estimates of n_c are subject to estimation errors as well as to stochastic errors because they used a parsimony method (see chapter 11).

A statistical study of variation in the rate of amino acid substitution was conducted by Ohta and Kimura (1971a). They examined the rates of amino acid substitution in hemoglobins and cytochrome c in various vertebrate lines and showed that the variance of the rate is 1.9 to 4.2 times higher than that expected under the Poisson process. This suggests that the rate is not strictly constant. Their test, however, was not very accurate because the correlation between different pairs of distances (d's) was ignored and some of the divergence times used were unreliable.

Kimura (1983a) devised an improved method for testing rate constancy. He considered the case where all species diverged at the same time. Suppose that s species are studied, and let x_i be the number of amino acid substitutions that occurred in the ith species after separation from the common ancestor. We assume that the number of amino acid substitutions between the ith and jth species is $d_{ij} = x_i + x_j$. The mean and variance of x_i are then given by $\bar{x} = \bar{d}/2$ and $V_x = (s+1)V_d/[2(s-1)]$, respectively, where $\bar{d}$ and V_d are the mean and variance of d_{ij} among $s(s-1)/2$ comparisons. Since $\bar{d}$ and V_d are observable quantities, we can estimate $\bar{x}$ and V_x. V_x can then be compared with the expected variance

$\sigma_x^2 = \bar{x}$ under the Poisson process. Application of this method to data on hemoglobin α and β chains, cytochrome c, myoglobin, and pancreatic ribonuclease from different orders of mammals (e.g., human, mouse, rabbit, dog, horse, bovine, etc.) has shown that V_x is significantly larger than σ_x^2 in two polypeptides out of the five examined [see Gillespie (1984) for a somewhat different type of analysis]. It should be noted, however, that even this method is not very accurate because the different species used (mammalian orders in the present case) may not have diverged at the same time.

The problem of inaccurate dating of divergence times can be avoided if we use only three species. Consider the three species A, B, and C in figure 4.4, and assume that A and B are known to be more closely related to each other than to C. We denote by d_{AB}, d_{AC}, and d_{BC} the numbers of amino acids substitutions between A and B, A and C, and B and C, respectively. We can then write $d_{AB} = x + y$, $d_{AC} = x + z$, $d_{BC} = y + z$, where x, y, and z are the numbers of substitutions for the branches given in figure 4.4. Therefore, if we know d_{AB}, d_{AC}, and d_{BC}, we can estimate x, y, and z. That is,

$$x = (d_{AB} + d_{AC} - d_{BC})/2 \qquad (4.12a)$$

$$y = (d_{AB} - d_{AC} + d_{BC})/2, \qquad (4.12b)$$

$$z = (-d_{AB} + d_{AC} + d_{BC})/2. \qquad (4.12c)$$

From figure 4.4, it is clear that if the rate of substitution is constant, the expectations of x and y are equal to each other. This method has been used by Sarich and Wilson (1973) and their colleagues to test whether or not the numbers of amino acid substitutions for the human and ape

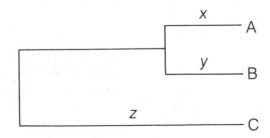

Figure 4.4.   Phylogenetic tree for three species. x y, and z are the numbers of amino acid substitutions.

lineage and the Old World monkey lineage (e.g., baboon, macaque) are the same. Using New World monkeys and other placental mammals as the external reference species (C), these authors examined data on hemoglobins α and β, fibrinopeptides A and B, carbonic anhydrase, cytochrome c, myoglobin, and lysozyme c (see Wilson et al. 1977a). However, they could not find a significant difference between the two groups of organisms. A similar test was also conducted for immunological distance data, which will be discussed in the next section, but the result obtained was the same.

A more complex statistical test of the molecular clock was conducted by Langley and Fitch (1974). In this test, no knowledge of evolutionary times is necessary, but we must know the branching pattern (topology) of the phylogenetic tree of the organisms used and the number of amino acid substitutions for each branch. For a given topology and a given set of observed numbers of amino acid substitutions for all branches, the expected branch lengths (expected numbers of amino acid substitutions) are estimated by using the maximum likelihood method, and the heterogeneity of the rate among different branches is tested with a χ^2 test (chapter 11). Langley and Fitch applied this method to data for hemoglobins α and β, cytochrome c, and fibrinopeptide A from 18 vertebrate species, showing that the rate heterogeneity is statistically significant.

Uzzell and Corbin (1971) studied the distribution of the number of amino acid substitutions in cytochrome c for a group of vertebrate species and showed that observed data fit the negative binominal distribution better than the Poisson distribution, the variance of the number of substitutions being about two times greater than the Poisson variance. This observation is in agreement with Langley and Fitch's results but suggests that the rate of substitution (λ) varies with amino acid site, following the gamma distribution [equation (9.57)]. Furthermore, in a statistical analysis of amino acid substitution data, Gillespie (1984, 1986) has suggested that the substitution rate varies randomly from time to time in each evolutionary lineage. If this is the case, the molecular clock may become episodic, with a cluster of substitutions being separated by periods with a few substitutions.

When extensive data on amino acid sequences are available and the times of divergence for the species are known, one can directly examine the relationship between evolutionary time and the number of amino acid substitutions. Figure 4.5 gives such a relationship for hemoglobins and myoglobins. The number of amino acid substitutions increases al-

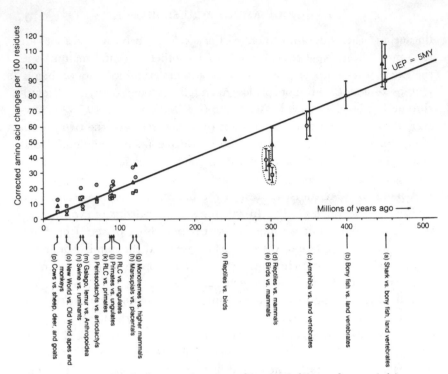

Figure 4.5.   Evolutionary changes of hemoglobin and myoglobin
chains. The number of amino acid substitutions (*d*) is plotted against
the divergence time obtained from paleontological records. Vertical
error bars for the older divergence points extend over ± 2 standard
deviations or to the 95 percent confidence level. Triangles represent
hemoglobin *α*, circles are hemoglobin *β*, and squares are myoglobin.
The straight line is the best linear regression fit to all data points. From
Dickerson and Geis (1983).

most linearly if we exclude the comparisons between birds and mammals. Particularly in the mammalian group, the linear relationship holds
very well, if we ignore the apparent random errors caused by stochastic
factors. However, careful examination suggests that the rate of amino
acid substitution for the period from 300 MY ago to 450 MY ago is
higher than that for mammalian species even if we exclude the bird-
mammal comparison. That is, the rate of substitution apparently deviates from constancy to some extent.

   It should be noted that not only the three globin chains but also
cytochrome c show a smaller number of amino acid substitutions in the
bird-mammalian comparison than expected from the linear relationship
(Dickerson and Geis 1983). At the present time, the mammalian line is

believed to have split from the reptile-bird line before these two groups
of organisms diverged (see figure 2.2). However, fossil records of birds
are notoriously poor, and the possibility that birds and mammals diverged after some group of reptiles (e.g., snakes) diverged cannot be
excluded (Dickerson and Geis 1983). If this is the case, the anomaly
concerning the amino acid substitution might disappear.

Studying the evolution of hemoglobins and myoglobins in vertebrates, Goodman et al. (1975) and Czelusniak et al. (1982) have concluded that the rate of amino acid substitution increased markedly following the gene duplication events separating hemoglobin and myoglobin
and α and β hemoglobins and that this increase was due to advantageous
mutations, improving the function of hemoglobin and myoglobin. Wilson et al. (1977a) and Kimura (1981a) have challenged this view and
contended that the increased substitution rate observed by Goodman and
his associates is probably caused by inaccurate dating of gene duplication
events. An increase in the rate of amino acid (nucleotide) substitution
was also observed by Li and Gojobori (1983) in duplicate globin genes
of the goat and the sheep. However, these authors concluded that it is
largely due to the relaxation of purifying selection (functional constraints) in duplicate genes rather than to advantageous mutations.
Nevertheless, it is likely that when a new gene is formed after gene
duplication its function is improved by nucleotide substitutions within
the gene (Dickerson 1971).

It is now clear that the rate of amino acid substitution is not strictly
constant but varies substantially with organism. Yet, if we consider a
long evolutionary time, it is approximately constant. As was noted by
Fitch (1976), the molecular clock does not run as regularly as the regular
clock or the isotope clock. It is quite "sloppy" but useful for obtaining
a rough idea of evolutionary time when fossil records are absent or unreliable. Note also that for constructing a phylogenetic tree for a group
of organisms the rate of amino acid substitution need not be constant.
As long as parallel or backward mutations are rare, one can reconstruct
a tree fairly easily (chapter 11).

Immunological Distance and Evolutionary Time

Although amino acid sequence data are useful for estimating evolutionary time and clarifying the genetic relationship of organisms, acqui-
sition of sequence data is time-consuming. As mentioned earlier, nucleotide sequencing is much easier than amino acid sequencing but, even

so, takes too much time. For taxonomic purposes, we need a simpler method. One such method is to use the intensity of immunological reaction between antigens and antisera prepared from different species. There are several methods for measuring the intensity of immunological reaction, but the simplest and most useful method seems to be that of quantitative microcomplement fixation of a purified protein, introduced by Sarich and Wilson (1966). The protein often used for this purpose is serum albumin (Champion et al. 1974). Briefly, the method is as follows. The antisera to be used are produced by immunization of rabbits with purified serum albumin from an organism of the group to be tested. The antisera produced strongly react with the albumin from this organism (homologous antigen) but less strongly with that from another species (heterologous antigen) for a given concentration of antisera. If the serum concentration is raised, however, the reaction with heterologous antigen increases to the level for homologous antigen. The degree of antigenic difference between two albumins is measured by the factor by which the antiserum concentration must be raised in order for a heterologous albumin to produce the same reaction as that with a homologous albumin. This factor is called the index of dissimilarity $(I.D.)$. The antigen-antibody reaction is measured by a method called quantitative complement fixation. Sarich and Wilson (1967) have shown that the logarithm of $I.D.$ is approximately linearly related to the time since divergence between the two organisms tested. They called $d_I = 100 \times \log_{10} I.D.$ the immunological distance. (They used the notation ID for d_I.) In many proteins (albumin, lysozyme, ribonuclease, etc.), this d_I is linearly related to the proportion of different amino acids between the two sequences compared (Prager and Wilson 1971; Benjamin et al. 1984). The reason why d_I should be a linear function of the proportion of different amino acids is not well understood. However, this empirical property of d_I is very useful for measuring the genetic distance between species, since the technique is much simpler than amino acid sequencing. According to Maxson and Wilson (1974), one unit of d_I corresponds to roughly one amino acid difference in albumin.

The relationship between d_I and the time (t) since divergence between two species may be written as

$$t = cd_I, \qquad (4.13)$$

where c is the proportionality constant and varies with the protein used and also to some extent with the group of organisms used. The c value that has been used for albumin is $(5.5-6) \times 10^5$ in mammals, reptiles, and frogs, and 1.9×10^6 in birds (Prager et al. 1974; Wilson et al. 1977a; Collier and O'Brien 1985). This suggests that the evolutionary rate of albumin is about three times slower in birds than in other organisms. However, the estimate of the evolutionary rate for birds may be erroneous, since the fossil records used are quite unreliable. If Alvarez et al.'s (1980) asteroid impact theory (chapter 2) is correct, most orders of birds might have evolved relatively recently (Wyles et al. 1983). If this is the case, the difference in the rate of albumin evolution between birds and the other groups of vertebrates might disappear.

For technical reasons, albumin cannot be used for microcomplement fixation in invertebrates. Beverley and Wilson (1984, 1985), therefore, used a larval hemolymph protein in the study of the phylogenetic relationship of various species of Drosophilidae and higher Diptera. Using information on fossils in amber, continental drift, island formation, etc., they estimated that the proportionality constant for this protein is $c = 8 \times 10^5$. This is of the same order of magnitude as that for albumin.

The estimate of t obtained by equation (4.13) is subject to four different types of errors. The first type is experimental error. According to Sarich and Wilson (1966), this error is generally less than 2 percent of the estimate, even if the estimate is relatively small. The second type of error arises when the antigen (protein) used is polymorphic in the species examined. When distantly related species are compared, however, this type of error is relatively small. The third type of error is generated by the fact that amino acid substitution in antigenic proteins occurs stochastically rather than deterministically. As mentioned earlier, this type of error will make the variance of d_I larger than the mean. Thus, it is much more significant than the first two types. The fourth type of error occurs because d_I is not strictly proportional to the number of amino acid substitutions (Champion et al. 1974). This type of error could be as important as the third type.

Nei (1977b) examined the (total) variance of d_I empirically by using d_I values obtained for various groups of organisms. He concluded that the variance of d_I is at least twice as large as the mean when the mean is small and that the ratio of the variance to the mean increases with increasing mean. Examining more extensive data for both albumin and larval hemolymph protein, Beverley and Wilson (1984) also showed that the variance of d_I is at least four times as large as the mean. Therefore, the immunological distance clock seems to be a little less accurate than the amino acid sequence clock.

Despite this inaccuracy, application of this method has brought about many interesting results in the study of evolution. Sarich and Wilson (1966, 1967) used this method to clarify the phylogenetic relationship of humans and apes. The results obtained were surprising. Unlike the prevailing view at that time, they showed that chimpanzees and gorillas are more closely related to humans, which belong to a different family (Hominidae), than to orangutans and gibbons, with which they form one family (Pongidae). Furthermore, their data indicated that humans, chimpanzees, and gorillas diverged about 5 MY ago, as mentioned earlier (chapter 2).

Another interesting finding is that different frog species belonging to the same genus often have large immunological distances similar to the values for different families or orders of mammals (Maxson et al. 1975; Post and Uzzell 1981; Maxson 1984). This suggests that frog species diverged a long time ago but retained similar morphological characters. Similarly, some pairs of *Drosophila* species apparently diverged a long

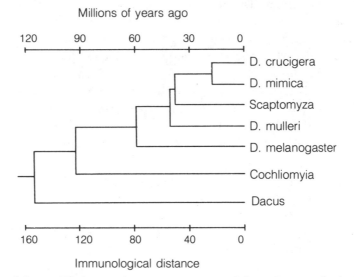

Figure 4.6.   Phylogenetic tree reconstructed from immunological distance data for seven fly lineages. *D. crucigera* and *D. mimica* are Hawaiian drosophilids, whereas *D. mulleri* and *D. melanogaster* are continental drosophilids. *Scaptomyza* is a fly genus closely related to *Drosophila*. *Drosophila, Cochliomyia,* and *Dacus* belong to different families. From Beverley and Wilson (1985).

time ago. Beverley and Wilson (1984, 1985) estimate that the species belonging to subgenus *Drosophila* (e.g., *D. virilis*, *D. mulleri*) diverged from the *D. melanogaster* species group about 60 MY ago, whereas Hawaiian *Drosophila* species diverged from subgenus *Drosophila* about 40 MY ago (figure 4.6).

The Hawaiian archipelago is known to be no older than 5–6 MY. Therefore, one might expect that all Hawaiian *Drosophila* species diverged relatively recently. Immunological data, however, suggest that the picture-winged group (e.g., *D. crucigera*) and the modified-mouth-parts group (e.g., *D. mimica*) of Hawaiian species diverged about 20 MY ago. Furthermore, the Hawaiian flies belonging to a different genus, *Scaptomyza*, are closer to Hawaiian drosophilids than to continental drosophilids, indicating that the Hawaiian fly fauna originated about 40 MY ago. This paradoxical finding is understandable if we consider the history of the Koko Seamount–Midway–Hawaii Archipelago, which has witnessed the sequential rise and erosion of many islands during the past 70 MY (see Beverley and Wilson 1985).

□ CHAPTER FIVE □

EVOLUTIONARY CHANGE OF NUCLEOTIDE SEQUENCES

In the study of evolution, DNA sequences are much more informative than protein sequences because a large part of DNA sequences are not encoded into protein sequences and there is degeneracy of the genetic code. Thus, the genetic differences in noncoding regions of DNA such as introns, flanking regions, and intergenic regions or silent nucleotide substitutions can only be studied by examining DNA sequences. Examination of DNA sequences also reveals detailed information on the mechanisms of deletion, insertion, unequal crossing over, transposition of genes, gene conversion, and even horizontal gene transfer.

The evolutionary change of DNA sequences is caused either by nucleotide substitution or by deletion and insertion (including inversion). In the following, I will consider these factors separately.

Nucleotide Substitution

When the number of nucleotide differences between two homologous DNA sequences is small, the number of nucleotide substitutions per nucleotide site is estimated by $\hat{p} = n_d/n$, where n_d and n are the number of different nucleotides between the two sequences and the total number of nucleotides compared, respectively. The statistical property of $\hat{p}$ is the same as that of equation (4.1), and its variance is given by (4.2). When $\hat{p}$ is appreciably large, however, this gives an underestimate of the number of nucleotide substitutions. To avoid this underestimation, various statistical methods have been proposed.

Jukes and Cantor's Method

A simple method of estimating the number of nucleotide substitutions was presented by Jukes and Cantor (1969). This method depends on the following mathematical model. We assume that nucleotide substitution occurs at any nucleotide site with equal probability and that at a given site a nucleotide changes to one of the three remaining nucleotides with a rate of λ per year. Consider two nucleotide sequences (X and Y) which diverged from a common ancestral sequence t years ago. We denote by q_t the proportion of identical nucleotides between X and Y and by $p_t = 1 - q_t$ the proportion of different nucleotides. The proportion of identical nucleotides (q_{t+1}) at time $t + 1$ (measured in years) can then be obtained in the following way. First, we note that a site which had the same nucleotide for X and Y at time t will remain the same at time $t + 1$ with probability $1 - 2\lambda$, neglecting the terms involving λ^2. Second, a site which had different nucleotides at time t will have the same nucleotide at time $t + 1$ with probability $2\lambda/3$, again neglecting the terms involving λ^2. Hence,

$$q_{t+1} = (1 - 2\lambda)q_t + \frac{2}{3}\lambda(1 - q_t)$$

If we write $q_{t+1} - q_t = dq_t/dt$ and drop subscript t for q_t, we therefore have

$$\frac{dq}{dt} = \frac{2\lambda}{3} - \frac{8\lambda}{3}q. \qquad (5.1)$$

Solution of this differential equation with the initial condition $q = 1$ for $t = 0$ gives

$$q = 1 - \frac{3}{4}(1 - e^{-8\lambda t/3}). \qquad (5.2)$$

Under the present model, the expected number of nucleotide substitutions per site (d) is given by $2\lambda t$. Therefore, d can be estimated by

$$\hat{d} = -\frac{3}{4}\log_e\left(1 - \frac{4}{3}p\right), \qquad (5.3)$$

where $p = 1 - q$ is the proportion of different nucleotides between X and Y (Jukes and Cantor 1969). The large-sample variance of $\hat{d}$ is given by

$$V(\hat{d}) = \left[\frac{\mathrm{d}d}{\mathrm{d}p}\right]^2 V(p) = \frac{9p(1-p)}{(3-4p)^2 n}, \tag{5.4}$$

where $V(p) = p(1-p)/n$ (Kimura and Ohta 1972).

Equation (5.3) gives a good estimate of d if the rate of nucleotide substitution is the same for all nucleotide pairs or if d is small. However, if these conditions are not satisfied, it gives an underestimate. For this reason, various other methods have been proposed.

Kimura's Two-Parameter Method

As discussed in chapter 3, the rate of transitional nucleotide substitution is often higher than that of transversional nucleotide substitution. Taking into account this observation, Kimura (1980) proposed a method for estimating the number of nucleotide substitutions per site. We again consider two DNA sequences that diverged t years ago and compare the nucleotides at each homologous site. Since there are four possible nucleotides (A, T, C, G) at each site, the total number of possible pairs of nucleotides is 16, as given in table 5.1. In this table, the first letter of each pair of nucleotide symbols stands for the nucleotide for sequence X and the second letter for sequence Y. Thus, the letter pairs in the first row represent the cases where the nucleotides of X and Y are identical, and the other letter pairs stand for the cases where they are different. When they are different, there are two different types of differences: type I and type II. Type I nucleotide differences are of the transition type and type II differences are of the transversion type. We denote by P and Q the frequencies of type I and type II differences, respectively.

Table 5.1   Sixteen different types of nucleotide pairs between sequences X and Y.

Identical nucleotides	AA	TT	CC	GG	Total
Frequency	R_1	R_2	R_3	R_4	R
Transition-type pair	AG	GA	TC	CT	Total
Frequency	P_1	P_1	P_2	P_2	P
Transversion-type pair	AT	TA	AC	CA	
Frequency	Q_1	Q_1	Q_2	Q_2	
	TG	GT	CG	GC	Total
Frequency	Q_3	Q_3	Q_4	Q_4	Q

Obviously, $R = 1 - P - Q$ represents the frequency of identical nucleotide pairs.

In Kimura's mathematical model, the rate of transitional nucleotide substitution per site per year (α) is assumed to be different from that of transversional substitution (2β) (see figure 3.4). The total substitution rate per site per year is therefore given by $\alpha + 2\beta$. Using this model, Kimura showed that the expected number of nucleotide substitutions per site between X and Y is given by

$$d \equiv 2\lambda t = 2\alpha t + 4\beta t$$

$$= -\frac{1}{2} \log_e[(1 - 2P - Q) \sqrt{1 - 2Q}]. \qquad (5.5)$$

This equation can be used for estimating d from information on P and Q. The variance of the estimate $(\hat{d})$ of d is given by

$$V(\hat{d}) = \frac{1}{n} [a^2 P + b^2 Q - (aP + bQ)^2], \qquad (5.6)$$

where

$$a = \frac{1}{1 - 2P - Q}, \qquad b = \frac{1}{2}\left[\frac{1}{1 - 2P - Q} + \frac{1}{1 - 2Q}\right].$$

Nucleotide Substitution Matrix

Although Kimura's two-parameter method is useful when the rates of transitional and transversional nucleotide substitutions are different, the actual pattern of nucleotide substitution is much more complicated than the two-parameter model, as can be seen from table 3.4. Even in mitochondrial DNA, where the rate of transitional mutation is much higher than the rate of transversional mutation, equation (5.5) does not seem to be applicable except for a relatively short evolutionary time (Aquadro et al. 1984). However, if we know the nucleotide substitution matrix for a unit evolutionary time, we can relate the proportion (p) of different nucleotides between a pair of sequences to the number of nucleotide substitutions, as in the case of amino acid substitution (chapter 4).

Let g_1, g_2, g_3, and g_4 be the frequencies of nucleotides A, T, C, and G, respectively, at a given evolutionary time. We denote by λ_{ij} the rate of change of the ith nucleotide to the jth nucleotide during one evolutionary time unit. The nucleotide frequencies in the next evolutionary time are then given by

$$g_1' = (1 - \lambda_1)g_1 + \lambda_{21}g_2 + \lambda_{31}g_3 + \lambda_{41}g_4, \qquad (5.7a)$$

$$g_2' = \lambda_{12}g_1 + (1 - \lambda_2)g_2 + \lambda_{32}g_3 + \lambda_{42}g_4, \qquad (5.7b)$$

$$g_3' = \lambda_{13}g_1 + \lambda_{23}g_2 + (1 - \lambda_3)g_3 + \lambda_{43}g_4, \qquad (5.7c)$$

$$g_4' = \lambda_{14}g_1 + \lambda_{24}g_2 + \lambda_{34}g_3 + (1 - \lambda_4)g_4, \qquad (5.7d)$$

where $\lambda_1 = \lambda_{12} + \lambda_{13} + \lambda_{14}$, $\lambda_2 = \lambda_{21} + \lambda_{23} + \lambda_{24}$, $\lambda_3 = \lambda_{31} + \lambda_{32} + \lambda_{34}$, and $\lambda_4 = \lambda_{41} + \lambda_{42} + \lambda_{43}$. If we use matrix algebra, (5.7) can be expressed as

$$\mathbf{g}' = \mathbf{Mg}, \qquad (5.8)$$

where $\mathbf{g}'$ and $\mathbf{g}$ represent the column vectors of g_1', g_2', g_3', g_4' and g_1, g_2, g_3, g_4, respectively, and $\mathbf{M}$ is the matrix

$$\mathbf{M} = \begin{bmatrix} 1 - \lambda_1 & \lambda_{21} & \lambda_{31} & \lambda_{41} \\ \lambda_{12} & 1 - \lambda_2 & \lambda_{32} & \lambda_{42} \\ \lambda_{13} & \lambda_{23} & 1 - \lambda_3 & \lambda_{43} \\ \lambda_{14} & \lambda_{24} & \lambda_{34} & 1 - \lambda_4 \end{bmatrix} \qquad (5.9)$$

If λ_{ij}'s remain constant, the nucleotide frequencies eventually reach the equilibrium values. These equilibrium values ($\tilde{g}_i$'s) are obtained by equating g_i' to g_i in (5.7) or by putting $\mathbf{g} = \mathbf{Mg}$. They become

$$\tilde{g}_i = A_{ii} / \sum_{j=1}^{4} A_{jj}, \qquad (5.10)$$

where A_{ij} is the cofactor of the element at the ith row and jth column in the following determinant (Tajima and Nei 1982).

$$A = \begin{vmatrix} \lambda_1 & -\lambda_{12} & -\lambda_{13} & -\lambda_{14} \\ -\lambda_{21} & \lambda_2 & -\lambda_{23} & -\lambda_{24} \\ -\lambda_{31} & -\lambda_{32} & \lambda_3 & -\lambda_{34} \\ -\lambda_{41} & -\lambda_{42} & -\lambda_{43} & \lambda_4 \end{vmatrix} \qquad (5.11)$$

We note that the nucleotide frequencies ($\mathbf{g}_t$) at time t are given by $\mathbf{g}_t = \mathbf{M}_t\mathbf{g}_0$, where $\mathbf{M}_t = \mathbf{M}^t$. This is identical in form with (4.7) for the case of amino acid substitution. Therefore, the proportion (p) of different nucleotides between a pair of homologous sequences is given by (4.8). On the other hand, the number of nucleotide substitutions per site (d) for t units of evolutionary time is given by

$$d = 2\left(\sum_{i=1}^{4} \bar{g}_i\lambda_i\right)t. \tag{5.12}$$

Thus, it is possible to estimate d from p if $\mathbf{M}$ is known. In practice, it is convenient to measure time in terms of the time length corresponding to 0.01 substitutions per site (one PAM).

One problem with this approach is that we need a reliable nucleotide substitution matrix, and to get this matrix we must have a large amount of substitution data. At the present time, we do not have such extensive data for any gene or gene family.

In the above computation of d, we used the iteration method. Theoretically, however, it is possible to derive the relationship between d and p (or more precisely nucleotide pair frequencies) analytically (Lanave et al. 1984). The only problem is the mathematical complication. This complication can be reduced if we use a smaller number of parameters for nucleotide substitution. For example, Kimura (1981b) considered the following substitution matrix.

$$\mathbf{M} = \begin{bmatrix} 1-(2\alpha+\alpha_1) & \beta_1 & \beta & \beta \\ \alpha_1 & 1-(2\alpha+\beta_1) & \beta & \beta \\ \alpha & \alpha & 1-(2\beta+\alpha_2) & \beta_2 \\ \alpha & \alpha & \alpha_2 & 1-(2\beta+\beta_2) \end{bmatrix}$$

This matrix involves 6 parameters rather than 12 parameters as required in (5.9). Therefore, d can be related to the observable quantities in table 5.1. A complete solution of this problem has been given by Gojobori et al. (1982a). A slightly different model involving four parameters has also been studied by Takahata and Kimura (1981). The results of these studies indicate that when d is smaller than 0.5, the Jukes-Cantor formula is quite accurate, but when d exceeds 0.5, the Jukes-Cantor formula tends to give underestimates of d in the presence of unequal rates of nucleotide substitution. In this case, Takahata and Kimura's and Gojobori et al.'s methods give better estimates. However, it should be noted that these methods are dependent on specific schemes of nucleotide substitution, and if actual nucleotide substitution does not follow these schemes, they are expected to give biased estimates. Furthermore, they are often inapplicable to actual data because of a negative argument in the logarithm of the formula used (Gojobori et al. 1982a).

Blaisdell (1985) recently introduced a least-squares method of estimation of the number of nucleotide substitutions. In this method, a theoretical substitution model such as Kimura's two-parameter model is fitted to observed data on P_i, Q_i, and R_i, and the substitution parameters that minimize the sum of squares of the differences between the observed and theoretical values of P_i, Q_i, R_i are estimated. Once all substitution parameters are estimated, it is possible to estimate the total number of nucleotide substitutions for the entire evolutionary process. One advantage of this method over others is that the numbers of nucleotide substitutions for the two evolutionary lineages involved can be estimated separately. The reader is advised to consult the original paper for the details.

Tajima and Nei's Method

Tajima and Nei (1984) developed another method which seems to be quite insensitive to various disturbing factors. This method is exact for Tajima and Nei's (1982) equal input model of nucleotide substitution, but for other types of unequal substitution rates it gives only an approximate estimate. [In the equal input model, the rate of substitution to the ith nucleotide is assumed to be the same, regardless of the original nucleotide, i.e., $\lambda_{1i} = \lambda_{2i} = \lambda_{3i} = \lambda_{4i} = a_i$ except for λ_{ii} in matrix (5.9).] Nevertheless, computer simulation has shown that it gives a quite reliable estimate of the number of nucleotide substitutions for a wide variety of substitution schemes. Here, we present only the final result of Tajima and Nei's formulation.

In this method, the number of nucleotide substitutions is estimated by the following formula.

$$\hat{d} = -b\log_e(1 - p/b), \tag{5.13}$$

where

$$b = \left(1 - \sum_{i=1}^{4} g_i^2 + p^2/c\right)/2. \tag{5.14}$$

Here, c is given by

$$c=\sum_{i=1}^{3}\sum_{j=i+1}^{4}x^2_{ij}/(2g_ig_j), \tag{5.15}$$

where x_{ij} $(i<j)$ is the relative frequency of nucleotide pair i and j when two DNA sequences are compared. The variance of the estimate of d obtained by (5.13) is approximately given by

$$V(\hat{d})=\frac{b^2p(1-p)}{(b-p)^2n}. \tag{5.16}$$

It is noted that when the rate of nucleotide substitution is the same for all nucleotide pairs, $b=3/4$, and equations (5.13) and (5.16) reduce to equations (5.3) and (5.4), respectively. In practice, b is usually smaller than 3/4 because of unequal rates of nucleotide substitution, and in this case equation (5.13) gives a larger value than the Jukes-Cantor formula.

EXAMPLE

Table 5.2 shows the numbers of 10 different nucleotide pairs between the DNA sequences for the coding regions of the rabbit α and mouse β^{maj} globin genes. The numbers at the first, second, and third nucleotide positions of codons are listed separately, the total number of homologous codons compared being 139. If we want to estimate d by using the Jukes-Cantor formula, we must know the proportion (p) of different nucleotide pairs between the two sequences compared. For the first nucleotide position, the number of nucleotide differences is 60. Therefore, $\hat{p}=60/139=0.432$. If we put this value into equation (5.3), we have $\hat{d}=0.643$. The variance of $\hat{d}$ is $V(\hat{d})=0.009819$ from (5.4).

Table 5.2 Observed numbers of the 10 different pairs of nucleotides between the DNA sequences for the rabbit α and mouse β^{maj} globin genes. The numbers at the first, second, and third nucleotide positions of codons are listed separately. The rabbit sequence is from Heindell et al. (1978), and the mouse sequence is from Konkel et al. (1978).

Position	Identical				Transitional		Transversional				
in codon	AA	TT	CC	GG	AG	TC	AT	AC	TG	CG	Total
First	18	8	19	34	15	9	8	10	8	10	139
Second	32	35	20	7	11	5	4	11	2	12	139
Third	1	4	35	27	5	30	2	3	12	20	139

Therefore, the standard error of $\hat{d}$ is 0.099. The $\hat{d}$ values for the second and third positions can be computed in the same way, and the results obtained are presented in table 5.3. It is clear that the number of nucleotide substitutions is highest at the third position and lowest at the second position, the first position number being intermediate.

In Tajima and Nei's method, we must compute the frequencies of all kinds of nucleotide pairs (x_{ij}). For example, for the first nucleotide position we have $x_{AA} = 0.1295$, $x_{AG} = 0.1079$, etc. Once the x_{ij}'s are obtained, the average frequency of the ith nucleotide for the two sequences under comparison (g_i) is given by $g_i = x_{ii} + \Sigma_{j \neq i} x_{ij}/2$, where $\Sigma_{j \neq i}$ indicates the summation of x_{ij} for all j's except $j = i$. Thus, we obtain $g_A = 0.248$, $g_T = 0.147$, $g_C = 0.241$, and $g_G = 0.363$ for the first position. We also have $p = \Sigma_{i<j} x_{ij} = 0.432$, $c = 0.243$, $p^2/c = 0.767$, and $b = 0.747$. Thus, the estimate of d is $\hat{d} = 0.644$ from (5.13). On the other hand, the variance of $\hat{d}$ becomes 0.009914 from (5.16). These values are essentially the same as those obtained from the Jukes-Cantor (JC) method. At the third position, however, Tajima and Nei's (TN) method gives a larger $\hat{d}$ (See table 5.3). Table 5.3 also includes the $\hat{d}$ values obtained by Kimura's two-parameter (K2P) method and Gojobori, Ishii, and Nei's (GIN) method. The former gives $\hat{d}$'s similar to those obtained by the JC method, whereas the latter method gives somewhat larger values even for the first and second positions. At the third position, the GIN method is inapplicable, because one of the arguments for the logarithms involved becomes negative.

In the above example, the JC method gives the same $\hat{d}$ value as that obtained by the K2P or TN method when $\hat{d}$ is equal to or smaller than

Table 5.3   Estimates $(\hat{d})$ of the number of nucleotide substitutions per site between the rabbit α and mouse β^{maj} globin genes for the first, second, and third nucleotide positions of codons. Four different methods are used.

Position in codon	JC Method	K2P Method	GIN Method	TN Method
First	0.64 ± 0.10	0.64 ± 0.10	0.68 ± 0.11	0.64 ± 0.10
Second	0.42 ± 0.07	0.42 ± 0.07	0.52 ± 0.12	0.42 ± 0.07
Third	0.88 ± 0.14	0.92 ± 0.16	—	1.03 ± 0.20

JC: Jukes-Cantor.   K2P: Kimura's 2-parameter.   GIN: Gojobori-Ishii-Nei.
TN: Tajima-Nei.   —: Inapplicable.

0.64. In general, the differences between the estimates of d obtained by the different methods are quite small when $\hat{d} < 0.5$. In these cases, therefore, there is no need to use a complicated method. However, if the substitution rate matrix is extremely skewed as in the case of mammalian mitochondrial DNA, the JC method may give an underestimate even if d is as small as 0.2.

Synonymous and Nonsynonymous Nucleotide Substitutions

It is clear from the above example that the rate of nucleotide substitution is much higher at the third position of codon than at the first and second positions. This is apparently due to the fact that many nucleotide substitutions at the third position are silent and do not change amino acids. However, not all substitutions at the third position are silent. Furthermore, some silent substitutions may also occur at the first position. It is, therefore, interesting to know the numbers of silent (synonymous) substitutions and amino acid altering (nonsynonymous) substitutions separately.

When the number of nucleotide substitutions between two DNA sequences is so small that there is no more than one nucleotide difference between any pair of homologous codons compared, the numbers of synonymous and nonsynonymous substitutions can be obtained simply by counting silent and amino acid altering nucleotide differences. However, when two or more nucleotide differences exist between a pair of codons, the distinction between synonymous and nonsynonymous substitutions is no longer simple. For this reason, Perler et al. (1980) developed a statistical method for estimating synonymous substitutions. More elaborate methods were also developed by Miyata and Yasunaga (1980) and Li et al. (1985b). The main difference between Perler et al.'s method and the latter two methods is that in the former an equal weight is given to two or more evolutionary pathways that are possible between a pair of codons, whereas in the latter a larger weight is given to an evolutionary pathway involving silent substitutions than to a pathway involving amino acid altering substitutions.

Unweighted Pathway Methods

In Perler et al.'s (1980) unweighted pathway method, nucleotide sites are divided into the groups of synonymous sites and nonsynonymous sites, each with three different categories. Nucleotide differences are also classified into synonymous and nonsynonymous differences, each with three categories. Therefore, the computation is quite complicated. Furthermore, their method is known to give underestimates of synonymous substitutions (Gojobori 1983). For this reason, Nei and Gojobori (1986) proposed another method, which is essentially an unweighted version of Miyata and Yasunaga's (1980) method. Since Nei and Gojobori's method is much simpler than Perler et al.'s and yet gives a better estimate, we shall discuss this method.

We first note that nucleotide sites in codons can be classified into four categories, i.e., the four-fold degenerate, three-fold degenerate, two-fold degenerate, and nondegenerate sites. The four-fold degenerate sites are those where all nucleotide changes result in synonymous substitutions. Thirty-two third positions of the 61 sense codons are of this type (see table 3.1). The three-fold degenerate sites are those in which two of the three possible changes are synonymous. Only the third positions of the three isoleucine condons belong to this category. The two-fold degenerate sites, where one of the three possible changes is synonymous, include 24 third positions of the 61 sense codons. They also include the first positions of four leucine codons (TTA, TTG, CTA, CTG) and four arginine codons (CGA, CGG, AGA, AGG). All other nucleotide sites are nondegenerate.

Let us compute the number of synonymous and nonsynonymous sites for each codon, considering the above properties. We denote by f_i the proportion of synonymous changes at the ith position of a codon ($i = 1$, 2, 3). The numbers of synonymous (s) and nonsynonymous (n) sites for this codon are then given by $s = \Sigma_{i=1}^{3} f_i$ and $n = 3 - s$, respectively. For example, in the case of codon TTA (Leu), $f_1 = 1/3$, $f_2 = 0$, and $f_3 = 1/3$. Thus, $s = 2/3$ and $n = 7/3$. For a DNA sequence of r codons, the total numbers of synonymous and nonsynonymous sites are, therefore, given by $S = \Sigma_{j=1}^{r} s_j$ and $N = 3r - S$, respectively, where s_j is the value of s for the jth codon.

We now compute the numbers of synonymous and nonsynonymous nucleotide differences between a pair of homologous sequences. We compare the two sequences codon by codon and count the number of nucleotide differences for each pair of codons compared. When there is only one nucleotide difference, we can immediately decide whether the difference is synonymous or nonsynonymous. For example, if the codon pairs compared are GTT (Val) and GTA (Val), there is one synonymous difference. We denote by s_d and n_d the numbers of synonymous and nonsynonymous differences per codon, respectively. In the present case, $s_d = 1$ and $n_d = 0$. When two nucleotide differences exist between the two codons compared, there are two possible ways to obtain the differences. For example, in the comparison of TTT and GTA there are two pathways to reach from one codon to the other. That is,

Pathway I   TTT (Phe) $\leftrightarrow$ GTT (Val) $\leftrightarrow$ GTA (Val)

Pathway II   TTT (Phe) $\leftrightarrow$ TTA (Leu) $\leftrightarrow$ GTA (Val).

Pathway I involves one synonymous and one nonsynonymous substitution, whereas pathway II involves two nonsynonymous substitutions. We assume that pathways I and II occur with equal probability. The numbers of synonymous and nonsynonymous differences then become $s_d = 0.5$ and $n_d = 1.5$, respectively. In some comparisons of codons, there is a pathway involving a termination codon. This type of pathway is eliminated from the computation.

When there are three nucleotide differences between the codons compared, there are six different possible pathways between the codons, and in each pathway there are three mutational steps. Considering all these pathways and mutational steps, one can again evaluate the number of synonymous and nonsynonymous differences in the same way as in the case of two nucleotide differences. It is now clear that the total numbers of synonymous and nonsynonymous differences can be obtained by summing up these values over all codons. That is, $S_d = \sum_{j=1}^{r} s_{dj}$ and $N_d = \sum_{j=1}^{r} n_{dj}$, where s_{dj} and n_{dj} are the numbers of synonymous and nonsynonymous differences for the jth codon, and r is the number of codons compared. Note that $S_d + N_d$ is equal to the total number of nucleotide differences between the two DNA sequences compared.

We can, therefore, estimate the proportions of synonymous (p_S) and nonsynonymous (p_N) differences by the equations

$$p_S = S_d/S, \qquad p_N = N_d/N, \qquad (5.17)$$

where S and N are the average numbers of synonymous and nonsynonymous sites for the two sequences compared. To estimate the numbers of synonymous (d_S) and nonsynonymous (d_N) substitutions per site, we use the Jukes-Cantor method [equation (5.3)] replacing p by p_S or p_N. This method, of course, gives only approximate estimates of d_S and d_N because the nucleotide substitution at the twofold and threefold degenerate sites does not really follow the Jukes-Cantor pattern, as noted by Perler et al. (1980). Despite this theoretical problem, computer simulations have shown that it gives reasonably good estimates of synonymous and nonsynonymous substitutions. Approximate variances of the estimates of d_S and d_N may be obtained by using equation (5.4).

Weighted Pathway Methods

In the method described above, we assumed that all evolutionary pathways between a pair of codons occur with equal probability. This assumption is unlikely to be correct. For example, in the comparison of TTT with GTA, one pathway involves one nonsynonymous and one synonymous substitution, whereas the other pathway involves two nonsynonymous substitutions. Since synonymous substitutions occur with a higher probability than nonsynonymous substitutions, it would be reasonable to assume that the first pathway occurs with a higher probability than the second. It should also be noted that when different nonsynonymous substitutions are involved in different pathways, the interchange of two similar amino acids would occur with a higher probability than the interchange of two dissimilar amino acids. Therefore, if we compute the number of nucleotide substitutions, giving different weights for different nucleotide substitutions, we would obtain a better estimate of silent substitutions. Using this idea, Miyata and Yasunaga (1980) developed a method of estimating synonymous and nonsynonymous substitutions.

In Miyata and Yasunaga's (1980) method, the weight given for each nucleotide substitution depends on whether or not the amino acids interchanged are biochemically similar. They first consider the following substitution coefficient *(v)* for each synonymous or nonsynonymous nucleotide substitution.

$$v = \begin{cases} 1 & \text{for } \delta = 0 \text{ (synonymous change)}, \\ 1 - \delta/3.5 & \text{for } 0 < \delta < 3.465, \\ 0.01 & \text{for } \delta > 3.465, \end{cases} \quad (5.18)$$

where δ is the degree of polarity and volume differences of the amino acids interchanged. It varies from 0.06 to 5.13, and the value for each amino acid pair is given by Miyata et al. (1979). The substitution coefficient for a pathway *(V)* is then given by the product of the substitution coefficients for all nucleotide changes in the pathway. For example, the coefficient for pathway TTT (Phe) ↔ GTT (Val) ↔ GTA (Val) is $V = 0.59 \times 1 = 0.59$, because $\delta = 1.43$ for the Phe ↔ Val change and $\delta = 0$ for a silent change (Miyata et al. 1979). Similarly, the coefficient of pathway TTT (Phe) ↔ TTA (Leu) ↔ GTA (Val) is $V = 0.82 \times 0.74 = 0.61$, since $\delta = 0.63$ for the Phe ↔ Leu change and $\delta = 0.91$ for the Leu ↔ Val change. The weights for the first and second pathways are then given by $w_1 = 0.59/(0.59 + 0.61) = 0.49$ and $w_2 = 0.61/(0.59 + 0.61) = 0.51$, respectively. In general, the weight for the ith pathway for a given pair of codons is given by

$$w_i = V_i / \sum_{k=1}^{l} V_k, \qquad (5.19)$$

where V_k is the value of V for the kth pathway and l is the number of pathways available. Once w_i is obtained, the numbers of synonymous and nonsynonymous differences are given by

$$s_d' = \sum_{i=1}^{l} w_i m_{Si}, \qquad n_d' = \sum_{i=1}^{l} w_i m_{Ni}, \qquad (5.20)$$

where m_{Si} and m_{Ni} are the numbers of synonymous and nonsynonymous substitutions for the ith pathway. For example, $m_{S1} = 1$ and $m_{N1} = 1$ for the first pathway of the TTT–GTA comparison, and $m_{S2} = 0$ and $m_{N2} = 2$ for the second pathway. Note that in the previous unweighted method, w_i is assumed to be $1/l$, so that s_d' and n_d' become equal to s_d and n_d, respectively. The total numbers of synonymous and nonsynonymous differences for the entire DNA sequence are then given by $S_d' = \sum_{j=1}^{r} s_{dj}'$ and $N_d' = \sum_{j=1}^{r} n_{dj}'$, respectively.

In Miyata and Yasunaga's method, the numbers of synonymous and nonsynonymous sites are also obtained by weighting different pathways. They consider the synonymous (s) and nonsynonymous (n) sites for all codons involved in a given pathway and take the averages ($\bar{s}$ and $\bar{n}$) over all codons in the pathway. They then compute the weighted averages of these averages to get the mean synonymous and nonsynonymous sites for the two homologous codons compared. That is,

$$s_A = \sum_{j=1}^{l} w_j \bar{s}_j, \qquad n_A = \sum_{j=1}^{l} w_j \bar{n}_j, \qquad (5.21)$$

where $\bar{s}_j$ and $\bar{n}_j$ are the values of $\bar{s}$ and $\bar{n}$ for the jth pathway for a given codon comparison. The total numbers of synonymous and nonsynonymous sites for the entire DNA sequence are given by $S' = \sum_{i=1}^{r} s_{Ai}$ and $N' = \sum_{i=1}^{r} n_{Ai}$, respectively, where i refers to the ith codon in the sequence. Therefore, the proportions of synonymous and nonsynonymous nucleotide differences are computed by $p_S = S'_d/S'$ and $p_N = N'_d/N'$, respectively. To estimate the number of nucleotide substitutions per site, we can again use the Jukes-Cantor method.

One might expect that Miyata and Yasunaga's method gives better estimates of synonymous and nonsynonymous substitutions than Nei and Gojobori's method. Nei and Gojobori's (1986) computer simulation, however, has shown that the two methods give very similar estimates and that the estimates obtained by the former method are not necessarily better than those obtained by the latter. The reason for this is that although different weights are given to different evolutionary pathways for a given pair of codons, the genetic code has such a property that the weights for different pathways are quite similar. We have already seen this property for the case of the TTT–GTA comparison, where the weights (0.49 and 0.51) for the two pathways are very close to each other despite the various values of δ involved. In some cases, the difference between the weights can be larger than this, but in those cases the probabilities of occurrence of nucleotide substitution are usually very small.

Li et al. (1985b) recently developed another weighted pathway method. They use a different weighting scheme based on the expected and observed frequencies of nucleotide substitutions. They also consider nondegenerate, twofold degenerate, and fourfold degenerate sites separately. Their method, however, again gives essentially the same estimates of d_S and d_N as those obtained by Nei and Gojobori's method.

Some Remarks

It should be noted that although there are four different methods of estimating the numbers of synonymous and nonsynonymous substitutions, all methods depend on a number of simplifying assumptions and give only approximate estimates. One problem in estimating these numbers is that synonymous and nonsynonymous sites are not fixed on the DNA sequence but vary from time to time. Therefore, as the number of nucleotide substitutions increases, the reliability of the estimates gradually declines.

Another way of examining the difference in the rate of synonymous and nonsynonymous substitutions is to compare the nucleotide substitutions at the three nucleotide positions of codons. Since about 72 percent of nucleotide substitutions at the third position are synonymous and most substitutions at the first and second positions are nonsynonymous, comparison of substitutions between these two groups of positions gives some idea about the relative rates of synonymous and nonsynonymous substitutions. In this case, the three nucleotide positions can be defined unambiguously so that statistical estimation is much simpler.

Pattern of Nucleotide Substitution

Synonymous and Nonsynonymous Substitution

As soon as nucleotide sequences of genes became available for related species, a conspicuous difference was noted between the rates of synonymous and nonsynonymous substitutions (Grunstein et al. 1976; Kafatos et al. 1977). Kimura (1977) showed that the rate of nucleotide substitution at the third position of codons in the histone gene is as high as the rate of amino acid substitution in fibrinopeptides, the highest rate known among polypeptides. He took this as evidence for the neutral mutation theory, since nucleotide substitutions at the third position are largely synonymous and synonymous substitutions would not be subject to purifying selection. Later studies have shown that the rate of synonymous substitution is much higher than that of nonsynonymous substitution in most functional genes.

Miyata et al. (1980) examined the rates of synonymous and nonsynonymous substitutions for various eukaryotic genes and concluded that while the rate of nonsynonymous substitution varies enormously from gene to gene, the rate of synonymous substitution is more or less the same for all genes. Li et al. (1985b) conducted a more extensive study on this problem, examining 42 eukaryotic genes. Their results are qualitatively the same as Miyata et al.'s, but unlike Miyata et al.'s conclusion, they showed that the rate of synonymous substitution varies considerably with gene (table 5.4). Indeed, the highest rate (11.8×10^{-9} for $\beta 2$ microglobulin) is about eight times higher than the lowest rate (1.4×10^{-9} for histone H4), though the standard errors of these estimates are quite large. Nevertheless, the synonymous rate is much more uniform than the nonsynonymous rate. Interestingly, the synonymous and nonsynon-

Table 5.4 Rates of synonymous and nonsynonymous substitutions in 16 eukaryotic genes. Most of the rates given here were obtained from comparison between different mammalian orders which are assumed to have diverged 80 MY ago. All rates are in units of 10^{-9} substitutions per site per year. From Li et al. (1985b).

Genes	Number of codons used	Nonsynonymous rate	Synonymous rate
Histone H4	101	0.004 ± 0.01	1.43 ± 0.29
Actin α	376	0.014 ± 0.01	3.67 ± 0.43
Gastrin	82	0.15 ± 0.01	3.52 ± 2.57
Insulin	51	0.16 ± 0.09	5.41 ± 1.98
Parathyroid hormone	90	0.44 ± 0.12	1.72 ± 0.51
Glycoprotein hormone, α	92	0.67 ± 0.28	6.23 ± 1.57
Growth hormone	189	0.95 ± 0.11	4.37 ± 0.56
Prolactin	195	1.29 ± 0.12	5.59 ± 0.92
α-globin	141	0.56 ± 0.09	3.94 ± 0.60
β-globin	144	0.87 ± 0.11	2.96 ± 0.46
Immunoglobulin V_H	100	1.07 ± 0.19	5.67 ± 1.36
$\beta2$ Microglobulin	99	1.21 ± 0.20	11.77 ± 9.91
Interferon $\alpha1$	166	1.41 ± 0.13	3.53 ± 0.61
Fibrinogen γ	411	0.55 ± 0.06	5.82 ± 0.67
Albumin	590	0.92 ± 0.07	6.72 ± 0.62
α-Fetoprotein	586	1.21 ± 0.08	4.90 ± 0.45
Average for 42 genes[a]		0.88	4.65

[a]Computed from the above 16 genes plus 26 other genes.

ymous rates are significantly correlated, the correlation coefficient being 0.55 (Graur 1985).

Most of the substitution rates presented in table 5.4 were obtained from comparisons of two or three organisms under the assumption of rate constancy. In some genes (e.g., the genes for insulin and globins), however, DNA sequences from several organisms are available so that the pattern of accumulation of mutations can be examined. Figure 5.1 shows the relationship between the number of nucleotide substitutions and evolutionary time for the insulin gene. The number of nonsynonymous substitutions increases almost linearly with evolutionary time. The number of synonymous substitutions also increases with evolutionary time, but the rate of increase does not seem to be constant. As shown by Perler et al. (1980), there is some tendency for the rate to decline with time, though the tendency is not as great as in Perler et al.'s result. This

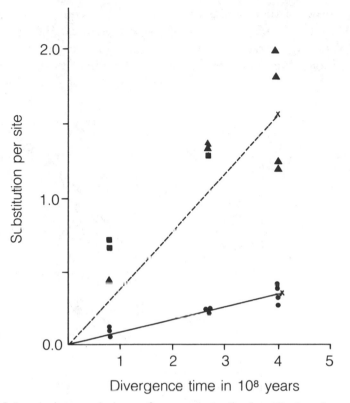

Figure 5.1. Accumulations of synonymous (broken line) and nonsynonymous (solid line) nucleotide substitutions in the insulin gene with time. From Li et al. (1985a). Copyright by the University of Chicago.

nonlinearity is caused mainly by comparisons involving the rat sequence (square marks in figure 5.1). According to Wu and Li (1985), the rate of synonymous substitution is about two times higher in mice and rats than in other mammalian orders. They speculate that this is caused by the comparatively shorter generation time in these organisms (see, however, Easteal 1985).

Influenza virus genes are known to evolve about 2 million times faster than eukaryotic nuclear genes, as mentioned earlier (chapter 3). Various strains of the influenza A virus have been isolated from the 1930s and kept in freezers. These strains can be used for studying the rate of nucleotide substitution, since the date of isolation of each strain is known.

Several genes of these strains have recently been sequenced (e.g., Blok and Air 1982; Krystal et al. 1983). Using these data, Hayashida et al. (1985) studied the rates of synonymous and nonsynonymous nucleotide substitutions. Figure 5.2 shows the results for the hemagglutinin (a),

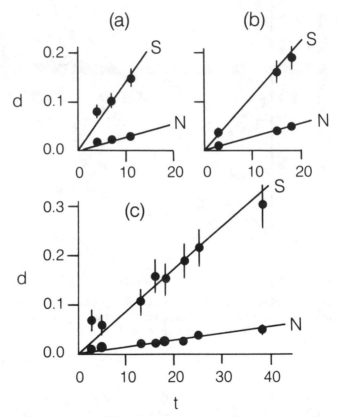

Figure 5.2.   Relationships between the number of nucleotide substitutions *(d)* and evolutionary time *(t)* in influenza virus genes. Evolutionary time represents the difference in the year of isolation between the two strains compared. S: Synonymous substitution. N: Nonsynonymous substitution. (a) Hemagglutinin. (b) Neuraminidase. (c) Nonstructural protein. The rates of synonymous substitutions per site per year for the hemagglutinin, neuraminidase, and nonstructural genes are 0.014, 0.011, and 0.009, respectively, whereas the corresponding rates for nonsynonymous substitution are 0.0029, 0.0028, and 0.0015. Vertical lines represent standard errors. From Hayashida et al. (1985). Copyright by the University of Chicago.

neuraminidase (b), and nonstructural protein (c) genes. Both synonymous and nonsynonymous substitutions increase almost linearly with evolutionary time, but the rate of substitution is much higher for synonymous substitutions than for nonsynonymous substitutions. This indicates that although the rate of nucleotide substitution in influenza virus genes is extremely high, some amino acids are evolutionarily conserved.

While the results from influenza virus genes indicate that both synonymous and nonsynonymous substitutions increase almost linearly with time, there is evidence that this is not always the case. I have already mentioned that the rate of synonymous substitution for rodents is apparently higher than that for other mammals. Examining the synonymous rates for various groups of organisms, Britten (1986) recently concluded that the rate is about five times lower in birds and higher primates than in rodents and sea urchins (see also Goodman et al. 1984). Although Britten's conclusion should be reexamined by using more extensive data, the rate of synonymous substitution seems to vary substantially with evolutionary lineage.

As is clear from table 5.4 and figures 5.1 and 5.2, the rate of synonymous substitution is about five times higher than the rate of nonsynonymous substitution. This suggests that synonymous substitutions are more useful as a clock than nonsynonymous substitutions when the time of divergence between two closely related species is to be estimated. In this case, the estimation of synonymous substitutions is relatively simple, and the error associated with the estimation is expected to be small. For estimating divergence times for distantly related organisms, however, the number of nonsynonymous substitutions (or amino acid substitutions) seems to be better than the number of synonymous substitutions. There are two reasons for this. First, the standard error of an estimate of the number of synonymous substitutions is quite large when the estimate is high (see table 5.4). Second, while the rate seems to be roughly constant within each group of organisms, it varies substantially among different groups, as mentioned above. By contrast, the number of nonsynonymous or amino acid substitutions seems to increase almost linearly for a larger period of time (figure 5.1 and chapter 4). For this purpose, the number of substitutions at the first and second positions of codons can also be used since the majority of the substitutions are nonsynonymous.

Rates of Substitution Among the Four Kinds of Nucleotides

In chapter 3, we discussed the relative mutation rates among the four different kinds of nucleotides using data from pseudogenes. In functional genes, however, the substitution rate depends on both mutation rate and selection intensity (chapter 13). Therefore, the relative substitution rates among the four nucleotides in functional genes are expected to be different from those of pseudogenes. These relative rates can be studied by comparing homologous functional genes. When the DNA sequences for a given gene are available for many different organisms (e.g., the human, mouse, rabbit, dog), it is possible to infer the ancestral nucleotide sequences (Fitch 1971a). Once this ancestral sequence is inferred, the relative nucleotide substitution rates among the four nucleotides can be obtained by comparing this ancestral sequence with sequences from extant species.

Table 5.5 shows the pattern of nucleotide substitution for functional genes obtained by this method. Transitional substitutions (45.5%) are again more frequent than expected under random substitution, but the C → T change is not the most frequent one, unlike the case in pseudogenes. Instead, the G → A change shows the highest frequency. The frequencies of different transversional changes are more uneven for functional genes than for pseudogenes. The difference in substitution patterns between pseudogenes and functional genes is clearly caused by natural selection that operates after the occurrence of mutation. Gojobori et al. (1982b) have shown that the pattern of nucleotide substitution for functional genes is consistent with Grantham's (1974) observation that substitutions between biochemically similar amino acids occur more frequently than those between dissimilar amino acids.

Table 5.5   Relative substitution rates among the four nucleotides A, T, C, and G in functional globin and ACTH genes. From Gojobori et al. (1982a).

Mutant nucleotide	Original nucleotide			
	A	T	C	G
A		5.0 ± 1.2	8.1 ± 0.6	20.9 ± 3.1
T	4.2 ± 2.0		9.4 ± 3.3	4.8 ± 0.8
C	6.3 ± 2.8	3.7 ± 1.6		11.5 ± 3.8
G	11.5 ± 1.4	1.7 ± 1.4	13.0 ± 1.5	

Rates of Nucleotide Substitution in Noncoding Regions

The noncoding regions of a gene include the 5′ flanking (untranscribed) region (5′FL), 5′ untranslated region (5′UT), 3′ untranslated region (3′UT), 3′ flanking (3′FL) region, and introns (INT) (see chapter 3). These regions undergo deletion and insertion more often than coding regions, so that it is difficult to obtain an accurate estimate of nucleotide substitutions. Nevertheless, a number of authors (e.g., Efstratiadis et al. 1980; Perler et al. 1980; Miyata et al. 1980; Miyata 1982; Li et al. 1985a) have studied the evolutionary change of these regions.

Table 5.6 shows the estimates obtained by Li et al. (1985a) from an analysis of more than ten genes. The substitution rate is lower in noncoding regions than in pseudogenes or synonymous sites. The 3′FL region shows a rate close to the synonymous rate. However, this rate is based on a small number of genes so that its reliability is low. The 5′UT and 3′UT regions show a rate that is less than a half of the synonymous rate. These regions are known to include highly conserved polynucleotides such as the TATA box (see chapter 3; Lomedico et al. 1979; Gojobori and Nei 1981; Proudfoot 1984). Apparently these sequences play an important role in protein synthesis, and probably because of this functional constraint, these regions have a low rate of substitution.

Table 5.6 includes the substitution rate for pseudogenes obtained by the method to be discussed in chapter 6. It is interesting to see that the synonymous rate is nearly the same as the pseudogene rate. This suggests that synonymous substitutions are subject to very little purifying selection. However, as noted by many authors (e.g., Sanger et al. 1977; Grantham 1980), synonymous codons for a particular amino acid are not

Table 5.6 Rates of nucleotide substitution (λ) for different regions of mammalian genes and for pseudogenes. Modified from Li et al. (1985a).

	Noncoding Region					Coding Region		
	5′FL[a]	5′UT	INT	3′UT	3′FL	SYN	NSY	Pseudogenes
λ[b]	2.36	1.74	3.70	1.88	4.46	4.65	0.88	4.85
R[c]	0.49	0.36	0.76	0.39	0.92	0.96	0.18	1.00

[a]FL: Flanking region. UT: Untranslated region. INT: Intron. SYN: Synonymous sites. NSY: Nonsynonymous sites.
[b]All rates are in units of 10^{-9} substitutions per site per year.
[c]Rate relative to the pseudogene rate.

used at random. In unicellular organisms such as *E. coli* and yeast, this nonrandom codon usage seems to be partly due to unequal concentrations of different tRNAs, as shown by Ikemura (1985). If this is the case, there must be a small degree of purifying selection even for synonymous substitutions (Kimura 1983a).

Mitochondrial DNA and Chloroplast DNA

Unlike nuclear genes, mitochondrial DNA (mtDNA) and chloroplast DNA (cpDNA) can be extracted relatively easily. Therefore, an increasing number of authors are using these DNAs for studying evolutionary problems.

Mitochondrial DNA in multicellular animals is about 16,000 base pairs (bp) long, and the length of the DNA is nearly the same for most animal species (Brown 1983). It is circular and encodes five known proteins and six unidentified reading frames (URF). There are no introns in these genes. It also contains genes for one 16S rRNA, one 12S rRNA, and 22 tRNAs. The evolutionary change of mtDNA occurs mainly through nucleotide substitution, although deletions and insertions are known to occur quite frequently in noncoding regions (Brown and Simpson 1981; Cann et al. 1984; Densmore et al. 1985). Plant mtDNA is much more complex than animal mtDNA and consists of several discrete units. For this reason, it is not as useful as animal mtDNA for evolutionary study.

The mtDNA in mammals is known to evolve much faster than nuclear DNA. Brown et al. (1982) has estimated that the rate of nucleotide substitution is $\lambda = 10^{-8}$ per site per year. [Nei (1985) obtained 7×10^{-9} from reanalysis of Brown et al.'s data.] This rate is about two times higher than the nuclear pseudogene rate and five to ten times higher than the average rate for nuclear DNA. This high rate of substitution is believed to be due to a high rate of mutation rather than to positive selection (Brown et al. 1979). Brown et al. (1982) have also shown that transitional nucleotide substitutions are much more frequent than transversional substitutions, the former accounting for 92 percent of the total changes (see also Aquadro and Greenberg 1983). Partly because of this high rate of transitional substitution, the proportion of nucleotide differences between two homologous sequences reaches a plateau relatively quickly (figure 5.3; Aquadro et al. 1984). For the first 10 million years, however, the number of nucleotide substitutions increases almost linearly. Therefore, this number can be used as a molecular clock within this time period.

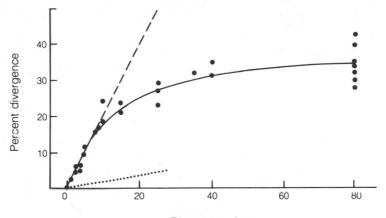

Figure 5.3. Relationship between percent sequence divergence (100*d*) and divergence time. The points represent estimates from pairwise comparisons of restriction endonuclease cleavage maps. The initial rate of mtDNA sequence divergence is shown by the dashed line and the rate of divergence of single-copy nuclear DNA by the dotted line. From Brown et al. (1979).

While the rate of nucleotide substitution in mtDNA is very high in mammalian species, this high rate may not apply to all kinds of organisms. Kikuno and Miyata (1986) have recently estimated the rates of synonymous substitutions for *Drosophila* and plant mtDNAs and suggested that the pattern of nucleotide substitution in mammalian mtDNA is exceptional and that the mitochondrial DNAs in nonmammalian or nonvertebrate species evolve in the same fashion as that of nuclear genes. The transition-transversion bias in insects and plants also does not seem as extreme as in the case of mammalian species (Wolstenholme and Clary 1985; Kikuno and Miyata 1986). However, since these conclusions are based on comparisons of relatively distantly related species of *Drosophila,* some more studies seem to be necessary (A. C. Wilson, personal communication).

Just like mtDNA, chloroplast DNA has a circular form but is much longer than animal mtDNA, the total length being about 150,000 bp. It encodes many proteins, the 23S, 16S, 5S, and 4.5S rRNAs and several tRNAs (Curtis and Clegg 1984). This DNA is quite conservative and shows a slow rate of evolution. Particularly conservative is the gene for ribulose-1, 5-biphosphate carboxylase/oxygenase (RBC), which is a key enzyme in the photosynthetic carbon metabolism of both prokaryotic and eukaryotic autotrophs. Comparison of the nucleotide sequences indicates that about 80 percent of the amino acids of a component of this protein are identical between cyanobacteria and higher plants, though nucleotides in silent sites are quite different (Curtis and Clegg 1984). Zurawski et al. (1984) examined the nucleotide sequences of three genes from barley and maize and obtained a rate of 1.1×10^{-9} per site per year. This is about ten times lower than the rate for mammalian mtDNA.

Because of this low rate of evolution, chloroplast DNA is not very suitable for studying the genetic relationship of closely related species. However, it is a useful material to clarify the evolutionary relationships of different genera, families, etc. For this purpose, the evolutionary change of this DNA is usually studied by using the restriction enzyme technique (Palmer and Zamir 1982; Bowman et al. 1983; Tsunewaki and Ogihara 1983).

Alignment of Nucleotide Sequences

So far we have assumed that the two homologous nucleotide sequences to be compared have no deletions or insertions and thus can be compared directly. In practice, however, the numbers of nucleotides involved in the two sequences compared are often different, and we must infer locations of deletions or insertions (gaps) and then align the two sequences. When the sequence difference is small, this alignment can be done relatively easily by inspection, particularly in the coding regions of DNA. In general, however, alignment of DNA sequences is not simple, and several methods have been developed for this purpose. Most of them can be used for both nucleotide and amino acid sequences.

Dot Matrix Method

To make our problem concrete, let us consider the following two nucleotide sequences.

$$\begin{array}{ll} \text{Sequence I} & \text{ATGCGTCGTT} \\ \text{Sequence II} & \text{ATCCGCGAT} \end{array} \qquad \text{(A1)}$$

Sequence I has ten nucleotides and sequence II has nine nucleotides. Therefore, at least one gap must be imposed in the alignment of these two sequences. A simple way of aligning these sequences is to make a two-dimensional comparison, as given in figure 5.4. In this comparison, dots are given when the nucleotides in sequences I and II are identical. Obviously, if the two sequences are identical, there will be a diagonal line of dots. If the sequences are identical except for a gap in one of the two sequences, the diagonal line will be shifted down or up in the middle of the line. Therefore, we can identify the gap. In practice, there are usually some nucleotide differences between the two sequences in addition to the gap, and this makes it difficult to identify the gap. Furthermore, the alignment of sequences produced in this way is quite subjective. For this reason, several mathematical methods have been developed to make an objective alignment.

Sequence Distance Method

One of the most popular methods of sequence alignment is that developed by Needleman and Wunsch (1970). In this method, the similarity between two sequences is measured by an index, and the alignment of the two sequences that maximizes the similarity is chosen. Later, Sellers (1974) developed another method in which the distance between two sequences is measured by an index and the alignment that minimizes this index is chosen. However, Smith et al. (1981) have shown

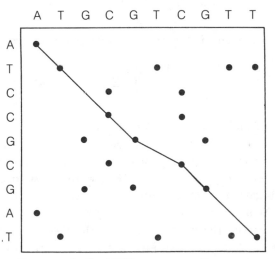

Figure 5.4.   Dot matrix for DNA sequence alignment.

that the two methods are essentially the same and give the same result for most cases. In the following, we consider the distance method rather than the similarity method.

Consider two DNA sequences A and B of length m and n, respectively, where $A = a_1 a_2 a_3 \cdots a_m$ and $B = b_1 b_2 b_3 \cdots b_n$. An alignment between two such sequences is defined as an ordered sequence of pairs, each pair containing one element from each sequence or an element of either sequence and the null element with the order of the original sequence preserved. Deletions or insertions (gaps) are indicated by alignment pairs containing a null element $(-)$. For example, the following alignment

$$\begin{array}{l} \text{ATGC--GTCGTT} \\ \text{AT--CCG--CGAT} \end{array} \qquad (A2)$$

contains three gaps of one element (length one), seven pairs of matched elements, and one pair of mismatched elements.

To measure the distance between two sequences, let α be the number of pairs of matched elements, β be the number of pairs of mismatched elements, and Δ_k be the number of gaps of length k. Since there are a total of $n + m$ nucleotides in the two sequences, we have

$$n + m = 2(\alpha + \beta) + \sum_k k\Delta_k.$$

The distance between the two sequences is then measured by

$$D = \text{Min}(\beta + \sum_k w_k \Delta_k), \qquad (5.22)$$

where $\text{Min}(\cdot)$ stands for the minimum value of $\beta + \sum_k w_k \Delta_k$ among all possible alignments. Here, w_k is a gap weight and usually given by $w_k = ku + v$, where u and v are positive constants. This means that a long gap contributes to the total distance D more than a short gap does.

The computation of D is quite complicated, but there are several algorithms. Waterman et al. (1976) proposed the following interative algorithm. That is, the distance (D_{ij}) for the first i nucleotides in one sequence and the first j nucleotides in the other sequence is computed by using the following recurrence equation.

$$D_{ij} = \text{Min}[D_{i-1,j-1} + \gamma(a_i, b_j), P_{ij}, Q_{ij}], \qquad (5.23)$$

where $\gamma(a_i, b_j) = 0$ if $a_i = b_j$, $\gamma(a_i, b_j) = \gamma$ if $a_i \neq b_j$, and

$$P_{ij} = \underset{1 \le k \le i}{\text{Min}} [D_{i-k,j} + w_k],$$

$$Q_{ij} = \underset{1 \le k \le j}{\text{Min}} [D_{i,j-k} + w_k].$$

Here, γ is a positive value, and $\text{Min}(x,y,z)$ is equal to the minimum value of x, y, and z.

The computation of D_{ij} is usually done by a computer. Gotoh (1982) has shown that when w_k is a linear function of k, i.e., $w_k = ku + v$, P_{ij} and Q_{ij} can be simplified to

$$P_{ij} = \text{Min}[D_{i-1,j} + w_1, P_{i-1,j} + u]$$

$$Q_{ij} = \text{Min}[D_{i,j-1} + w_1, Q_{i,j-1} + u].$$

This simplification saves computer time tremendously when long sequences are aligned. In the computation of D_{ij}, one may set $D_{00} = P_{00} = Q_{00} = 0$, $D_{i0} = Q_{i0} = w_i$ $(1 \le i \le m)$ and $D_{0j} = P_{0j} = w_j (1 \le j \le n)$. The distance ($D$) between the two sequences is equal to D_{mn}.

Let us illustrate the computation of D_{ij} using the two sequences in (A1). We use $w_k = 2k + 2$ for the weight for Δ_k and $\gamma(a_i, b_j) = 3$ for $a_i \neq b_j$. In the computation of D_{ij}, it is convenient to consider the $(m+1) \times (n+1)$ matrix, letting i and j start from 0. We set $D_{1,0} = w_1 = 4$, $D_{2,0} = 6, \cdots, D_{10,0} = 22$, and $D_{0,1} = 4$, $D_{0,2} = 6, \cdots, D_{0,9} = 20$ (see figure 5.5). We also set $P_{0i} = Q_{i0} = w_i$ and $D_{0,0} = 0$. $D_{1,1}$ is then obtained by using equation (5.23). In the present case, $a_1 = b_1$, so that $\gamma(a_1, b_1) = 0$. $P_{1,1} = \text{Min}[D_{0,1} + w_1, P_{0,1} + u] = 6$. Similarly, $Q_{1,1} = 6$. Therefore, $D_{1,1} = \text{Min}[0, 6, 6] = 0$. This value is given in figure 5.5. In the case of $D_{2,1}$, we have $D_{1,0} = 4$, $\gamma(a_2, b_1) = 3$, $P_{2,1} = 4$, and $Q_{2,1} = 8$. Thus, $D_{2,1} = \text{Min}[7, 4, 8] = 4$. Similarly, we have $D_{1,2} = 4$. This process will be continued until all D_{ij}'s are obtained. These values are given in figure 5.5. The distance between the two sequences is equal to $D = D_{10,9}$, which is 10.

The computation of D_{mn} is not sufficient for finding the alignment of sequences. The alignment associated with this D_{mn} is obtained by tracing

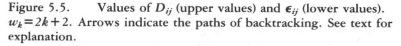

		A	T	G	C	G	T	C	G	T	T
	0	4	6	8	10	12	14	16	18	20	22
A	4	0 1	4 2	6 2	8 2	10 2	12 2	14 2	16 2	18 2	20 2
T	6	4 4	0 1	4 2	6 2	8 2	10 3	12 2	14 2	16 3	18 3
C	8	6 4	4 4	3 1	4 1	8 2	10 2	10 1	14 2	16 2	18 2
C	10	8 4	6 4	7 5	3 1	7 3	9 2	10 3	13 3	15 2	17 2
G	12	10 4	8 4	6 1	7 4	3 1	7 2	9 2	10 3	13 2	15 2
C	14	12 4	10 4	10 4	6 1	7 4	6 1	7 1	11 2	13 3	15 2
G	16	14 4	12 4	10 1	10 4	6 1	10 7	9 1	7 1	11 2	13 2
A	18	16 5	14 4	14 4	12 4	10 4	9 1	13 7	11 4	10 1	14 3
T	20	18 4	16 5	16 4	14 4	12 4	10 1	12 1	13 4	11 1	10 1

Figure 5.5.   Values of D_{ij} (upper values) and ϵ_{ij} (lower values). $w_k = 2k + 2$. Arrows indicate the paths of backtracking. See text for explanation.

back through the D matrix from D_{mn} to find how D_{ij} was incremented at each stage of iteration to get the minimum distance D_{mn}. This is usually done by using a computer. In this respect, Gotoh's (1982) computer algorithm has a nice feature. (N. Saitou has written a computer program.) It computes a direction matrix $[\epsilon_{ij}]$, which is associated with the D matrix, and from this direction matrix one can easily find the alignment. This direction matrix consists of elements 1, 2, 3, 4, 5, 6, and 7. Elements 1, 2, and 4 indicate "go back diagonally, i.e., from cell (ij) to $(i-1, j-1)$," "go back horizontally, i.e., from (ij) to $(i-1, j)$," and "go back vertically, i.e., from (ij) to $(i, j-1)$," respectively. Elements 3, 5, and 6 are combinations of executions 1 and 2, 1 and 4, and 2 and 4, respectively. Element 7 is a combination of executions 1, 2, and 4. These combinations of different executions produce different alignments for the same D_{mn} value. However, to minimize the number of gaps, one should follow the previous direction when there are alternatives.

In figure 5.5, the elements of the direction matrix are given in italics underneath the D_{ij} value. If we follow the direction given by these elements starting from the cell (10, 9), we obtain the following alignment

$$\text{ATGCGTCGTT}$$
$$\text{ATCCG-CGAT} \qquad \text{(A3)}$$

This pair of sequences has one gap of one element and two mismatches. Therefore, the distance is $w_1 + 2\gamma = 10$, which agrees with the value obtained above. By contrast, the alignment given in (A2) has a distance of $3w_1 + \gamma = 15$. Therefore, alignment (A3) is considered to be better than (A2). Incidentally, alignment (A3) is the same as that obtainable from the dot matrix in figure 5.4.

The choice of alignment depends considerably on the relative values of w_k and γ. If we use a small value of w_k relative to γ, we will have an alignment with many gaps and few mismatched nucleotides. However, since deletions and insertions occur less frequently than nucleotide substitutions, such an alignment would not be realistic. Therefore, it is generally advisable to make w_1 larger than γ. For a detailed discussion of this problem, see Smith et al. (1981).

In the above treatment of sequence alignment, I have implicitly assumed that there is no deletion or insertion at the beginning or the end of the sequences. Theoretically, it is possible to make a sequence alignment without this assumption, but the alignment procedure is quite complicated (Smith et al. 1981). In practice, however, this problem occurs relatively infrequently because a pair of DNA sequences can usually be divided into several segments, each of which is bounded by strongly conserved regions, and alignment can be done for each segment separately. Strongly conserved regions can usually be identified by inspection. However, if this is difficult, Karlin et al.'s (1985) statistical method may be used. It should also be noted that for the alignment of coding regions amino acid sequences are generally more informative than DNA sequences because the former are conserved more often than the latter.

When one wants to compare several sequences with each other, alignment should be done simultaneously for all sequences. This problem is also quite complicated, and the reader may refer to Sankoff and Cedergren (1983) and Waterman (1984).

Evolutionary Changes Due to Deletion and Insertion

Recent data on nucleotide sequences of related genes indicate that a substantial proportion of evolutionary change of DNA arises from deletion and insertion of nucleotides, particularly in noncoding regions of DNA. It seems that most deletions and insertions are relatively short and occur with an appreciable frequency (e.g., Efstratiadis et al. 1980; Langley et al. 1982; Cann and Wilson 1983). It is therefore possible to conduct a statistical study of the effects of these deletions and insertions on DNA divergence. There are two methods.

If we have two homologous DNA sequences, it is possible to identify deletions and insertions by using the alignment algorithm mentioned above and count the number of nucleotides involved in each deletion or insertion (gap). Deletions and insertions can also be identified by examining the pattern of the restriction site map for each DNA sequence. If we know the number of nucleotides involved in all gaps, the effect of deletions and insertions on DNA divergence can be measured by the number of gap nucleotides (nucleotides in the gaps) per nucleotide site. Let m_g be the number of nucleotides involved in all gaps and m_T be the total number of nucleotides compared. The number of gap nucleotides per site is then given by

$$g_m = m_g/m_T. \qquad (5.24)$$

Equation (5.24) seems to be appropriate when the number of occurrences of deletion and insertion is small. When this number is high, however, deletion and insertion may have occurred at the same DNA position. In this case, g_m gives an underestimate of the total number of nucleotides involved in deletions and insertions. To make a correction for this effect, Tajima and Nei (1984) proposed the following method.

Consider two homologous nucleotide sequences (X and Y) that diverged from a common ancestral sequence t evolutionary time units ago (e.g., years or 1,000 years). We assume that the length of a deletion or insertion is short compared with the total length of the DNA sequence and that deletion and insertion occur independently. Let α be the proportion of DNA that is deleted during unit evolutionary time, i.e., $\alpha = m_D/m_T$, where m_D is the number of nucleotides deleted and m_T is the total number of nucleotides before deletion. Note also that α is the number of nucleotides deleted per nucleotide site and is usually a very small quantity. Similarly, we denote by β the proportion of nucleotides that are inserted during unit evolutionary time, i.e., $\beta = m_I/m_T$, where m_I is the number of nucleotides inserted. For mathematical convenience, we assume that deletion occurs first and insertion follows and that m_T remains more or less constant because of the compensating effects of deletion and insertion. In practice, α and β may vary with evolutionary time, and we denote the values of α and β for the ith evolutionary time unit by α_i and β_i, respectively. If we assume that deletion and insertion occur independently in sequences X and Y, the total number of nucleotides that are deleted or inserted per nucleotide site over the entire evolutionary time (t) is given by

$$g = 2 \sum_{i=0}^{t-1} (\alpha_i + \beta_i)$$

$$= 2(\bar{\alpha} + \bar{\beta})t, \tag{5.25}$$

where $\bar{\alpha}$ and $\bar{\beta}$ are the averages of α_i and β_i over the evolutionary time, respectively. Note that g measures only the DNA divergence due to deletion and insertion and no consideration is given to the DNA changes due to nucleotide substitution.

The value of g can be estimated in the following way. We first consider the evolutionary change of the number of nucleotides (m) in the lineage of X. Let $m_X(t)$ be the total number of nucleotides at time t in this lineage. We then have

$$m_X(t) = m_X(t-1)(1 - \alpha_{t-1})(1 + \beta_{t-1})$$

$$= m_X(0) \prod_{i=0}^{t-1} (1 - \alpha_i)(1 + \beta_i)$$

$$\simeq m_X(0)e^{-\Sigma\alpha_i + \Sigma\beta_i}, \tag{5.26}$$

where $m_X(0)$ is the initial number of nucleotides. A similar expression can be obtained for m for Y, i.e., $m_Y(t)$, with $m_Y(0) = m_X(0)$. On the other hand, the proportion of homologous DNA segments shared by X and Y is given by

$$m_{XY}(t) = m_{XY}(t-1)(1 - \alpha_{t-1})^2$$

$$\simeq m_X(0)e^{-2\Sigma\alpha_i}, \tag{5.27}$$

since insertions do not create any homologous DNA segments. Therefore, we have

$$P = \frac{m_{XY}}{\sqrt{m_X m_Y}} = e^{-\Sigma_{i=0}^{t-1}(\alpha_i + \beta_i)} \qquad (5.28)$$

where m_X, m_Y, and m_{XY} are the observed values of $m_X(t)$, $m_Y(t)$, and
$m_{XY}(t)$. Thus, g in (5.25) can be estimated by

$$g = -2 \log_e P. \qquad (5.29)$$

To see the pattern of accumulation of DNA changes due to dele-
tion/insertion, Tajima and Nei (1984) computed the evolutionary distance given by (5.29) for the 5′ flanking region (including about 120
bp upstream starting from the cap site), 5′ leader region (about 50 bp
between the cap site and the initiation codon), intron I (about 130 bp),
and 3′ noncoding region (about 130 bp) of globin genes as well as for
the coding region (about 438 bp or 146 codons). They used Efstratiadis
et al.'s (1980) data for the noncoding region and Hunt et al.'s (1978)
data for the coding region. Their results indicate that the coding regions
of the globin genes are strongly conserved and even the comparison between the human and shark α globin genes gives a value of $g = 0.084$.
Furthermore, the increase of g with evolutionary time was nonlinear. By
contrast, the noncoding regions showed a much more rapid change due
to deletions and insertions, though the rate of change varied considerably among the four different noncoding regions examined. The 3′ noncoding region showed the highest rate and the 5′ leader region the lowest. The initial rate of increase of g was approximately 5×10^{-9} per site
per year for the former and 2×10^{-10} per site per year for the latter,
though the rate of increase of g was not really constant. The higher rate
for the former than the latter occurred apparently because the 5′ leader
region plays an important role for mRNA processing and translation,
and thus the DNA sequence is not flexible. The nonlinear relationship
between g and t is mainly due to the fact that a deletion or insertion
occasionally involves a large number of nucleotides.

Estimation of DNA Divergence from Restriction-Site Data

A restriction endonuclease recognizes a specific sequence of nucleotide
pairs, generally four or six pairs in length, and cleaves it. Therefore, if

a circular DNA such as mitochondrial DNA has m recognition (restriction) sites, it is fragmented into m segments after digestion by this enzyme. The number and locations of restriction sites vary with nucleotide sequence. The higher the similarity of the two DNA sequences compared, the closer the cleavage patterns. Therefore, it is possible to estimate the number of nucleotide substitutions between two homologous DNAs by comparing the locations of restriction sites. Similarly, the number of nucleotide substitutions may be estimated from the proportion of DNA fragments that are shared by the two DNA sequences. This problem was first studied by Upholt (1977), but his mathematical formulation was not very accurate. Later, a number of authors (Nei and Li 1979; Kaplan and Langley 1979; Gotoh et al. 1979; Kaplan and Risko 1981; Nei and Tajima 1983) have studied this problem in more detail. In the following, I first consider the evolutionary change of restriction sites and then discuss the estimation of nucleotide differences from restriction site data.

Evolutionary Change of Restriction Sites

We assume that the four types of nucleotides (A, T, C, G) are randomly arranged in the DNA sequence under investigation, and all nucleotide sites have the same probability of substitution. This assumption does not always hold, but since the restriction enzyme method is used only when the number of nucleotide substitutions per site (d) is relatively small (say, $d<0.3$), violation of the above assumption does not introduce serious errors (Tajima and Nei 1982). At any rate, under this assumption the expected number of restriction sites (m) for a restriction enzyme with a recognition sequence of r nucleotides (usually $r=4$ or 6) is given by

$$E(m) = m_T a, \qquad (5.30)$$

where m_T is the total number of nucleotides and a is the probability that a sequence of r nucleotides in the DNA is a restriction site.

Under our assumption, a is given by

$$a = g_A^{r_A} g_T^{r_T} g_C^{r_C} g_G^{r_G}, \qquad (5.31)$$

where g_A, g_T, g_C, and g_G are the frequencies of nucleotides A, T, C, and G in the DNA sequence, respectively, and r_A, r_T, r_C, and r_G are the

numbers of A, T, C, and G in the recognition sequence, respectively
$(r_A + r_T + r_C + r_G = r)$. For example, the recognition sequence of enzyme
EcoRI is GAATTC, so that $r_A = 2$, $r_T = 2$, $r_C = 1$, and $r_G = 1$. When a
restriction enzyme identifies more than one type of recognition sequence
(e.g., HaeI), a is given by a somewhat different formula, as will be
discussed later. Usually, a is much smaller than 1. For example, in the
human mitochondrial DNA, $m_T = 16,569$, $g_A = 0.25$, $g_T = 0.31$,
$g_C = 0.14$, and $g_G = 0.30$, so that $a = 0.00025$ and $E(m) = 4.2$ for EcoRI.
This expected number of restriction sites per DNA sequence agrees quite
well with the observed number (3.8) (Tajima 1983b). Obviously, a is
generally greater for four-base restriction enzymes than for six-base enzymes. However, when the nucleotide frequencies deviate considerably
from 1/4, a can be small even for a four-base enzyme.

Let us now consider two DNA sequences (X and Y) that diverged t
years (or generations) ago and compare all possible restriction sites of the
two sequences. We note that there are m_T possible restriction sites in a
circular DNA of m_T nucleotides. In a linear DNA, the possible number
of restriction sites is $m_T - r + 1$, but m_T is usually much larger than r,
so that the possible number is again approximately m_T. In the comparison of restriction sites between two DNA sequences, there are four different cases. A sequence of r nucleotides at a particular location of the
DNA is a restriction site (1) for both X and Y, (2) for X but not for Y,
(3) for Y but not for X, or (4) for neither X nor Y. Let m_X and m_Y be
the numbers of restriction sites for DNA sequences X and Y, and m_{XY}
be the number of restriction sites shared by the two sequences. The
numbers of observations for the above four different cases are then given
by m_{XY}, $m_X - m_{XY}$, $m_Y - m_{XY}$, and $m_T - m_X - m_Y + m_{XY}$, respectively.

Let us now derive the probabilities of having these four events, considering a restriction enzyme with a unique recognition sequence. Let w_i
be the probability that a sequence of r nucleotides at a location of the
DNA is different from the recognition sequence by i nucleotides, P be
the probability that a site of a sequence of r nucleotides which was a
restriction site at time 0 becomes a nonrestriction site at time t, and Q_i
be the probability that a sequence of r nucleotides which was originally
different from the recognition sequence by i nucleotides becomes a restriction site at time t. Since the expected number of restriction sites
remains constant over time, we have the relationship

$$aP = \sum_{i=1}^{r} w_i Q_i. \qquad (5.32)$$

The probability that a sequence of r nucleotides in the DNA is a restriction site for both X and Y is then given by $a(1-P)^2+\Sigma_i w_i Q_i^2$. The first term in this expression represents the probability that a restriction site that existed in the ancestral sequence still remains a restriction site, whereas the second term represents the probability of having a new restriction site by mutation. The sum of the two terms may be written as aS, where $S=(1-P)^2+\Sigma_i w_i Q_i^2/a$ is the probability that X and Y share the same recognition sequence at a given site. The probability that a sequence of r nucleotides is a restriction site for X but not for Y is $aP(1-P)+\Sigma_i w_i Q_i(1-Q_i)=a[1-\{(1-P)^2+\Sigma_i w_i Q_i^2/a\}]=a(1-S)$. The probability of having the third case is the same as that for the second case. The probability that a sequence of r nucleotides is not a restriction site for both X and Y is $aP^2+\Sigma_i w_i(1-Q_i)^2=1-a[2-(1-P)^2-\Sigma_i w_i Q_i^2/a]=1-a(2-S)$ since $\Sigma_i^r w_i=1-a$.

In these probabilities, S may be written as

$$S=(1-p)^r, \tag{5.33}$$

where p is the probability that sequences X and Y have different nucleotides at a given nucleotide position. This p is related to the expected number of nucleotide substitutions per site (d) by $p=(3/4)[1-\exp(-4d/3)]$ in (5.3). If the rate of nucleotide substitution per site per year is λ, d is given by $d=2\lambda t$. Therefore, it is possible to estimate d if we know S.

We also note that S may be approximated by $(1-P)^2$, since $\Sigma_i w_i Q_i^2/a$ is usually much smaller than $(1-P)^2$ (Nei and Li 1979). This approximation amounts to neglecting shared restriction sites that have newly arisen by independent mutations in X and Y. In practice, an even more approximate formula for S, i.e.,

$$S=e^{-2r\lambda t}$$

may be used as long as $d<0.25$ (Nei and Li 1979; Li 1981a; Kaplan and Risko 1981). In this case, $P=1-e^{-r\lambda t}$.

Estimation of Nucleotide Substitutions

Let us now consider how to estimate the number of nucleotide substitutions per site (d) from actual data. Since the probabilities of having the four cases are a function of d, we can use the maximum likelihood method to estimate d. The maximum likelihood method for estimating

d was first studied by Kaplan and Langley (1979), Gotoh et al. (1979), and Kaplan and Risko (1981). Later, Nei and Tajima (1983) simplified it considerably. We first consider the likelihood of having the observed values of m_X, m_Y, and m_{XY} with the parameters of S and a. It can be written as

$$L = C(aS)^{m_{XY}}[a(1-S)]^{m_X+m_Y-2m_{XY}}$$
$$\times [1 - a(2-S)]^{m_T-m_X-m_Y+m_{XY}}, \qquad (5.34)$$

where C is a constant.

Previously, we defined a in terms of nucleotide frequencies. In practice, we do not know nucleotide frequencies, but a can be estimated simultaneously with S. The maximum likelihood estimates ($\hat{a}$ and $\hat{S}$) of a and S are given by solving the following equations.

$$\frac{\partial \log_e L}{\partial a} = \frac{m_X + m_Y - m_{XY} - m_T a(2-S)}{a[1-a(2-S)]} = 0. \qquad (5.35)$$

$$\frac{\partial \log_e L}{\partial S} = \frac{m_{XY}-(m_X+m_Y-m_{XY})S}{S(1-S)} + \frac{(m_T-m_X-m_Y+m_{XY})a}{1-a(2-S)} = 0, \qquad (5.36a)$$

$$= \frac{2m_{XY}-(m_X+m_Y)S}{S(1-S)(2-S)} = 0. \qquad (5.36b)$$

Therefore,

$$\hat{a} = \frac{m_X+m_Y-m_{XY}}{m_T(2-S)}, \qquad (5.37)$$

$$\hat{S} = m_{XY}/\hat{m}, \qquad (5.38)$$

where $\hat{m} = (m_X+m_Y)/2$. Note that (5.36b) is obtained by putting (5.37) into (5.36a). Note also that if we substitute $\hat{S}$ for S, (5.37) becomes

$$\hat{a} = \frac{m_X+m_Y}{2m_T}. \qquad (5.37a)$$

The variances [$V(\hat{S})$ and $V(\hat{a})$] of $\hat{S}$ and $\hat{a}$ are

$$V(\hat{S}) = \frac{S(1-S)(2-S)}{2m_T a}, \qquad (5.39)$$

$$V(\hat{a}) = \frac{a(1+S)}{2m_T}. \qquad (5.40)$$

Once $\hat{S}$ is obtained, the proportion of nucleotide differences (p) can be estimated by

$$\hat{p} = 1 - \hat{S}^{1/r}, \qquad (5.41)$$

and d by equation (5.3). When $d < 0.25$, $S = e^{-2r\lambda t}$ approximately. Therefore, $d \equiv 2\lambda t$ can also be estimated by

$$\hat{d}_1 = [-\log_e \hat{S}]/r \qquad (5.42)$$

(Nei and Li 1979). The variances of these estimates are given by

$$V(\hat{p}) = \frac{(1-p)^2(2-S)(1-S)}{2r^2 \bar{m} S}, \qquad (5.43)$$

$$V(\hat{d}) = \frac{9(1-p)^2(2-S)(1-S)}{2r^2 \bar{m}(3-4p)^2 S}, \qquad (5.44)$$

$$V(\hat{d}_1) = \frac{(2-S)(1-S)}{2r^2 \bar{m} S}, \qquad (5.45)$$

where $\bar{m}$ is equal to $m_T a$ and can be estimated by $\hat{m} = (m_X + m_Y)/2$.

In the above formulation, we considered restriction-site data obtained by a single restriction enzyme. However, the above formulation applies to data from many different restriction enzymes as long as they have the same r value. The only thing one has to do in this case is to take the summation of m_X, m_Y, m_{XY}, and m_T for all enzymes.

Restriction Enzymes with Multiple Recognition Sequences

Most restriction enzymes used for evolutionary studies recognize a unique sequence of 4 or 6 nucleotides. However, there are enzymes that recognize multiple sequences of a given number of nucleotides. For example, *Hind*II recognizes the sequence GTPyPuAC, where Py is either T or C and Pu is either A or G. In this case, a is given by $(g_A + g_G)$ $(g_T + g_C)g_A g_T g_C g_G$, whereas $S = (1-p)^4(1-2p/3)^2$. S can be approximately written as $(1-p)^{16/3}$ for $p \leq 0.3$. Therefore, if we redefine r as

Table 5.7   Examples of various types of type II restriction enzymes, expected frequencies of restriction sites (a), and the r values. One example from each type is given.

Enzyme	Recognition sequence	a	r
Four-base			
*Hae*III	GGCC	$g_C^2 g_G^2$	4
*Mnl*I[a]	CCTC	$g_T g_C^3 + g_A g_G^3$	4
Five-base			
*Hinf*I	GANTC	$g_A g_T g_C g_G$	4
*Eco*RII	CC(^A_T)GG	$(g_A + g_T) g_C^2 g_G^2$	14/3
*Mbo*II[a]	GAAGA	$g_A^3 g_G^2 + g_T^3 g_C^2$	5
Six-base			
*Eco*RI	GAATTC	$g_A^2 g_T^2 g_C g_G$	6
*Hae*I	(^A_T)GGCC(^A_T)	$(g_A + g_T)^2 g_C^2 g_G^2$	16/3
*Hind*II	GTPyPuAC	$(g_A + g_G)(g_T + g_C) g_A g_T g_C g_G$	16/3
Seven-base			
*Eca*I	GGTNACC	$g_A g_T g_G^2 g_C^2$	6

[a] *Mnl*I recognizes double strand sequences $\frac{CCTC}{GGAG}$ and $\frac{GAGG}{CTCC}$ whereas *Mbo*II recognizes $\frac{GAAGA}{CTTCT}$ and $\frac{TCTTC}{AGAAG}$. However, these enzymes are used very rarely.

$r = 16/3$, $S = (1-p)^r$ still holds. Another six-base enzyme with multiple recognition sequences is *Hae*I, which recognizes (^A_T)GGCC(^A_T). In this case, $a = (g_A + g_T)^2 g_C^2 g_G^2$, and $S \simeq (1-p)^{16/3}$. Namely, r is the same as that for *Hind*II. The values of a and r for nine different types of restriction enzymes are given in table 5.7. Most (type II) restriction enzymes belong to one of these groups, so that we can compute a and r.

Some Remarks

It should be noted that our formulation depends on a number of assumptions. One of them is that the rates of substitution between the four different nucleotides are more or less the same. This assumption does not necessarily hold, but its effect is not serious, since the present method is intended to be applied to the case of relatively small values of d, and in this case the effect of rate heterogeneity is not important, as mentioned earlier. The second assumption we have implicitly made is that the frequencies of the four nucleotides are more or less the same.

This assumption also does not necessarily hold, but its effect on the estimate of d again seems to be small. Kaplan and Langley (1979) and Kaplan and Risko (1981) developed a method of estimating d when this assumption does not hold, but their method gives essentially the same result as that of equation (5.42) unless d is larger than 0.3 (Kaplan 1983).

In the above formulation, we have considered a given region of DNA of which the total number of nucleotides is m_T. The model used is most appropriate for mitochondrial DNA or chloroplast DNA. In the study of evolutionary divergence of nuclear DNA, however, all restriction fragments which hybridize with a given DNA probe are studied. Namely, the entire DNA of an organism is first digested by a given restriction enzyme, and DNA fragments of different lengths are separated by electrophoresis. The fragments which are homologous to a given DNA probe are then identified by the Southern blot method. In this case, any restriction fragment which is partially or wholly homologous to this probe can be detected, as can be seen from figure 10.1. It is clear from this figure that one restriction site almost always exists on each of the right- and left-hand sides of the probe region. If we consider only the DNA region which is covered by the DNA probe, the mathematical theory presented above obviously applies (Page et al. 1984). In this case, however, information on the evolutionary change of DNA outside the probe region is lost. Actually, it can be shown that the same theory is applicable to the DNA region that includes the restriction sites existing outside the probe, as long as d is relatively small. The reason for this is that S is approximately $\exp(-2r\lambda t)$ and this S can still be estimated by the proportion of shared sites, including the outside restriction sites (M. Nei, unpublished).

Estimation of Nucleotide Substitutions from Many Different Enzymes

When different kinds of enzymes with different r values are used, a somewhat complicated maximum likelihood method is required to estimate p or d. The likelihood corresponding to (5.34) is

$$L = C \prod_{i=1}^{k} (a_i S_i)^{m_{XYi}} [a_i(1 - S_i)]^{m_{Xi} + m_{Yi} - 2m_{XYi}}$$

$$\times [1 - a_i(2 - S_i)]^{m_T - m_{Xi} - m_{Yi} + m_{XYi}}, \tag{5.46}$$

where i refers to the ith type of enzymes with r_i recognition nucleotides and k is the number of different kinds of restriction enzymes. In the present case, it is better to estimate p or d directly rather than S_i, since $S_i = (1-p)^{r_i}$ is a function of p. We then have

$$\frac{\partial \log_e L}{\partial p} = \Sigma_i r_i \frac{[2m_{XYi} - (m_{Xi} + m_{Yi})(1-p)^{r_i}]}{[1-(1-p)^{r_i}][2-(1-p)^{r_i}]} = 0, \tag{5.47}$$

$$\frac{\partial \log_e L}{\partial a_i} = \frac{1}{a_i}\left[m_{Xi} + m_{Yi} - m_{XYi} - \frac{a_i(m_T - m_{Xi} - m_{Yi} + m_{XYi})(2-S_i)}{1-a_i(2-S_i)} \right] = 0. \tag{5.48}$$

From (5.48), we obtain

$$a_i = \frac{m_{Xi} + m_{Yi} - m_{XYi}}{m_T(2-S_i)}. \tag{5.49}$$

Equation (5.47) does not yield a simple expression for p. However, p can be estimated easily if we use the following iteration method that can be obtained from (5.47).

$$\hat{p} = \hat{p}_1 \frac{\Sigma_i r_i (\hat{m}_i - m_{XYi})/[\{1-(1-\hat{p}_1)^{r_i}\}\{2-(1-\hat{p}_1)^{r_i}\}]}{\Sigma_i r_i \hat{m}_i /[2-(1-\hat{p}_1)^{r_i}]}, \tag{5.50}$$

where $\hat{p}_1$ is a trial value of $\hat{p}$ (Nei and Tajima 1983). When $\hat{p} = \hat{p}_1$, $\hat{p}$ is the maximum likelihood estimate of p. In practice, we can use $\hat{p}_1 = 1 - S^{1/r}$ for a particular kind of restriction enzymes as a trial value. Usually, four or five cycles of iterations are sufficient for obtaining the maximum likelihood estimate.

Once $\hat{p}$ is obtained, d is estimated by (5.3). The variance of $\hat{d}$ thus obtained is given by

$$V(\hat{d}) = 1/\sum_{i=1}^{k} [1/V(d_i)], \tag{5.51}$$

where $V(d_i)$ is the value of (5.44) for the ith type of restriction enzymes when $S = (1-\hat{p})^r$ is used. The variance of $\hat{p}$ can also be obtained in the same way by using (5.43).

EXAMPLE

Gotoh et al. (1979) have compiled restriction-site data for rat and mouse mitochondrial DNAs. In the comparison of two strains of the rat, A and B, four six-base enzymes, one four-base enzyme and two multiple-sequence enzymes with $r = 16/3$ were used. The numbers of restriction sites identified are given in table 5.8. We first compute $\hat{p}_1$ by using the data for six-base enzymes, where $\hat{S} = 22/23$. $\hat{p}_1$ then becomes $1 - (22/23)^{1/6} = 0.0073813$. If we use (5.50), we have $\hat{p}_2 = 0.0097470$ as the second estimate. Further iterations give $\hat{p}_3 = 0.0097983$ and $\hat{p}_4 = 0.0097983$. We can therefore take $\hat{p} = 0.00980$ as the maximum likelihood estimate. The estimate of d can be obtained by using equation (5.3), and it becomes $\hat{d} = 0.00986$. To compute the variance of $\hat{d}$, we must first determine $V(d_i)$ in (5.51). Substituting $(1 - 0.0098)^r$ for S and m_i for $\bar{m}$ in (5.44), we have $V(d_i) = 0.00003912$ for $r = 6$, 0.00005966 for $r = 4$, and 0.000015429 for $r = 16/3$. We therefore have $V(\hat{d}) = 0.00002049$ from (5.51), and the standard error becomes 0.00453. A similar set of restriction-site data exists for the comparison of the rat and mouse (table 5.8). In this case, the data for six-base enzymes give $\hat{p}_1 = 0.247327$, and after a few cycles of iterations we obtain $\hat{p} = 0.250$. Using equations (5.3) and (5.51), we therefore have $\hat{d} = 0.303 \pm 0.073$.

Table 5.8   Numbers of restriction sites observed in the comparisons of mitochondrial DNA between rat strains A and B and between the rat and mouse.

Enzymes used	r	$\dfrac{m_X + m_Y}{2}$	m_{XY}
Rat(A)−Rat(B)			
Six-base[a]	6	23	22
Four-base[b]	4	22	21
Multiple-sequence[c]	16/3	6.5	6
Rat−Mouse			
Six-base[d]	6	16.5	3
Four-base[e]	4	7	2
Multiple-sequence[f]	16/3	4	1

[a] *Bam*HI, *Eco*RI, *Hind*III, *Hpa*II. [b] *Hae*III. [c] *Hae*II, *Hind*II. [d] *Bam*HI, *Eco*RI, *Hind*III, *Hpa*I, *Pst*I. [e] *Hha*I. [f] *Hae*II, *Hind*II.

Restriction Fragment Data

In the above theory, we assumed that all restriction sites can be mapped on the DNA sequence so that shared and nonshared restriction sites can be determined. Although mapping of restriction sites is not difficult, it is often too laborious to be used for a large-scale population survey. In the case of mitochondrial or chloroplast DNA, where the total length is fixed, there is a simpler method. That is, comparing the electrophoretic patterns of DNAs digested by a restriction endonuclease between the two DNA sequences in question, one can estimate d. The rationale for this is that the degree of genetic divergence between two DNA sequences is correlated with the proportion of DNA fragments shared by them. Three different methods (Upholt 1977; Nei and Li 1979; Engels 1981a) have been developed for estimating d from this information. According to Kaplan's (1983) computer simulation, however, all of them give essentially the same results. In the following, let us discuss Nei and Li's method because it is simple and mathematically a little more rigorous than the others.

Nei and Li have shown that the expected proportion of shared DNA fragments (F) can be expressed by the following approximate formula

$$F \simeq G^4/(3 - 2G), \qquad (5.52)$$

where G is $e^{-r\lambda t}$. Therefore, $d \equiv 2\lambda t$ can be estimated from F by the above equation. To estimate d, we first note that F can be estimated by

$$\hat{F} = 2m_{XY}/(m_X + m_Y), \qquad (5.53)$$

where m_X and m_Y are the numbers of restriction fragments in DNA sequences X and Y, respectively, whereas m_{XY} is the number of fragments shared by the two sequences. If $\hat{F}$ is obtained, we can estimate G by the iteration formula,

$$\hat{G} = [\hat{F}(3 - 2\hat{G}_1)]^{1/4}, \qquad (5.54)$$

where $\hat{G}$ is an estimate of G and $\hat{G}_1$ is a trial value of $\hat{G}$. This iterative computation is done until $\hat{G} = \hat{G}_1$ is obtained. Usually a few cycles of iterations are sufficient. I suggest that $\hat{F}^{1/4}$ be used as the first trial value of $\hat{G}_1$. Once $\hat{G}$ is obtained, d can be estimated by

$$\hat{d} = -(2/r)\log_e\hat{G}. \qquad (5.55)$$

Suppose that one obtains $m_X = 40$, $m_Y = 36$, and $m_{XY} = 20$ for a pair of mtDNAs by using 10 six-base restriction enzymes. We then have $\hat{F} = 0.5263$ and $\hat{G}_1 = \hat{F}^{1/4} = 0.8517$. Putting these into equation (5.54), we have $\hat{G} = 0.9089$. Now regarding this value as $\hat{G}_1$ in (5.54), we have $\hat{G} = 0.8816$. Repeating this process a few more times, we obtain $\hat{G} = 0.8938$ as the final estimate. We therefore have $\hat{d} = 0.037$ from equation (5.55).

It should be noted that the above method is useful only for the case of small d because F rapidly declines as d increases. When F is small, it is affected by random errors so much that the estimate of d is unreliable. Kaplan's (1983) computer simulation suggests that (5.55) gives an underestimate of d when $d \geq 0.1$. In general, the estimate obtained by this method (length-difference method) has a larger variance than that obtained by the previous site-difference method. Furthermore, estimates of d obtained by the length-difference method are more susceptible to the errors caused by undetectable restriction fragments or fragment length differences than those obtained by the site-difference method (Tajima 1983b). In usual electrophoresis of DNAs digested by a restriction enzyme, small fragments are often undetected. For example, in Brown et al.'s (1979) study of mitochondrial DNA, a DNA fragment less than about 150 nucleotides was not detected. In the study of nuclear DNA by using the Southern blot method, even larger fragments are often undetected. Two DNA fragments with similar but not identical lengths are also indistinguishable in usual electrophoresis.

Nevertheless, the length-difference method is simple to use and gives a fairly accurate estimate of d when d is smaller than 0.05. Therefore, in the estimation of d for a pair of highly homologous DNA sequences, it is a useful method. Avise et al. (1979), Kessler and Avise (1985), and Saunders et al. (1986) took advantage of this method in their studies of the evolutionary relationships of polymorphic mitochondrial DNAs in vertebrate organisms and horseshoe crabs. Terachi et al. (1984) also used this method (Engels' method) in their study of the evolutionary relationship of chloroplast DNAs from wheat and its related species.

DNA Hybridization

Another method of quick determination of the extent of nucleotide differences between a pair of DNA sequences is DNA hybridization. In this method, hybrid DNA from the single strands of two different species is formed, and the extent of nucleotide differences is estimated from the thermostability of the hybrid DNA. Thermostability decreases as the proportion of mismatched nucleotides increases. This technique has been used extensively by Britten and Kohne (1968), Walker (1968), Kohne (1970), and other authors in the study of the organization of eukaryotic genomes. Indeed, it was this technique that led Britten and Kohne (1968) to discover the existence of a large amount of repetitive DNA in the genome of higher organisms. Ironically, repetitive DNAs are a nuisance in the study of evolutionary differentiation of genomes, and they are usually eliminated from the study.

The basic procedure of this method is as follows. (1) DNAs obtained from the nuclei of tissue cells are sheared by sonication into fragments with an average length of several hundred nucleotides, and repetitive (multiple-copy) DNAs are eliminated. (2) The remaining "single-copy" DNAs are denatured into single strands. (3) Single-strand DNAs from one species are hybridized with those from the other species. (4) The thermal stability of the hybrid DNAs is measured by the temperature in Celsius at which 50 percent of the hybrid DNAs are dissociated into single strands. (5) This 50 percent melting temperature for the hybrid DNAs from different species is compared with that for the hybrid DNAs from the same species. The difference between the two temperatures (T_D) is known to be approximately linearly related to the proportion of nucleotide differences between the two DNA sequences examined; the 50 percent melting temperature declines as this proportion (p) increases. Namely, p is related to T_D by

$$p = cT_D, \tag{5.56}$$

where c is the proportionality constant and believed to be about 0.01 (Britten et al. 1974). For more detailed aspects of the experimental method, the reader may refer to Britten et al. (1974), Sibley and Ahlquist (1981), and Hunt et al. (1981). The details vary with the author and the organism used.

It should be noted that equation (5.56) is an empirical formula and its theoretical basis is not well established. The c value also apparently varies with the experimental condition. Kohne et al. (1972) have suggested $c = 0.015$ instead of $c = 0.01$. For this reason, Sibley and Ahlquist (1981) do not want to convert T_D into p. Rather, they use T_D as a measure of the extent of DNA divergence. Note also that the separation of repetitive and nonrepetitive DNAs is somewhat arbitrary, and even the so-called "single-copy" DNAs are expected to include a substantial proportion of genes of low copy number such as those of globin genes. Since the separation of repetitive and nonrepetitive DNAs is not very rigorous, the same homologous set of "single-copy" DNAs may not be obtained from the two species under investigation. Probably for this and some additional reasons (McCarthy and Farquhar 1972), the experimental error of T_D seems to be substantial. Sibley and Ahlquist (1984) suggest that five to six replicate observations be made for the same pair of species. When closely related species are studied, the effect of DNA polymorphism within species should also be taken into account. Another problem with this method is the effect of deletions and insertions of nucleotide sequences. The proportion of deletions and insertions is expected to increase as the evolutionary time increases, but their effect on T_D is not well understood. Since a large proportion of the DNAs used for hybridization experiment are noncoding regions (Britten 1986), this method will be affected more seriously by deletions and insertions than the sequence comparison method for coding regions.

Despite these problems, DNA hybridization seems to be a useful technique for studying the genetic relationship of relatively closely related species. Recently, Sibley and Ahlquist (1981, 1984, and references therein) have used this technique extensively for clarifying the evolutionary relationships of bird and hominoid species. Table 5.9 shows the estimates of T_D for the pairwise comparisons of seven species of primates. The evolutionary tree obtained by UPGMA (see chapter 11) from these data is given in figure 2.4 (chapter 2). This suggests that the two

Table 5.9   Estimates of T_D for the pairwise comparisons of seven species of primates. From Sibley and Ahlquist (1984).

	Human	Chimp	Pygmy chimp	Gorilla	Orangutan	Gibbon
Chimpanzee	1.8 ± 0.1					
Pygmy chimp	1.9 ± 0.1	0.7 ± 0.1				
Gorilla	2.4 ± 0.1	2.1 ± 0.1	2.3 ± 0.1			
Orangutan	3.6 ± 0.1	3.7 ± 0.1	3.7 ± 0.1	3.8 ± 0.1		
Gibbon	5.2 ± 0.1	5.1 ± 0.1	5.6 ± 0.2	5.4 ± 0.1	5.1 ± 0.1	
Baboon	7.7 ± 0.1	7.7 ± 0.1	8.0 ± 0.2	7.5 ± 0.2	7.6 ± 0.1	7.4 ± 0.2

species of the chimpanzee are closer to man than to the gorilla, which is in turn closer to man than to the orangutan or the gibbon.

Figure 2.4 includes the evolutionary time scale as well as the T_D value. The evolutionary time scale in that figure was obtained by assuming that the orangutan diverged from the human and African ape group 13 MY ago. The relationship between evolutionary time (t) and T_D obtained under this assumption is $t = 3.5 \times 10^6 \, T_D$. This calibration suggests that the human and chimpanzee lines diverged about 7 MY ago. This is somewhat higher than Sarich and Wilson's estimate (about 5 MY ago), but since both estimates are dependent on a number of assumptions, it is difficult to make any definite conclusion at the present time. Note that if the relationship $t = 3.5 \times 10^6 \, T_D$ is correct and if we assume $p = 0.01 \, T_D$, then the average rate of nucleotide substitution for "single-copy" DNAs is given by $\lambda = 0.01 \times 3.7/(2 \times 13 \times 10^6) = 1.4 \times 10^{-9}$ per nucleotide site per year from the comparison of the orangutan and the human and African ape lines. This seems to be a little lower than what is expected from the average rate for the coding and noncoding regions of DNAs for the genes so far studied (table 5.3).

According to Britten (1986), the λ value varies with taxonomic group. He estimates that λ is 1.3×10^{-9} for higher primates and birds and 6.6×10^{-9} for rodents and sea urchins. The reason for this difference is not known, but he speculates that the mutation rate in higher primates and birds is lower than that in rodents and sea urchins, probably because the former group of organisms has developed a better DNA repair system. At any rate, if this is the case, there will be no universal molecular clock. However, note that even this group-dependent molecular clock is very useful for the study of evolution.

CHAPTER SIX

GENOMIC EVOLUTION

The evolutionary change of DNA can be divided into two different categories, i.e., (1) nucleotide substitution in genes and (2) change in DNA content or the total number of nucleotides in the genome. In the preceding two chapters, we were mainly concerned with the first class of change. Until recently, the second class of change was thought to be relatively unimportant except when long-term evolution was considered. Recent studies, however, indicate that the total number of nucleotides in the genome changes quite frequently and thus the second class of change plays an important role in evolution (Dover and Flavell 1982; Nei and Koehn 1983). In this chapter, I would like to present a summary of recent studies in this area.

Evolutionary Change of Genome Size

During about three billion years of evolution from bacteria to mammals, genome size has increased enormously, the DNA content for mammals being about 1,000 times higher than that of bacteria. Some viruses such as φX174 and F1 have a genome consisting of only 6 to 8 genes (about 6,000 nucleotides), although these viruses would not represent the oldest form of organisms. This increase in genome size was clearly important for organisms evolving from simple to complex form. For a highly developed, complex organism to maintain its life, a large number of genes are required. In fact, there are many genes that exist only in higher organisms. For example, the genes for fibrinogen, haptoglobin, and immunoglobulins exist only in vertebrate animals.

However, a close examination of the genomic sizes of various organisms shows that DNA content is not necessarily correlated with the complexity of the organism (Sparrow et al. 1972; Hinegardner 1976). For example, a species of lungfish has a DNA content about 40 times higher than that of mammals (table 6.1). Many amphibians also have a larger amount of DNA than mammals. Thus, a high DNA content alone is not sufficient to produce a complex organism. For a complex organism

Table 6.1   DNA contents of various organisms.

Organism	Nucleotide pairs per genome	Organism	Nucleotide pairs per genome
Mammals	3.2×10^9	Fruitfly	0.1×10^9
Birds	1.2×10^9	Maize	7×10^9
Lizards	1.9×10^9	Neurospora	4×10^7
Frogs	6.2×10^9	E. coli	4×10^6
Most bony fish	0.9×10^9	T_4 phage	2×10^5
Lungfish	111.7×10^9	λ phage	1×10^5
Echinoderm	0.8×10^9	ϕX174	6×10^3

to be produced, there must be a sufficiently large number of different kinds of genes in the genome. A large fraction of DNA in lungfish and salamanders is heterochromatic and thus seems to be nonfunctional (Hinegardner 1976).

Mechanisms of Increase in DNA Content

The high DNA contents of higher organisms are believed to have occurred largely by gene duplication. There are two types of gene duplication. One is chromosome duplication, and the other is the duplication of a small segment of chromosome *(tandem duplication)* by unequal crossover. A common type of chromosome duplication is genome duplication. As shown in table 6.1, mammalian DNA is about 1,000 times larger than *Escherichia coli* DNA. If the increase in DNA content is entirely due to genome duplication, there must have been about ten genome duplications from bacteria to mammals ($2^{10} \simeq 1,000$). If bacteria evolved about 3×10^9 years ago (chapter 2), genome duplication must have occurred on the average once in every 300 million years (Nei 1969). On the other hand, if DNA content increases continuously by unequal crossover at a rate of k per year, the DNA content (n_t) at time t is given by $n_t = n_0 (1 + k)^t \simeq n_0 \exp(kt)$, where n_0 is the initial DNA content. From bacteria to mammals, DNA content increased 1,000 times in about 3×10^9 years. Therefore, k is estimated to be 2.3×10^{-9}. This means that the DNA content comparable to that of mammals would increase by an average of seven nucleotide pairs per year.

In plant evolution, genome duplication or polyploidization played an important role, as documented by Stebbins (1950). It also seems to have been quite important in the evolution of animals. Cytological and biochemical studies indicate that genome duplication has occurred several times in fish and amphibians (Ohno 1970; Ferris and Whitt 1979). It seems that genome duplication was quite common in animal evolution before sex chromosomes were differentiated. Once the differentiation of sex chromosomes was completed in the mammalian, avian, and reptilian lineages, genome duplication seems to have disrupted the mechanism of sex determination, and thus the resulting tetraploid almost never spread through the population. In most fish and amphibians, sex chromosomes have not been established, and tetraploid males and females can be maintained without much difficulty (Ohno 1967).

Ohno (1970) has argued that genome duplication is generally more important than tandem gene duplication because the latter may duplicate only parts of the genetic system of structural genes and regulatory genes and may disrupt the function of duplicate genes, whereas the former duplicates the entire genetic system with little harmful effect. Recent molecular studies of genome organization in eukaryotes, however, indicate that most genes do not exist as single copies in the genome but rather in clusters. Therefore, tandem duplication apparently played an important role in evolution. Good examples of clustered genes are the rRNA genes and tRNA genes. Table 6.2 gives the number of rRNA

Table 6.2 Numbers of rRNA genes and tRNA genes per haploid genome in various organisms.

Gene	Organism	Number	Genome size (base pairs)
rRNA	Human mitochondrial genome	1	16,600
	Mycoplasma capricolum	2	1×10^6
	Escherichia coli	7	4×10^6
	Saccharomyces cerevisiae	140	5×10^7
	Drosophila melanogaster	130–250	1×10^8
	Xenopus laevis	400–600	8×10^9
	Human	300	3×10^9
tRNA	Human mitochondrial genome	22	16,600
	E. coli	100	4×10^6
	S. cerevisiae	320–400	5×10^7
	D. melanogaster	750	1×10^8
	X. laevis	7,800	8×10^9
	Human	1,300	3×10^9

and tRNA genes per haploid genome for a variety of organisms. The mitochondrial genome of mammals has only one set of 12S and 16S rRNA genes. The mycoplasmas are the smallest self-replicating prokaryotes, and *Mycoplasma capricolum* contains only two sets of rRNA genes. The number of rRNA genes for *E. coli* is four or five times larger than that for *M. capricolum*. The numbers of rRNA genes in yeast, fruitflies, *Xenopus,* and mammals are even larger. Thus, the number of rRNA genes has gradually increased in the evolutionary process, apparently because a higher organism requires more rRNAs for protein synthesis. A similar tendency is observed for the number of tRNA genes (table 6.2).

Tandem duplication is important not only for increasing the number of genes with the same function but also for generating genes with new functions. For example, the human β globin gene family has five functional genes (figure 6.1). One of them (ϵ) is functional only in the early embryonic stage, whereas $^{G}\gamma$ and $^{A}\gamma$ produce globin polypeptides only in the fetal stage. The remaining two functional genes (δ and β) are responsible for adult hemoglobins. However, the nucleotide sequences of these genes have a high degree of similarity, so that all of them were apparently produced by tandem duplication (see the following section).

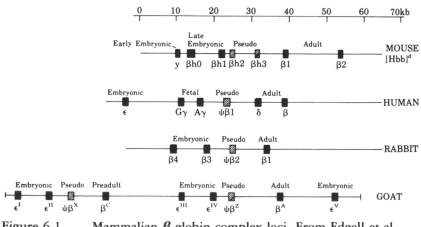

Figure 6.1.   Mammalian β globin complex loci. From Edgell et al. (1983).

Formation of New Genes

Duplication of Complete Genes

If two duplicate genes are produced from a gene, one of them may mutate drastically and acquire an entirely different function. The simplest way to determine whether a pair of genes have descended from a common ancestor is to examine the similarity of the nucleotide sequences of the genes or the amino acid sequences of the proteins encoded by the genes. In fact, by examining the amino acid sequences of the α, β, and γ chains of hemoglobins in man, Ingram (1963) was able to show that the genes responsible for the three chains of human hemoglobin were produced by gene duplication. Comparison of the three chains indicates that the proportion of amino acids shared by the α and β chains is as high as 41 percent, while that for the β and γ chains is even higher (73 percent) (table 6.3). These similarities are so high that the probability that they have occurred by chance is negligible. Ingram further showed that myoglobin also originated from the same common ancestor as that of the three chains of hemoglobins.

After Ingram's study, many examples of formation of new genes by gene duplication were discovered. Table 6.3 gives some examples. The approximate time of divergence for each pair of homologous proteins was computed by using the amino acid substitution matrix method discussed in chapter 4. The functions of some pairs of homologous proteins such as hemoglobin and myoglobin are considerably different, while other homologous proteins such as hemoglobin β and δ chains still maintain essentially the same function. Human β and δ chains are apparently interchangeable, since the proportion of hemoglobin $\alpha_2\delta_2$ in adults varies considerably among individuals without any noticeable effect. It is also noted that the pairs of homologous proteins between which the amino acid sequences differ by more than 50 percent generally have differentiated functions. In general, there is little functional differentiation between a pair of proteins for which the sequence divergence is less than 15 percent.

Under certain conditions, however, a gene with new function may be formed through a relatively small number of mutational steps. This occurs particularly when the substrates of the original and mutant enzymes are closely related. The normal strain of the bacteria *Pseudomonas aerugi-*

Table 6.3 Extents of divergence and functional differences between proteins derived from gene duplications. Chemical activities include differences in catalytic action and in binding to substrates, inhibitors, antigens, etc. From Dayhoff and Barker (1972).

Proteins	Amino acid diff. (%)	Divergence time (MY)	Chemical activities	Aggregation properties	Action sites
Hemoglobin-myoglobin	77	1100	−	+ +	+
Growth hormone-prolactin	75	200	+	−	+
Immunoglobulin heavy and light chains	75	400	+ +	+	−
Immunoglobin μ and γ chain C regions	70	350	+	+	+
Trypsin-thrombin	65	1500	+	−	+
Lactalbumin-lysozyme	63	350	+ +	−	+
Immunoglobulin κ and λ chain C regions	62	300	−	−	−
Hemoglobin α and β chains	59	600	−	+	−
Hemoglobin β and γ chains	27	130	−	−	−
Protamines, salmine AI and AII	22	100	−	−	−
Chymotrypsin A and B	21	270	−	−	−
Growth hormone-lactogen	15	23	+	−	+
Hemoglobin β and δ chains	8	40	−	−	−
Alcohol dehydrogenase E and S chains	2	10	+	−	−

+ + Very different.   + Different.   − Similar.

nosa uses acetamide and propionamide as sources of nitrogen but not valeramide and phenylacetamide. By exposing this strain to mutagenic agents and by conducting artificial selection, however, Betz et al. (1974) produced a number of mutant strains which can utilize valeramide or phenylacetamide. Studies on the biochemical properties of the new enzymes produced have suggested that only a few steps of mutational changes were involved in the formation of the new genes.

Gene Elongation

Genes with improved function may also arise by elongation of genes. Gene elongation is usually caused by duplication of a gene or part of a gene. Many proteins of present-day organisms have internal repeats of amino acid sequences, and these repeats often correspond to separate functional or structural domains of the protein. This suggests that the genes coding for these proteins were formed by internal gene duplication. Indeed, it seems that most eukaryotic genes were produced by duplication and elongation of primordial genes or minigenes that existed in the early stage of gene evolution (Darnell 1978; Doolittle 1978; Ohno 1981; Blake 1985).

This view is supported by the nucleotide sequences of the ovomucoid genes in birds. Ovomucoid is a protein present in egg whites and is responsible for the trypsin inhibitory activity of egg whites. The ovomucoid polypeptide consists of three functional domains (figure 6.2). Each of these domains is capable of binding to one molecule of trypsin or other serine protease. The amino acid similarities between domains I and II, I and III, and II and III are 46, 33, and 30 percent, respectively. These high degrees of similarity suggest that the ovomucoid gene was produced by triplication of a primordial domain gene (Kato et al. 1978). Interestingly, the coding regions of the gene for the three domains are separated by introns (Stein et al. 1980), suggesting that the introns are remnants of the ancient intergenic regions which were duplicated simultaneously with the coding regions. However, the coding region corresponding to each protein domain is composed of two exons rather than one, another intron existing between them. This suggests that each domain region itself is a product of duplication of a small primordial gene, though the similarity between the two exons is no longer high.

Another striking example of gene elongation by duplication is the $\alpha2$ type I collagen gene [$\alpha2(I)$] from chickens. This gene has a length of approximately 38 kb and contains more than 50 exons. Twenty-one of these exons have been sequenced [see Li (1983) for the literature], and two of them contain 45 base pairs (bp), twelve 54 bp, four 99 bp, and three 108 bp—all in multiples of the 9 bp that code for the triplet Gly·X·Y, where X and Y are often prolines. It is possible that all these exons were derived by multiple duplications and recombinations.

There are many other examples of gene elongation by duplication (Li 1983: Doolittle 1985). Many of them involve one or more domain repetitions, and the duplication event can easily be inferred from sequence similarities. In some proteins, it is difficult to identify functional domains, and thus the correspondence between exons and functional domains cannot be revealed easily. In these cases, examination of protein

DOMAIN I

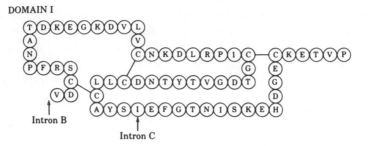

DOMAIN II

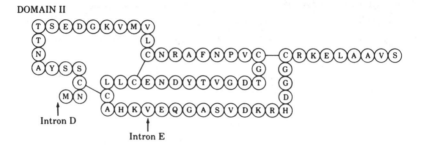

DOMAIN III

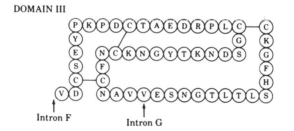

% HOMOLOGY		
DOMAINS	A.A.	N.A.
I/II	46	66
II/III	30	42
I/III	33	50

Figure 6.2.   Three functional domains of the secreted ovomucoid. The homologies between domains at the amino acid (A.A.) level and the mRNA (N.A.) level are shown at the bottom of the figure. From Stein et al. (1980).

folding structure often reveals the ancient functional domains or *modules* of a protein (Gō 1981). It is possible that each of these modules repre-sents a primordial gene in ancient organisms. Ohno (1984) argued that primordial genes were very short and that most present-day genes were produced by several rounds of duplication of these primordial genes.

Gene elongation seems to occur occasionally even at the present time. Haptoglobin in man is composed of two α chains and two β chains. There are two types of α chains in human haptoglobin, α^1 and α^2. Furthermore, two polymorphic forms of haptoglobin α^1 are known, called fast (F) and slow (S). The difference between these two forms is attributable to the amino acid at position 54, i.e., lysine for F and glutamic acid for S. Studies of amino acid sequences have shown that the α^2 (143 amino acids) is nearly twice as long as the α^1 chain (84 amino acids) and consists of portions of the F and S forms of the α^1 chain. From this observation, Smithies et al. (1962) predicted that the gene for α^2 is a product of unequal crossover which occurred between the F and S allelic genes in a heterozygote. This prediction was recently confirmed by Maeda et al. (1984), who showed that the unequal crossover occurred within two different introns of the gene. Since the α^2 gene is present only in man with the possible exception of artiodactyls (Bowman and Kurosky 1982) and no amino acid difference is observed between the homologous parts of α^1 and α^2 chains, this unequal crossover must have occurred very recently. The current gene frequency of α^2 is 30–70 percent in human populations (Mourant et al. 1976). Many human geneticists believe that the α^2 chain gene has selective advantage over the α^1 chain gene. However, the biological basis for this is unknown, though several suggestions have been made (see Bowman and Kurosky 1982). At any rate, if the α^2 chain replaces the α^1 chain, man will have a longer gene for the α chain than other primate species.

As mentioned in chapter 3, prokaryotic genes do not have introns, and this makes it difficult to explain the origin of introns. Some authors (see Flavell 1985 for the literature) argue that introns were inserted into the genes, whereas others maintain that the introns in prokaryotes were eliminated in the process of evolution. At the present time, it is difficult to know which hypothesis is correct. However, whatever the origin is, split genes with exons and introns seem to have helped in producing genes with new functions.

Hybrid Genes

Unequal crossover may occur in a DNA region including two genes. This may produce a new gene containing parts of the two genes. An example of this type of new gene is the Lepore hemoglobin gene in man. This gene is composed of a 5′ region of the δ chain gene and a 3′ region of the β chain gene (Baglioni 1962; Lewin 1985). This type of unequal crossover seems to occur rather frequently, since there are already several different types of Lepore hemoglobins reported. This high frequency of unequal crossover in the β and δ gene region is, of course, attributable to the close linkage and high similarity of the β and δ genes, the nucleotide sequence similarity being 93 percent. It has long been known from the study of the Bar locus in *Drosophila* that duplicate gene regions are quite unstable, probably because the homology between the genes occasionally disturbs a proper chromosomal (DNA) pairing in meiosis.

In practice, however, hybrid genes seem to have some deleterious effect, unless the original genes are retained together with the hybrid genes. Thus, the Lepore hemoglobin genes are kept in low frequency in populations. On the other hand, if the original genes are retained, the hybrid gene may evolve into a new gene. One possible example is the clupeine Z gene in the herring, which probably arose through a crossover between the clupeine YI and YII genes. Fitch (1971b) has shown that the probability that these three genes arose by simple duplication and subsequent amino acid substitution is very small. Examining the amino acid sequences of many blood proteins in vertebrates, Doolittle (1985) has shown that seemingly unrelated proteins often have partial sequence homology and argued that this is caused by exon shuffling among different genes. He regards this as an important mechanism of producing new genes (see also Gilbert 1978).

Horizontal Gene Transfer

It is well known that bacteriophages can transfer host genes from one bacterial strain to another by the process of transduction. The possibility that similar horizontal gene transfer occurs in higher organisms has been debated for many years (Wilson et al. 1977a). Recently, strong evidence suggesting that this occurs even in eukaryotes has been provided by Busslinger et al. (1982). These authors noticed that the genome of the sea urchin *Psammechinus miliaris* contains a histone H3 gene variant distinguishable from the predominant H3 gene. (Note that histone genes form a multigene family.) Sequencing of these genes showed that the variant gene is quite different from the predominant type in nucleotide sequence (about 11%) but is virtually identical with the histone gene from another sea urchin species (*Strongylocentrotus dröbachiensis*) belonging to a different family (sequence difference of 0.2%). By contrast, the predominant type gene was similar to that of a species (*P. lividus*) belonging to the same genus *Psammechinus*, and the gene from *S. dröbachiensis* was similar to that from its closely related species *Strongylocentrotus purpuratus*. Since the genus *Strongylocentrotus* diverged from *Psammechinus* about 65 MY ago and the genes from these two genera are quite different (about 11%) except the rare variant gene of *P. miliaris*, this observation suggests that the *P. miliaris* variant gene was recently transferred from *S. dröbachiensis*.

If this hypothesis of horizontal gene transfer is correct, it probably occurred by the aid of some virus. It is known that viruses such as retroviruses are capable of extricating chromosomal genes and incorporating them into the host genome (Bishop 1981) and that they are able to cross species boundaries (Benveniste and Todaro 1974, 1976). However, it should be noted that the frequency of occurrence of horizontal gene transfer in eukaryotes is apparently extremely low because most genes so far studied do not show abnormal patterns of interspecific similarity analogous to the histone H3 gene in sea urchins. Note also that the functionality of the variant H3 gene in *P. miliaris* has not been confirmed.

Repetitive DNA and Multigene Families

The genome of eukaryotes is known to contain various types of *repetitive DNA* (Britten and Kohne 1968). Repetitive DNA is any piece of nucleotide sequence which is repeated several to many times in the genome. The proportion of repetitive DNA varies greatly with organism. In lower eukaryotes, it amounts to 10 to 20 percent of the total genome, and repetition is only moderate. In higher animals, up to 50 percent of the DNA is moderately or highly repetitive (figure 6.3). In higher plants, the proportion of repetitive DNA may be even higher, reaching 80 percent. The function of repetitive DNA is largely unknown, and some classes of repetitive DNA seem to be nonfunctional ("junk DNA"; Ohno 1972). However, certain groups of repetitive DNA have clearly defined functions. Groups of repetitive DNAs with known functions are called *multigene families*.

The distribution of the number of copies of repetitive DNAs or repetition frequency can be studied by DNA hybridization experiments (Britten and Kohne 1968). In mammals, the repetition frequency varies widely from several copies to several million copies (figure 6.3). Al-

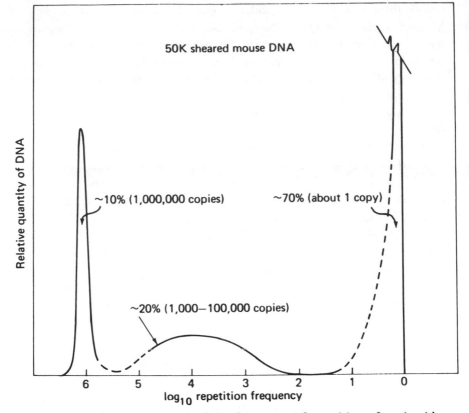

Figure 6.3.   Spectrogram of the frequency of repetition of nucleotide sequences in the DNA of the mouse. The relative quantity of DNA is plotted against the logarithm of the repetition frequency. The dashed segments of the curve represent regions of considerable uncertainty. From Britten and Kohne (1968), reprinted by permission, The American Association for the Advancement of Science, copyright 1968.

though the distribution of copy number is generally continuous, it is convenient to classify repetitive DNA into three classes, i.e., highly repetitive DNA, moderately repetitive DNA, and mildly repetitive DNA. Highly repetitive DNA is usually composed of a large number of copies of a short DNA sequence. This class of DNA is sometimes called satellite DNA since it often forms a satellite band when the total DNA is fractionated on the basis of nucleotide composition by using CsCl centrifugation.

The satellite DNA of the mouse is composed of approximately one million copies of identical or slightly variant nucleotide sequences of about 234 bp (Southern 1975). The prototype sequence of this repeating unit is given in figure 6.4. A quick glance at this figure indicates that there are several levels of subunits in this sequence. Hörz and Altenburger (1981) have shown that this sequence can be broken down into subunits of 1/2, 1/4, and 1/8 the total length. When the total sequence is divided into two subunits of 117 bp, there is a similarity of 81 percent between the subunits. When the prototype sequence is divided into four subunits of 58 bp, the similarity becomes 60 percent. Actually, a close examination of the prototype reveals that it is composed of eight related sequences which are designated as $\alpha_1, \alpha_2, \alpha_3, \alpha_4$, $\beta_1, \beta_2, \beta_3$, and β_4 (see Hörz and Altenburger 1981). The lengths of these repeats are not the same. The α subunit is about 28 bp long, whereas the β subunit is about 30 bp long, and they alternate. Interestingly, the α and β subunits consist of three consensus sequences related to each other, i.e., GA_5TGA, GA_6CT, and GA_5CGT. Thus, it is possible that the entire mouse satellite DNA was derived from a single ancestral sequence of 9 bp (possibly GAAAAATGT) by repeated gene duplication.

Highly repetitive satellite DNA exists in most higher eukaryotes, and it usually consists of a large number of copies of a short nucleotide sequence. For example, *Drosophila virilis* has several different classes of repetitive DNA, and three dominant classes are composed of the repeats of the sequences ACAAACT, ATAAACT, and ACAAATT. These se-

GGACCTGGAATATGGCGAGAAAACTGAAAATCACGGAAAATGAGAAATACACACTTTA

GGACGTGAAATATGGCGAG^GAAAACTGAAAAAGGTGGAAAATT^TAGAAATGTCCACTGTA

GGACGTGGAATATGGCAAGAAAACTGAAAATCATGGAAAATGAGAAACATCCACTTGA

CGACTTGAAAAATGACGAAATCACTAAAAAACGTGAAAAATGAGAAATGCACACTGAA

Figure 6.4. Prototype sequence of the repeat unit of mouse satellite DNA. After Hörz and Altenburger (1981).

quences are repeated over one million times (Gall et al. 1974). These highly repetitive DNAs are usually located in the heterochromatic regions of chromosomes and appear to have no biological function.

Another class of highly repetitive DNA, which is predominant in the human genome, is the so-called *Alu* family. There are about 300,000 members in this family, and each member is about 300 bp long. A majority of the members have one *Alu*I restriction site and are scattered throughout the genome at intervals of approximately 6,000 bp. Because of this property, the members of this family were once thought to be controlling elements of DNA replication (Jelinek et al. 1980). It now seems that this family also represents nonfunctional DNA, as will be discussed later.

Moderately repetitive DNAs are those which are repeated a dozen to several thousand times in the genome. They include rRNA genes, tRNA genes, and immunoglobulin heavy chain variable region genes. In the human genome, there are about 300 copies of rRNA genes, each of which consists of about 30,000 bp. These genes are located in clusters on five different chromosomes. Needless to say, these genes are of vital importance in cell physiology since they are required for producing rRNA. Transfer RNA is also produced by a multigene family which has more than one thousand members. The genomic organization of this gene family has not been well studied, but the genes are apparently clustered as in the case of rRNA genes (Lewin 1985). Similarly, 5S RNA genes form a gene family consisting of many tandem duplications (Tartof 1975). Another well-known multigene family is that of immunoglobulin heavy chain variable region genes. This family belongs to the superfamily of immunoglobulin genes and is responsible for producing heavy chain variable regions of immunoglobulins. There seem to be about 200 members in the mouse genome and 80 members in the human genome. It is clear from these examples that duplication of genes is an effective way to produce a large quantity of the RNAs or proteins that are essential for the survival of an individual.

In addition to the above two classes of repetitive DNA, there are mildly repetitive DNAs or multigene families. Most of them have well-defined biological functions, and the number of members in a family is usually a dozen or less. Typical examples are the β globin, the α globin, and the immunoglobulin constant chain gene families. As is clear from figure 6.1, most mammalian genomes have four to nine β or β-like globin genes, including pseudogenes. At least one of them is responsible for adult hemoglobins, whereas at least two of them participate in the production of embryonic or fetal hemoglobins. In the case of the human globins, the β and δ genes produce the adult β and δ globins, respectively, whereas the $^A\gamma$ and $^G\gamma$ code for fetal gamma globin. The ϵ gene is expressed only in the early embryonic stage. It is interesting to note that the globin genes are arranged in the order of gene expression from the 5' side of the gene cluster in the human, mouse, and rabbit genomes. In the goat genome, however, this is not the case. Note also that the arrangement and function of the first four genes of the goat β globin family are very similar to those of the next four genes (figure 6.1). Apparently one set of the four genes was produced by block gene duplication from the other set (Cleary et al. 1981; Li and Gojobori 1983).

If we note that many new genes with new functions in higher organisms were produced by gene duplication, we would expect that some of the current gene families are evolutionarily related. This is indeed the case with many gene families. A well-known example is the evolutionary relationship between the α and β globin gene families in mammals. These two families are now located on different chromosomes in most mammals, but they form a superfamily of globin genes, together with the myoglobin gene.

Another example of a superfamily of genes is that of the immunoglobulin genes. Immunoglobulins are macromolecules determining the immunity to various antigens in vertebrates. An immunoglobulin consists of two identical heavy chains and two identical light chains (figure 6.5). There are five types of heavy chains (μ, δ, γ, ϵ, and α) and two types of light chains (κ and λ). Both the heavy chains and light chains are composed of a variable (V) and a constant (C) region (figure 6.5). These immunoglobulin chains are produced by three different gene families, i.e., the heavy chain, the kappa chain, and the lambda chain gene families (see figure 6.6). Each family has the genes for both variable and constant regions of immunoglobulins and the genes for the joining (J) segment. In the case of the heavy chain gene family, there is an additional subfamily of diversity (D) genes (see Hunkapiller et al. 1982). The genes in the three gene families show substantial amounts of sequence similarities and are thus considered to have descended from a common ancestral gene (figure 6.7). That is, the three gene families form a superfamily of genes. Actually, this superfamily includes not only the three gene families but also the gene families for the major histocompatibility complex (MHC), T cell receptors, etc. (figure 6.7).

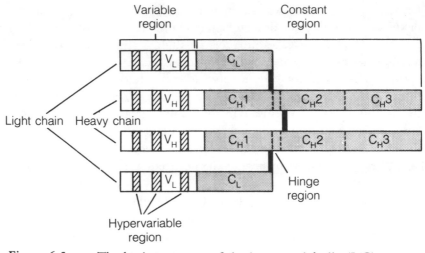

Figure 6.5.   **The basic structure of the immunoglobulin (IgG) molecule.**

Clarification of the developmental process of the generation of immunoglobulin or antibody diversity is one of the most important achievements in the recent history of molecular biology (Leder 1982; Tonegawa 1983). Since there are a large number of different antigens entering into an animal, the repertoire of antibodies must be very large. The antigen specificity of an immunoglobulin is determined by the V region, and thus immunoglobulin diversity is caused mainly by variability of the amino acid sequence in the V region. Variability of the V

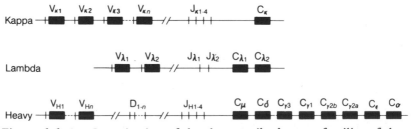

Figure 6.6.   Organization of the three antibody gene families of the mouse. Exon-intron organization of heavy-chain constant-region genes is not shown. V, D, J, and C denote variable, diversity, joining, and constant region genes, respectively. From Hunkapiller et al. (1982).

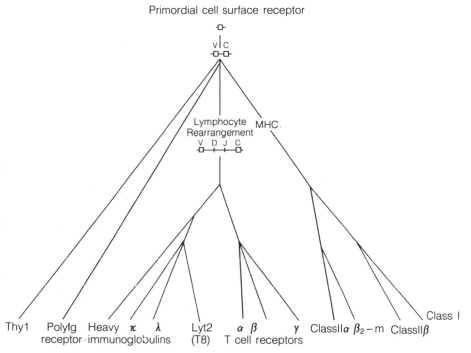

Figure 6.7.   Evolutionary relationship of multigene families for immunoglobulins, T cell receptors, major histocompatibility complex (MHC), Thy-1 polypeptide, and others. From Hood et al. (1985).

region is generated by genetic variation of the V, J, and D region genes. In the mouse H chain gene family, there are about 200 variable region genes (V_H), four J region genes (J_H), and ten D region genes (D_H). In the process of immunoglobulin production, one of the V_H genes, one of the J_H genes, and one of the D_H genes are randomly combined (DNA rearrangement) to produce a rearranged V gene, and this gene produces one immunoglobulin V region (figure 6.8). Therefore, the DNA rearrangement in the heavy chain gene family alone produces $200 \times 4 \times 10 = 8,000$ different V region sequences. Furthermore, the V-D-J joining often introduces insertion and deletion of nucleotides at the joining points, and this factor would double the number of different amino acid sequences in the V region. Similarly, the kappa and lambda gene families are expected to produce about 2,000 and 8 different L chain sequences, respectively. Therefore, germline gene variation alone

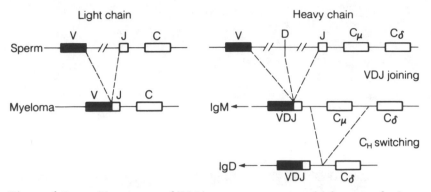

Figure 6.8.   Two types of DNA rearrangements which occur during the differentiation of antibody-producing cells. In light and heavy chain genes V-J or V-D-J joining juxtaposes the gene segments encoding the V_L and V_H genes. Subsequently in heavy-chain genes class, or C_H, switching may occur. The class switch leads to the expression of different immunoglobulin classes. Sperm indicates DNA undifferentiated with regard to antibody function, whereas myeloma denotes DNA in which one or more DNA rearrangements have occurred. From Hunkapiller et al. (1982).

is capable of producing $16,000 \times (2,000 + 8) \simeq 3.2 \times 10^7$ different immunoglobulins. This number seems to be more than sufficient for protecting vertebrate organisms from foreign antigens. In practice, somatic mutation (amino acid change) is also known to occur with a high frequency (about 3 percent per amino acid site) in the variable region genes (Kim et al. 1981; Tonegawa 1983), and this seems to be a major factor contributing to immunoglobulin diversity (Gojobori and Nei 1986). The immunoglobulin gene superfamily is a remarkable genetic system, in which gene duplication is effectively used for generating great diversity of amino acid sequences. Essentially the same genetic system exists for the T-cell receptor gene family, which is evolutionarily related to the immunoglobulin family (Hood et al. 1985).

Evolution of Multigene Families

The basic process of evolution of multigene families is gene duplication. Through gene duplication, the number of genes in a family may increase step by step. Once a gene duplicates into two, a further increase in the number of genes is facilitated by unequal crossover. However, to explain the evolution of highly repetitive DNA, simple gene duplication is not always sufficient. There are several hypotheses for explaining the evolution of repetitive DNA.

Saltatory Replication

The idea of saltatory replication was first presented by Britten and Kohne (1968) to explain the evolution of highly repetitive DNA. They suggested that highly repetitive DNA was generated by a rapid multiplication of a prototype sequence in a relatively short period of evolutionary time, and sequence variation among them was generated by the accumulation of random mutations that occurred later. This idea was used more forcefully by Southern (1975). Noting that the mouse repetitive DNA consists of several levels of hierarchical repeats, he proposed that there were several saltatory replications. The first stage was the multiplication of the basic sequences GA$_5$TGA, GA$_6$CT, and GA$_5$CGT to produce the α and β sequences discussed earlier. The second stage of multiplication was to produce a repeat of the 58 bp given in figure 6.4, and the third stage was to produce a sequence of 234 bp. The final stage of saltatory replication was the amplification of the 234 bp sequence into a million copies. According to Southern's estimation, these saltatory replications occurred quite a long time ago, i.e., 3 to 20 MY ago.

There seem to be some problems with this explanation. First, the mechanism of saltatory replication is unknown. We can assume that at certain times in the past gene duplication occurred many times very rapidly by unequal crossover, but why did it occur only at certain times, and how did it occur? We have no answers to these questions at the present time. Second, according to this hypothesis, amplification of a large number of duplicate DNAs (a million copies) occurred only at the final stage. Why was this stage so special? Third, when two closely related species have the same class of repetitive DNA, the repetitive sequences from one species are often more similar to each other than to those from the other species, as will be discussed below. This is difficult to explain in terms of saltatory replication.

Concerted Evolution

The ribosomal RNA gene family in the African toads *Xenopus laevis* and *X. mulleri* consists of 400 to 600 members. Each member consists of the 18S and 28S RNA transcription units and a nontranscribed spacer DNA (figure 6.9). Brown (1973) and his colleagues have shown that the nucleotide sequences of spacer DNA are virtually the same among member genes of the same species but differ by about 10 percent between X. *laevis* and X. *mulleri*. This observation cannot be explained by the hypothesis that the duplication of rRNA genes occurred before divergence of the two species, because according to this hypothesis we expect that the differences in nucleotide sequence between different repeats of the same species are similar to those between repeats of different species. The explanation becomes harder if we note that the nucleotide sequences of the 18S and 28S coding regions are identical between X. *laevis* and X. *mulleri*. Actually, the 18S and 28S coding regions are very similar even among distantly related organisms; the coding regions of higher plants are closer in sequence to those of *Xenopus* than spacer sequences of X. *laevis* are to those of X. *mulleri*.

This puzzling observation can be explained by the theory of *coincidental* or *concerted evolution* originally proposed by Smith et al. (1971) and Smith (1974). According to this theory, unequal crossover that occurs randomly between nonhomologous genes or chromosome repeats plays an important role, and because of this factor the number of genes in the cluster may increase or decrease depending on chance. Unequal crossover also has an effect to homogenize genes in the cluster, as shown in figure 6.10. That is, in the absence of mutation all genes in the cluster will eventually become identical. In reality, of course, mutation always occurs, so that a gene family is expected to have some variant genes.

It is now clear that when a species diverges into two and the gene cluster in each descendant species evolves independently, the cluster within

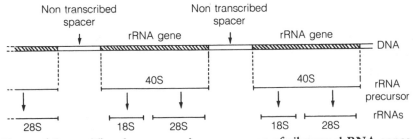
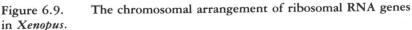

Figure 6.9.   **The chromosomal arrangement of ribosomal RNA genes in *Xenopus*.**

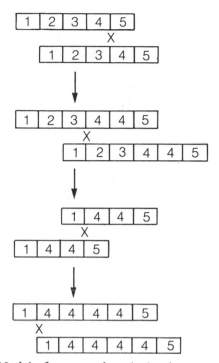

Figure 6.10.   Model of concerted evolution by unequal crossover. Different numbers show different genes. In the presence of unequal crossover, the genes on the same chromosome are gradually homogenized. Adapted from Ohta (1980).

each species tends to have similar gene copies because of unequal cross-over, whereas the genes belonging to the cluster of the two species gradually diverge by mutation. This is exactly what we have observed in the spacer regions of the rRNA gene family in *Xenopus*. I also mentioned that the 18S and 28S coding regions are identical between *X. laevis* and *X. mulleri,* as well as between different copies of the same species. This identity has apparently been maintained by strong purifying selection that operates for the coding regions. Thus, we can explain the entire observation about the rRNA gene family in *Xenopus* in terms of unequal crossover, mutation, and purifying selection. In addition to these factors, gene conversion also seems to play an important role in the evolution of multigene families. The role of gene conversion is similar to that of unequal crossover. The only difference between unequal crossover and gene conversion is that the former may increase or decrease the number of repetitive genes, whereas the latter does not change the number. Particularly in the evolution of the major histocompatibility complex (MHC), gene conversion seems to play an important role (Mellor et al. 1983).

At any rate, if we consider unequal crossover, gene conversion, mutation, and purifying selection, most observations about multigene families can be explained. Of course, the relative importance of these factors would vary from gene family to gene family. For example, in the case of rRNA genes unequal crossover or gene conversion seems to play a very important role. In many organisms, the spacer regions of the rRNA gene show polymorphism with respect to restriction sites (e.g., Coen et al. 1982; Williams et al. 1985; Suzuki et al. 1986). Curiously, this polymorphism exists mainly among different local populations rather than within populations. This suggests that homogenization of genes in the rRNA gene cluster occurs very rapidly. It is quite possible that some nonrandom process (e.g., biased gene conversion) speeds up this homogenization (Ohta and Dover 1984).

By contrast, unequal crossover seems to occur with a very low frequency in the evolution of the immunoglobulin heavy-chain variable-region genes (V_H). As mentioned earlier, there are about 200 V_H genes in the mouse genome and about 80 V_H genes in the human. This gene family also seems to be subject to concerted evolution (Smith et al. 1971; Hood et al. 1975). If turnover of genes occurs quickly by means of unequal crossover, we would expect the genes from the same species to be closely related, whereas the genes from distantly related organisms such as the mouse and the human should be quite different. (The human and the mouse diverged about 75 MY ago.) To see whether this is the case, Gojobori and Nei (1984) studied the evolutionary relationship of 16 V_H genes from the mouse and five V_H genes from the human, examining the nucleotide differences among them. Contrary to the expectation, however, their results have shown that many genes in the same species are distantly related to each other and that some of the mouse genes are more closely related to some human genes than to other mouse genes (figure 6.11). Furthermore, it has been estimated that all 21 genes studied originated from a common ancestral gene about 300 MY ago, i.e., the time when mammals diverged from reptiles.

This observation suggests either that the number of V_H genes in mammals has steadily increased for the last 300 MY or that if concerted evolution occurred, the rate of occurrence of unequal crossover was very

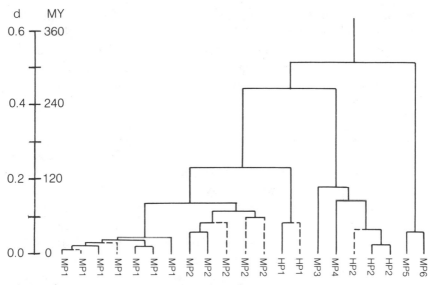

Figure 6.11.   Phylogenetic tree for 16 mouse (M) and five human (H) V_H germline genes. Dashed lines show pseudogenes. From Gojobori and Nei (1984). Copyright by the University of Chicago.

low. However, since the genome of the shark *Heterodontus francisci,* which diverged from the mammalian line about 400 MY ago, apparently has many V_H genes (Litman et al. 1985), this gene family seems to have been subjected to concerted evolution for a long time. If this is the case and slow concerted evolution generated the current mammalian V_H gene families, with the total number of repetitive genes (n) being kept more or less constant, we can obtain a rough estimate of the rate of unequal crossover (r) using the following approximate formula.

$$r = \frac{n-1}{\bar{t}} \left(1 - \frac{1}{s} \right), \qquad (6.1)$$

where s is the number of genes sampled, and $\bar{t}$ is the *coalescence time* for sampled genes, i.e., the time at which all sampled genes trace back to a single ancestral gene (F. Tajima in Gojobori and Nei 1984; see also Ohta 1983b). In the case of the mouse V_H genes, $n = 160$, $s = 16$, and $\bar{t} = 300$ MY from figure 6.14 below. Therefore, $r = 5 \times 10^{-7}$ per year per gene. In the case of human V_H genes, $n = 80$, $s = 5$, and $\bar{t} = 270$ MY, so that $r = 2.3 \times 10^{-7}$ per year per gene. This is quite a slow process. Equation (6.1) depends on the assumption that "gene transfer" occurs at random irrespective of the position of the locus. Ohta (1984) has shown that if "gene transfer" occurs more often between neighboring loci than between distantly located loci, r can be as high as 10^{-6}. Even this rate is much lower than that ($r = 10^{-2} - 10^{-4}$) for rRNA genes (Strachan et al. 1985).

In recent years, a number of authors have developed mathematical or computer models of concerted evolution. In this case, gene turnover occurs both in the genome and in the population. Therefore, a rigorous treatment of the problem is quite complicated. The reader who is interested in this problem may refer to Smith (1974), Ohta (1980, 1983a), Ohta and Dover (1984), Nagylaki (1984), and the papers cited therein.

In some multigene families, the number of genes seems to be kept rather constant by natural selection. For example, most human haploid genomes have two α globin genes, i.e., α_1 and α_2, which code for the same polypeptide. In Southeast Asia, however, there are many individuals who carry chromosomes with one or no α gene (Na-Nakorn and Wasi 1970). They show a clinical syndrome called α-thalassemia. There are also individuals who carry chromosomes with three α genes, though the frequency is quite low (Goossens et al. 1980). A restriction site analysis of the α globin family in the chimpanzee, gorilla, orangutan, and gibbon has indicated that these organisms also have two α globin genes that are virtually identical within species (Zimmer et al. 1980). (The chimpanzee seems to have a rather high frequency of chromosomes with triple α genes.) Closely related nonallelic α-chain genes also seem to exist in monkeys, horses, goats, deer, mice, dogs, and chickens. From these observations, Zimmer et al. (1980) proposed that the DNA region coding for the adult hemoglobin α chains has been in the duplicate form for at least 300 MY and that concerted evolution has continually kept the duplicate genes from diverging greatly. Apparently, strong selection is operating in favor of chromosomes with two α globin genes.

A similar type of concerted evolution also seems to be operating in the β-δ globin cluster, the γ globin cluster of two genes, and others (Zimmer et al. 1980; Hardison 1984). In these clusters, however, selection for maintaining two genes does not seem as strong as in the case of α globin genes. A number of authors (Ohta 1981; Wills and Londo 1981; Yokoyama 1983) have developed population genetics models for studying the maintenance of the number of copies of genes and polymorphic alleles for these cases.

Vesuvian Model of Generating Repetitive DNA

We have already mentioned that the *Alu* gene family in the human genome has about 300,000 members which are dispersed on the entire genome and that these member genes were once thought to play important cell functions such as the control of DNA replication. Recent studies, however, suggest that the *Alu* family might be a gigantic class of pseudogenes. This idea of pseudogenes is due to Ullu and Tschudi (1984), who have studied the nucleotide sequence of 7SL RNA genes from *Drosophila melanogaster* and *Xenopus laevis*. 7SL RNA is an abundant cytoplasmic RNA which functions in protein secretion as a component of the signal recognition particle. Probably because of its important function in cell physiology, the nucleotide sequence of the 7SL RNA gene has been conserved. Ullu and Tschudi have shown that sequence similarity is 87 percent between *Xenopus* and man and 64 percent between *D. melanogaster* and man. Furthermore, they have shown that the 5' and 3' portions of the 7SL RNA gene are homologous to the human *Alu* sequence. If we note that there is no *Alu* sequence in the fruitfly genome, this suggests that the *Alu* sequence was derived from the 7SL RNA gene by deletion of the middle portion of the gene after fruitflies diverged from vertebrates. Deletion of the middle portion seems to have occurred by (1) an aberrant RNA joining or splicing event similar to the processing of mRNA molecules, (2) reverse transcription of the "processed RNA," and (3) insertion of the reverse-transcribed DNA into the genome by the aid of retroviruses or some other mechanism.

However, there are several different classes of *Alu* or *Alu*-like sequences in the genome of mammalian species, and some of them are apparently derived from tRNA genes rather than from 7SL RNA genes (Rogers 1985). The proportions of 7SL-derived and tRNA-derived sequences seem to vary considerably with organism.

Why are there so many copies in the *Alu* or *Alu*-like families? It is possible that the 7SL RNA or tRNA genes keep producing *Alu* or *Alu*-like sequences for some reason. This is not unusual, since many nuclear genes seem to produce so-called *processed pseudogenes*. For example, the mouse genome has many globin pseudogenes, and at least some of them lack introns (Leder et al. 1981). These pseudogenes are considered to have arisen by reverse transcription of processed mRNAs and subsequent insertion of the reverse-transcribed DNA into the genome. This process seems to occur with a relatively high frequency. P. Leder (see Lewin

1981) likened this process of the "pumping out" of pseudogenes from a locus to the generation of lava by a volcano, calling it the Vesuvian model. This Vesuvian model seems to apply for many genes in addition to the α globin genes.

In the case of *Alu* and *Alu*-like sequences, however, there is a possibility that the sequences themselves generate more copies even if they are nonfunctional. This is because many *Alu* or *Alu*-like sequences retain the promoter region inside the sequence and are thus transcribed. Transcribed sequences may again be inserted into the genome through the process of reverse transcription. This cascade effect (or cascade replication) seems to have greatly enhanced the chance of dispersion of *Alu*-like sequences throughout the genome. Processed pseudogenes of ordinary genes cannot be transcribed, so that they do not generate any pseudogene.

In addition to short, *Alu*-like sequences, some mammalian genomes contain 10^4-10^5 copies of long interspersed sequences (LINES) with 6–7 kb. One of them is the rodent L1 repeat family studied by Hardies et al. (1986). This family seems to include a dozen or more functional genes with a long open reading frame, but the majority of the repeat sequences are apparently processed pseudogenes. In this family, all pseudogenes seem to have been derived from functional genes because they are apparently nontranscribable.

Movable Genetic Elements and Retroviruses

Until recently, all genes were considered to be located on fixed points on a chromosome and to stay there forever unless they were moved by occasional translocation or inversion. Some time ago, McClintock (1951) showed that certain genes or genetic elements in maize were movable within and between chromosomes, but the general significance of the discovery was not recognized. In recent years, many *movable genetic elements* or *transposable elements* have been discovered in both prokaryotes and eukaryotes (Shapiro 1983). These elements usually exist as multiple copies in the genome and sometimes transpose other genes as well as themselves.

Movable elements may be divided into two classes according to their state prior to insertion into the genome. Some elements (most temperate viruses) replicate extrachromosomally but can also become inserted into the genome and replicate passively as part of the genome. Such elements are called *episomes*. Other elements (such as the bearers of many antibiotic resistance genes in bacteria) cannot replicate separately and are found only in the inserted state. These elements are called *transposons*. In prokaryotes, transposons can be further divided into two classes, i.e., *true transposons* and *insertion sequences,* according to whether or not the movable elements carry genes unrelated to transposition (Campbell 1983). The transposons in eukaryotes are not as well characterized as those in prokaryotes, and they are often called *transposable elements* or simply *movable elements*.

Transposons in Prokaryotes

Among movable elements, there is a range of size and complexity (figure 6.12). The simplest bacterial elements are insertion sequences

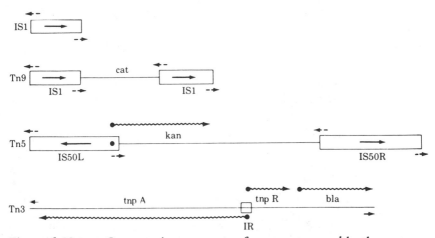

Figure 6.12.   Comparative structure of some transposable elements of bacteria. IS1 is 768 base pairs long, with an imprecise inverted repeat (arrows with dashed tails) in the terminal 34 nucleotides. Tn9 consists of two IS1 elements bracketing a gene for chloramphenicol acetyltransferase. In Tn5, two oppositely oriented, independently transposable, 1,500 base pair sequences (IS50) bracket a gene for kanamycin resistance. Tn3, 4,957 nucleotides in length, has a precise inverted repeat (arrows with solid tails) of 38 bases at the termini. From Campbell (1983).

like IS1 (768 bp long), and their functions are concerned only with the ability to transpose. An insertion sequence (IS) consists of three segments, i.e., a central protein-coding region and two short terminal repeats (figure 6.12). In the case of IS1, each of the two terminal repeats is 34 bp long. The terminal repeats are either in the same direction (direct repeats) or in the reverse direction (reverse repeats). The protein-coding region includes a few genes coding for transposases.

True transposons in bacteria are usually much larger in size than insertion sequences and consist of a central protein-coding region and two terminal modules (figure 6.12). The protein-coding region includes marker genes such as those for drug resistance, whereas the two terminal modules are direct repeats or reverse repeats of one of the IS or IS-like elements. For example, transposon Tn5 is about 5,700 bp long and includes genes for kanamycin resistance. The modules for this transposon are reverse repeats of insertion sequence IS50. Modules are responsible for the movement of transposons from one site to another in the genome. When a transposon moves, the original site usually retains the transposon. Therefore, the number of transposons in a genome increases. However, there is some mechanism to reduce the number, so that the total number of transposons in the genome cannot be very large. Since some modules may exist in the genome as independent IS elements, it is believed that transposons evolved by an association of two originally independent modules with the genes in the central region (Berg et al. 1984). In principle, any gene may be integrated into a transposon so that it may move from one location to another in the genome.

Movable Genetic Elements in Eukaryotes

Movable genetic elements in eukaryotes are similar in structure to bacterial transposons and again consist of a central protein-coding region and two flanking terminal repeats. There are many different kinds of movable elements in the genome of higher organisms, and each family contains a large number of copies. For example, the Ty (transposon in yeast) element in yeast is 6.3 kb long, and the 330 bp at each end constitute direct repeats, called δ. The number of Ty elements in the yeast genome is about 30, though it varies considerably among different strains (Lewin 1985). In addition to these, there are about 100 independent δ elements called solo δ's.

The movable elements of D. melanogaster have been studied extensively in recent years. There are several different types of movable elements. So far the best-characterized family is that of *copia*. The *copia* element is about 5 kb long and has direct terminal repeats of 276 bp and inverted repeats of 17 bp at the terminal ends. The number of copies of *copia* in the genome varies among strains, usually from 20 to 60, and copies are widely dispersed (Montgomery and Langley 1983). The *copia* family has many related families such as those of 412, 297, 17.6, etc. Each of these families has 10 to 100 members. Therefore, the *copia* and *copia*-like families account for almost 5 percent of the *D. melanogaster* genome (Rubin et al. 1981). These families code for abundant mRNA, but their function is not well known. Some *copia*-like elements, such as the 17.6 element, have a molecular structure similar to that of retroviruses (Saigo et al. 1984).

Recently, an intensive study has been made about the P element in *D. melanogaster*. The normal element consists of 2,900 bp and has inverted terminal repeats of 31 bp (O'Hare and Rubin 1983). In addition to these, there are many smaller heterogeneous P elements which have apparently arisen by internal deletion from the 2.9 kb P elements. Most *D. melanogaster* strains (P strains) recently established from natural populations of North America have about 50 copies of this element per genome, but the laboratory strains (M strains) established before 1960 lack this element. P elements are unique in manifesting so-called *hybrid dysgenesis* when a male from a P strain is crossed with a female from an M strain (Kidwell 1983a). Hybrid dysgenesis is a term used to denote various dysgenic traits that appear in the offspring from crosses of certain strains (Kidwell et al. 1977). The dysgenic traits include a high rate of deleterious mutations, chromosomal aberration, and sterility.

Hybrid dysgenesis occurs only when the P elements exist together with the cytoplasm from the M strain. It seems that the P elements code for a transposase that is inactive in the M cytoplasm. Whether or not this view is correct, the transposability of P elements has been used for transferring nuclear genes from one strain to another strain (Rubin and Spradling 1982).

The origin of the P element is virtually unknown. Kidwell (1979) proposed the hypothesis that this element invaded the natural population of *D. melanogaster* around the mid-1950s and quickly spread through the population because of its transposability. If this hypothesis is correct, the conversion of the M genome into the P genome in the United States populations must have occurred within 15 years (Kidwell 1983b;

Uyenoyama 1985). Engels (1981b) maintains the view that the P element has existed for a long time and that the absence of the P element from old laboratory strains occurred by the effect of inbreeding or random genetic drift. Recently, Lansman et al. (1985) have shown that many species of the genus *Drosophila* have this element and that the *melanogaster* P element is less homologous to that of its closely related species than to that of more distantly related species such as *D. willistoni* group species. From this observation, they speculated that the P element might occasionally be transferred from one species to another.

In recent years, a number of mathematical models have been developed to explain the maintenance of transposable elements in eukaryotic populations (e.g., Brookfield 1982; Charlesworth and Charlesworth 1983; Langley et al. 1983; Kaplan and Brookfield 1983). However, to understand the real mechanism of maintenance of transposable elements in populations, we need more basic genetic data on the rules of excision and transposition of transposable elements.

The evolutionary significance of movable genetic elements has been discussed by a number of authors (e.g., Campbell 1983; Berg et al. 1984; Syvanen 1984). At the present time, the discussion is largely speculative, but it is generally believed that movable elements increase the mutation rate of the genes that exist near the elements and also cause various chromosomal mutations. These genic and chromosomal mutations may be useful for increasing the chance of survival of the organism (Chao et al. 1983; Hartl et al. 1983). In bacteria, transposons may enter into plasmids or episomes, so that they may become vectors for transferring genes from one species to another (horizontal gene transfer).

Some authors (e.g., Bingham et al. 1982; Ginzburg et al. 1984) have speculated that movable elements like P elements may play an important role in the generation of reproductive isolation between species. Their argument is as follows. Suppose that one population is split into two at an evolutionary time and the two resultant populations (X and Y) are geographically isolated. We assume that transposons similar to P elements invade one of the two populations, say X. Then, the hybrid between the two populations will show hybrid dysgenesis. This single invasion of a movable element may not be sufficient for developing a sterility barrier, but one can assume that several invasions of different types of movable elements in both populations may develop complete reproductive isolation.

There are problems in this argument. First, there is no evidence that interspecific hybrid inviability or sterility is caused by movable elements or similar factors. Second, it is not clear why one population is invaded by one type of movable element, while the other is not. Third, even if only one population is invaded at a certain evolutionary time, the other population will also be infected later when the two populations come into contact. Ginzburg et al. (1984) have argued that if movable elements cause almost complete sterility, this dilemma can be avoided. However, if they cause complete sterility, the elements cannot spread through even the first population. This is true even with the hypothesis of repeated invasions. The only way one can resolve this dilemma is to assume that the transposable elements that infected one population have synergistic effects when combined with the transposable elements from the other population. At the present time, however, there seems to be no evidence that such synergistic effects really exist.

Retroviruses

Certain RNA viruses are known to be reverse-transcribed into double-stranded DNA at the time of reproduction, and the DNA transcribed are used for producing the next generation of RNA viruses. These viruses are called *retroviruses*. Retroviruses contain single-stranded RNA molecules, and each virus carries two copies of the genome. Thus, they are effectively diploid. These viruses were first discovered because of their ability to produce tumors in animals. Examples are the Rous sarcoma virus, Moloney murine leukemia virus, and C-type viruses in birds and mammals.

A typical retrovirus contains three genes, i.e., *gag, pol,* and *env* (figure 6.13). The *gag* gene codes for the protein components of the nucleoprotein core of the virus. The *env* gene is responsible for components of the envelope of the virus, whereas the *pol* gene specifies the enzyme reverse transcriptase. Reverse transcriptase converts the RNA into a double-stranded DNA, and this DNA is integrated into the host genome. The integrated DNA is transcribed by the host machinery to produce viral RNAs, which serve both as mRNA and as genomes for the virus. Integration is a normal part of the life cycle and is necessary for transcription. The integrated DNA *(proviral DNA)* may also replicate as a part of the host genome.

The retroviral genome has direct repeats, designated R, at both ends and other repeat components U3 and U5 (figure 6.13). The proviral

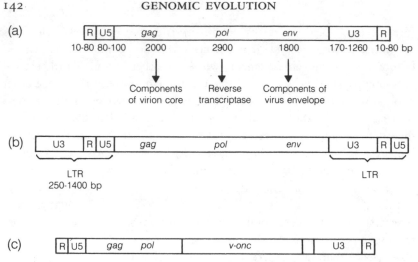

Figure 6.13.   Retrovirus. (a) RNA genome. (b) Proviral DNA integrated into the host DNA genome. (c) Replication-defective viral genome. The *env* gene is replaced by an oncogene *(v-onc)*.

DNA has longer direct repeats called *long terminal repeats* (LTRs). An LTR consists of three components, U3, R, and U5. LTRs are generated at the time of reverse transcription. Part of the viral genome may occasionally be replaced by a host cellular gene. Figure 6.13(c) shows such a virus, where the *env* gene is replaced by the *v-onc* gene. The resulting virus is defective and cannot sustain an infective cycle by itself. However, it can be propagated with the aid of a *helper virus* that provides the missing viral functions.

Many retroviruses carry oncogenes *(onc)*. Oncogenes are responsible for producing tumorigenic cells. The *onc* genes carried by retroviruses are called *v-onc,* and they transform a certain type of host cells into tumor cells which grow with unrestricted cell division. Host cellular genes homologous to *v-onc* genes are called cellular oncogenes *(c-onc)*. All *v-onc* genes are considered to have been derived from *c-onc* genes through DNA transduction. The *c-onc* genes are believed to play an important cell function in normal conditions. There are many different types of oncogenes, and because of their oncogenic propensity they are being studied intensively.

Another important class of retroviruses is the AIDS virus, which is responsible for the acquired immunodeficiency syndrome (AIDS). The AIDS and AIDS-related retroviruses also have the same general genomic organization as that of other retroviruses and contain the *gag*, *pol*, and *env* genes as well as LTRs. In addition to them, however, there are two open reading frames (ORF) called *A* and *B* (Rabson and Martin 1985). The proteins coded for by these ORFs are not known but apparently are responsible for infecting and killing T lymphocytes, which in turn causes human immunodeficiency. ORF A is located between the *pol* and *env* genes, whereas ORF B is located between the *env* gene and the right-hand LTR. Interestingly, the DNA sequence of the AIDS virus genome varies considerably from strain to strain, indicating the existence of a large amount of DNA polymorphism (Shaw et al. 1984; Luciw et al. 1984).

Viral DNA integrates into the host genome at randomly selected sites. Therefore, retroviruses may acquire any host gene. However, the gene acquired by the virus usually lacks introns. This is because when viral RNA genomes are transcribed from a proviral DNA which includes a host gene, the transcript goes through the process of intron splicing. Despite the elimination of introns, the acquired gene is expressed at a high level as part of the viral transcription unit.

As will be discussed in the next section, the genome of higher eukaryotes harbors many pseudogenes that lack introns. They are usually dispersed on different chromosomes. As indicated by Leder et al. (1981), these pseudogenes might have been dispersed by the aid of retroviruses. A functional gene carried by a proviral DNA is expected to lose its introns when viral genomes are formed. A reverse-transcript DNA from these viral genomes may then be inserted into the host genome to become a pseudogene.

Nonfunctional DNA

Nonfunctional DNA and Pseudogenes

As mentioned earlier, a large proportion of eukaryotic DNA is apparently nonfunctional in the sense that it is not essential for RNA transcription or even if RNA is produced, it is not functional. A large part of nonfunctional DNA is accounted for by intergenic regions and introns. Highly repeated DNA also seems to be largely nonfunctional. In addition to these classes of DNA, there are many pseudogenes. A *pseudogene* is a DNA segment with high homology with a functional gene but containing nucleotide changes that prevent its expression. Nonfunctionality of a gene is caused by nonsense or frameshift mutation in the coding region or by mutational damage of the gene regulatory system, intron excision system, etc. When a gene is duplicated into two, one of them may become nonfunctional as long as the other gene is functional. Pseudogenes may also arise by reverse transcription of mRNAs or other RNAs and integration of the DNAs produced into the genome by the aid of retroviruses as discussed above. These pseudogenes lack introns and often contain poly A's near the 3' end and direct or reverse repeats at both ends. They are usually called *processed pseudogenes* or *retroposons*.

It should be noted that mutations accumulate with a high frequency in pseudogenes so that their identity will gradually disappear. These pseudogenes would then exist simply as nonfunctional DNA. Some of these pseudogenes or nonfunctional DNA may be eliminated through natural selection, but the majority of them seem to stay in the genome simply because they do not interfere with cell physiology. Note that some amphibians have a large amount of DNA in the heterochromatic region, which is possibly nonfunctional. Yet this excessive DNA does not seem to particularly lower the fitness of this group of organisms.

Evolutionary Change of Pseudogenes

The evolutionary change of pseudogenes is of special interest because they can be used for testing the neutral mutation theory. Under the neutral theory, the rate of nucleotide substitution is expected to be higher for functionally less important genes or parts of genes than for functionally more important genes, since the latter would be subject to stronger purifying selection (chapter 14). On the other hand, selectionists believe that most nucleotide substitutions are caused by positive Darwinian selection (Clarke 1970; Milkman 1976), and in this case the rate of nucleotide substitution in functionally less important genes or parts of genes is expected to be relatively lower because the mutations in these regions of DNA would not produce any significant selective advantages. Since pseudogenes are nonfunctional, they are suited for testing the neutral theory. For this test, however, we must estimate the rate of nucleotide substitution for pseudogenes.

Let us consider this problem using figure 6.14. This figure shows the plausible phylogenetic tree for mouse globin pseudogene $M\psi\alpha3$, its functional counterpart mouse $M\alpha1$, and rabbit functional gene $R\alpha$. T_d and T_n represent the time since duplication of $M\psi\alpha3$ and $M\alpha1$ and the

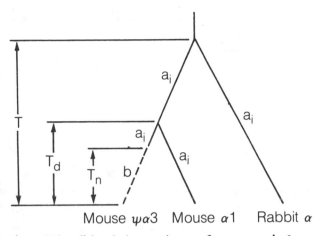

Figure 6.14. **Plausible phylogenetic tree for mouse *ψα*3, mouse *α*1, and rabbit *α*. T denotes the divergence time between mouse and rabbit, T_d the time since divergence of mouse *ψα*3 and *α*1, and T_n the time since nonfunctionalization of mouse *ψα*3. a_i denotes the rate of nucleotide substitution per site per year at the *i*th position of codons in the normal globin genes, and *b* the rate of substitution for mouse pseudogene *ψα*3. From Li et al. (1981).**

time since Mψα3 became nonfunctional, respectively, whereas T stands for the time since divergence between the mouse and the rabbit. T is known to be about 80 MY. a_i denotes the rate of nucleotide substitution per site per year at the *i*th nucleotide position of codon in the normal globin genes, and *b* the rate of substitution for the mouse pseudogene. Since there is no selective constraint, *b* is assumed to be the same for all the three nucleotide positions. From this figure, we can set up the following equations.

$$d_{ABi} = 2a_iT_d + (b - a_i)T_n, \tag{6.2}$$

$$d_{ACi} = 2a_iT + (b - a_i)T_n, \tag{6.3}$$

$$d_{BCi} = 2a_iT, \tag{6.4}$$

where d_{ABi}, d_{ACi}, and d_{BCi} are the numbers of nucleotide substitutions at the *i*th nucleotide position between Mψα3 and Mα1, between Mψα3 and Rα, and between Mα1 and Rα, respectively. The values of d_{ABi}, d_{ACi}, and d_{BCi} are estimated by equation (5.3) from the proportions of different nucleotides. Since T is known, we can easily estimate a_i by $d_{BCi}/(2T)$. We also note that

$$y_i \equiv d_{ACi} - d_{BCi} = (b - a_i)T_n. \qquad (6.5)$$

Therefore, T_d can be estimated by

$$T_d = (\Sigma d_{ABi} - \Sigma y_i)/(2\Sigma a_i). \qquad (6.6)$$

To estimate T_n and b, we can use equation (6.5) and apply the standard least-squares method, since there are two unknowns and three equations. However, essentially the same results are obtained by the following simple formulas.

$$T_n = (y_{12} - y_3)/(a_3 - a_{12}), \qquad (6.7)$$

$$b = (a_3 y_{12} - a_{12} y_3)/(y_{12} - y_3), \qquad (6.8)$$

where $y_{12} = (y_1 + y_2)/2$ and $a_{12} = (a_1 + a_2)/2$. In practice, T_n obtained by (6.7) may be larger than T_d. In this case, we set $T_n = T_d$, because by definition $T_n \leq T_d$. When $T_n = T_d$, equation (6.5) gives

$$b = (\Sigma a_i T_d + \Sigma y_i)/(3T_d). \qquad (6.9)$$

Li et al. (1981) applied the above method to data for three globin pseudogenes, i.e., mouse $\psi\alpha 3$, human $\psi\alpha 1$, and rabbit $\psi\beta 2$. Table 6.4 shows their estimates of T_d, T_n, b, a_1, a_2, and a_3 together with those for the goat globin pseudogenes obtained by Li (1983). It is clear that the rate of nucleotide substitution for pseudogenes is much higher than that for functional genes. Even compared with the rate for the third positions of codons, it is about two times higher. Similar results have been obtained by Miyata and Yasunaga (1981) for $M\psi\alpha 3$ and by Gojobori and Nei (1984) for two immunoglobulin V_H pseudogenes in mice. These results clearly indicate that functionally less important genes evolve faster than functionally more important genes and thus support the neutral mutation theory.

Our estimates of T_n suggest that pseudogenes have existed in the genome for tens of millions of years. The persistence of pseudogenes for a long period of time is also supported by the existence of the same processed pseudogene for immunoglobulin ϵ chains in hominoids and mon-

Table 6.4   Times since gene duplication (T_d), times since nonfunctionalization (T_n), and rates of nucleotide substitution per site per year (b) for various globin pseudogenes. a_1, a_2, and a_3 denote the rates of nucleotide substitution for the first, second, and third nucleotide positions of codons in functional genes, respectively. From Li et al. (1981) with additional data.

Gene	T_d (MY)	T_n (MY)	b ($\times 10^{-9}$)	a_1 ($\times 10^{-9}$)	a_2 ($\times 10^{-9}$)	a_3 ($\times 10^{-9}$)
Mouse $\psi\alpha 3$	27	23	5.0	0.75	0.68	2.65
Human $\psi\alpha 1$	49	45	5.1	0.75	0.68	2.65
Rabbit $\psi\beta 2$	44	43	4.1	0.94	0.71	2.02
Goat $\psi\beta^x$ and ψ^z	46	36	4.4	0.94	0.71	2.02
Average			4.7	0.85	0.70	2.34

keys which diverged about 30 MY ago (Ueda et al. 1985). This indicates that individuals carrying pseudogenes are not necessarily less fit in the populations. Indeed, the genome of higher organisms is known to carry numerous pseudogenes (Temin 1985).

Reactivation of Pseudogenes

Once a gene becomes nonfunctional, it is generally difficult to revive it. This is because while there are many ways to silence a gene, a back mutation that revives a pseudogene must occur at exactly the same site where the first forward mutation occurred, and the chance for this to occur is very small. Furthermore, once a gene becomes nonfunctional, many more destructive mutations may occur in the gene, and thus the probability of reviving a pseudogene would decrease with evolutionary time.

However, there are two possible exceptions to this rule. First, in bacteria there are genes that are nonfunctional under normal conditions but mutate to functional genes under certain circumstances. A well-studied example is the phospho-β-glucosidase system of the family Enterobacteriaceae. The β-glucosides are metabolized by a complex system of permeases and phospho-β-glucosidases with different substrate specificities. *Klebsiella* species possess a complete set of permeases and hydrolases and consequently metabolize the aryl β-glucosides arbutin and salicin and the disaccharide cellobiose. Wild-type strains of *Salmonella* and *Esche-*

richia coli are unable to utilize any β-glucoside sugars, apparently because of the nonfunctionality of some genes responsible for the phospho-β-glucosidase system. However, both *Salmonella* and *E. coli* can produce the β-glucoside positive phenotypes by mutation. This suggests that the nonfunctional genes in *Salmonella* and *E. coli* can be reactivated by mutation. Hall et al. (1983) called these apparently nonfunctional genes *cryptic genes*. According to these authors, bacterial genomes contain many cryptic genes.

Hall et al. (1983) speculated that these cryptic genes are maintained in bacterial populations because they are occasionally reactivated and this reactivation is essential for survival when no other sugars are available for metabolism. They visualize three states for the β-glucosidase genes, i.e., functional state, cryptic state, and truly pseudogene state, and hypothesize that a cryptic gene can mutate to both the functional state and the true pseudogene state, but a true pseudogene cannot be activated anymore. In this hypothesis, functional genes and cryptic genes are not required in normal environments, but cryptic genes never disappear from the population. This is because they are capable of supplying functional genes when these are needed and thus have selective advantage over true pseudogenes. This problem has been studied mathematically by Hall et al. (1983) and Li (1984).

Second, some nonfunctional genes in a multigene family may be reactivated by unequal crossover or gene conversion. That is, when a nonfunctional region of a pseudogene is replaced by the functional nucleotide sequence through gene conversion or unequal crossover, the pseudogene may be reactivated. For example, the immunoglobulin V_H gene family in mice is known to contain many pseudogenes (see figure 6.11), but some of them are caused by one or two nucleotide changes. These genes would easily be reactivated if gene conversion or recombination occurred at the right place.

□ CHAPTER SEVEN □
GENES IN POPULATIONS

In the last four chapters, we have been concerned mainly with long-term evolution, neglecting genetic polymorphism within species. When we consider the evolutionary change of genes or populations for a relatively short period of time, however, we must consider polymorphism. It is important to realize that genetic polymorphism and long-term evolution are merely two different phases of the same evolutionary process. Therefore, if one wants to propose any reasonable theory of evolution, he must be able to explain both polymorphism and long-term evolution with the same principle. To study the extent of polymorphism within species, however, we need to know the basic properties of gene and genotype frequencies in populations. In this chapter, I shall discuss this problem. The problem of measuring the extent of genetic variation within and between populations will be discussed in the following three chapters.

Gene Frequencies and Hardy-Weinberg Equilibrium

The frequency of a particular allele in a population is called the *gene* or *allele frequency*. It is a fundamental parameter in the study of evolution, since the genetic change of a population is usually described by the change in gene frequencies.

Consider a locus with two alleles, A_1 and A_2. In diploid organisms there are three possible genotypes at this locus, i.e., A_1A_1, A_1A_2, and A_2A_2. Let N_{11}, N_{12}, and N_{22} be the numbers of genotypes A_1A_1, A_1A_2, and A_2A_2 in a population, respectively, and $N_{11} + N_{12} + N_{22} = N$. The relative frequencies of A_1A_1, A_1A_2, and A_2A_2 are then given by $X_{11} = N_{11}/N$, $X_{12} = N_{12}/N$, and $X_{22} = N_{22}/N$, respectively. On the other hand, the (relative) *gene frequency* (or *allele frequency*) of A_1 is given by

$$x_1 = (2N_{11} + N_{12})/(2N)$$

$$= X_{11} + X_{12}/2. \tag{7.1}$$

Obviously, the gene frequency of A_2 is $x_2 = 1 - x_1$.

Theoretically, the *genotype frequencies* X_{11}, X_{12}, and X_{22} can take any nonnegative value between 0 and 1 with the restriction of $X_{11} + X_{12} + X_{22} = 1$. In many organisms, however, mating occurs at random, and genotypes are produced by random union of male and female gametes. In this case, the genotype frequencies are given by the expansion of $(x_1 + x_2)^2$. That is,

$$X_{11} = x_1^2, \quad X_{12} = 2x_1 x_2, \quad X_{22} = x_2^2. \tag{7.2}$$

This property was first noted by Hardy (1908) and Weinberg (1908) independently, so that it is called the *Hardy-Weinberg* principle. It should be noted that the above genotype frequencies are obtained by a single generation of random mating as long as the gene frequency is the same for male and female gametes.

The relationship between gene and genotype frequencies when more than two alleles exist at a locus is essentially the same as that for the case of two alleles. Let m be the number of alleles existing at a locus and denote the ith allele by A_i. There are m possible homozygotes and $m(m-1)/2$ possible heterozygotes, the total number of genotypes being $m(m+1)/2$. We represent the frequency of genotype $A_i A_j$ by X_{ij}. Then, the frequency of the ith allele is given by

$$x_i = X_{ii} + \frac{1}{2} \sum_{j \neq i} X_{ij}, \tag{7.3}$$

where $\sum_{j \neq i}$ indicates the summation of X_{ij} over all j's except for $j = i$. For example, when there are three alleles, the frequency of A_1 is given by $x_1 = X_{11} + (X_{12} + X_{13})/2$.

The genotype frequencies under random mating are given by the expansion of $(x_1 + x_2 + \cdots + x_m)^2$. Therefore, the frequencies of homozygote $A_i A_i$ and heterozygote $A_i A_j$ are

$$X_{ii} = x_i^2, \quad X_{ij} = 2x_i x_j. \tag{7.4}$$

These are the Hardy-Weinberg proportions for multiple alleles.

Estimation of Gene Frequencies in a Random Mating Population

When population size is large, it is not easy to examine the genotypes of all individuals in order to determine the gene frequencies. Therefore, we must sample a certain number of individuals from the population and estimate the population gene frequencies from this sample. Surely the accuracy of the estimates obtained depend on the sample size, but it is also affected by dominance, mating system, and the gene frequencies themselves. This problem has been discussed in many textbooks (e.g., Elandt-Johnson 1970; C. C. Li 1976; Spiess 1977). In the following, we consider only simple cases.

Codominant Alleles

When all genotypes are identifiable, the gene frequencies can be estimated by counting the number of genes in the sample. Suppose that there are m alleles at a locus and n individuals are sampled from the population. Let n_{ij} be the number of individuals for genotype A_iA_j, the total number of individuals being n, i.e., $\Sigma\, n_{ij} = n$. The gene frequency of A_i is then estimated by

$$\hat{x}_i = \left(2n_{ii} + \sum_{j \neq i} n_{ij} \right) / (2n). \qquad (7.5)$$

It can be shown that this is the maximum likelihood estimate of x_i. If we assume that the sampling of genes is multinomial, the variance $[V(\hat{x}_i)]$ of $\hat{x}_i$ and the covariances $[Cov(\hat{x}_i, \hat{x}_j)]$ of $\hat{x}_i$ and $\hat{x}_j$ are given by

$$V(\hat{x}_i) = \hat{x}_i(1 - \hat{x}_i)/(2n), \qquad (7.6a)$$

$$Cov(\hat{x}_i, \hat{x}_j) = -\hat{x}_i\hat{x}_j/(2n). \qquad (7.6b)$$

EXAMPLE

Most protein polymorphisms detectable by electrophoresis are controlled by codominant alleles. At the red-cell acid phosphatase locus in man, there are three major alleles, A, B, and C, all being codominant. Hopkinson and Harris (1969) examined the genotypes of 880 individuals in England and obtained the results given in table 7.1. From these results we can obtain the estimates ($\hat{x}_1$, $\hat{x}_2$, and $\hat{x}_3$) of allele frequencies of A, B, and C. They become $\hat{x}_1 = 0.373$, $\hat{x}_2 = 0.570$, and $\hat{x}_3 = 0.057$. On the other hand, the variances are given by $V(\hat{x}_1) = (0.373 \times 0.627)/1760 = 0.0002658$, $V(\hat{x}_2) = 0.0002785$, and $V(\hat{x}_3) = 0.0000611$. Therefore, the standard error of $\hat{x}_1$ is $[V(\hat{x}_1)]^{1/2} = 0.016$. This is less than one-tenth of $\hat{x}_1$. Similarly, the standard errors of $\hat{x}_2$ and $\hat{x}_3$ become 0.017 and 0.018, respectively.

Table 7.1 Observed and expected numbers of genotypes at the red cell acid phosphatase locus in man in a survey of an English population. Data from Hopkinson and Harris (1969).

Genotype	AA	BB	CC	AB	AC	BC	Total
Observed number	119	282	0	379	39	61	880
Expected number	122.4	285.9	2.9	374.2	37.4	57.2	880

The expected numbers of genotypes under random mating are obtained by replacing x_i's in (7.4) by their respective estimates $\hat{x}_i$'s and multiplying X_{ii} and X_{ij} by n. The results obtained are presented in table 7.1. The agreement between the observed and expected frequencies can be tested by the χ^2 test. The general formula of χ^2 is

$$\chi^2 = \Sigma \frac{(O-E)^2}{E},$$
(7.7)

where O and E stand for the observed and expected frequencies, and Σ indicates summation over all genotypes. In the present case, we have $\chi^2 = 3.43$. The number of degrees of freedom for the χ^2 is 3, since there are five independent observations (genotype frequencies) and we have estimated two independent parameters (gene frequencies). This χ^2 is not significant at the 5 percent level, implying that the agreement between the observed and expected frequencies is satisfactory.

In the above example, we computed the expected genotype frequencies by using (7.4). This is perfectly all right as long as sample size is large. When sample size is small, however, more accurate values of the expected numbers of genotypes are obtained by

$$\hat{n}_{ii} = \frac{n\hat{x}_i(2n\hat{x}_i - 1)}{2n-1}, \quad \hat{n}_{ij} = \frac{4n^2\hat{x}_i\hat{x}_j}{2n-1}$$
(7.8)

(Levene 1949).

Dominant Alleles

Let A and a be the dominant and recessive alleles at a locus, respectively. In a randomly mating population, the frequency of recessive homozygote aa is given by x_2^2, where x_2 is the frequency of allele a.

Therefore, if n_R is the number of recessive homozygotes in a sample of n individuals, x_2 may be estimated by

$$\hat{x}_2 = \sqrt{n_R/n}. \tag{7.9}$$

The estimate of the frequency (x_1) of A is, of course, given by $\hat{x}_1 = 1 - \hat{x}_2$. The sampling variance of $\hat{x}_2$ is

$$V(\hat{x}_2) = (1 - \hat{x}_2^2)/(4n). \tag{7.10}$$

When multiple alleles exist, dominance relationship may vary from case to case. The ideal procedure of estimation of gene frequencies for these cases is the maximum likelihood method, which is known to be most efficient statistically at least in large samples. Usually this method requires an iterative solution of the maximum likelihood equation (e.g., Elandt-Johnson 1970).

Deviations from Hardy-Weinberg Proportions

It is known that the Hardy-Weinberg equilibrium holds for many polymorphic loci in outbreeding organisms. However, it can be disturbed by a number of factors such as inbreeding, assortative mating, and natural selection. In this section, I shall discuss this problem without going into the details. The reader who is interested in the details may refer to Crow and Kimura (1970) or C. C. Li (1976).

A Pair of Alleles

When there are two alleles at a locus, any deviation from Hardy-Weinberg proportions may be measured by a single parameter (F) called the *fixation index* (Wright 1951, 1965). If we use this index, the genotype frequencies are given by

$$X_{11} = (1 - F)x_1^2 + Fx_1, \tag{7.11a}$$

$$X_{12} = 2(1 - F)x_1x_2, \tag{7.11b}$$

$$X_{22} = (1 - F)x_2^2 + Fx_2. \tag{7.11c}$$

The fixation index F can be positive or negative, depending on the case. Some important factors that cause a non-zero value of F are as follows.

Inbreeding.   Inbreeding increases the frequency of homozygotes, and
if there is no other factor, F is equal to Wright's (1969) inbreeding
coefficient. In this case, F is always positive. In self-fertilizing organisms, the frequency of heterozygotes is reduced by half every generation,
and eventually all genotypes become homozygous with $F = 1$. Some organisms reproduce by both selfing and outbreeding with certain probabilities. When the probability of outcrossing is t, the equilibrium value
of F is $(1 - t)/(1 + t)$.

Assortative Mating.   Another type of nonrandom mating is assortative
mating, where mating occurs selectively between individuals having similar
phenotypic characters. For example, the statures of husband and wife in
human populations are highly correlated. People with deaf-mutism also
tend to mate with each other. Assortative mating has an effect similar
to that of inbreeding and increases the frequency of homozygotes. Unlike inbreeding, however, the effect is limited only to the loci concerned
with the character with which assortative mating occurs (Karlin 1969;
Wright 1969).

Subdivision of a Population.   It is often difficult to know the breeding
structure of a population, and the population under investigation may
include several mating units or subpopulations. In this case, even if
Hardy-Weinberg equilibrium holds in each subpopulation, the genotype
frequencies in the entire population may deviate from Hardy-Weinberg
proportions.
    Consider a population divided into s subpopulations, in each of which
Hardy-Weinberg equilibrium holds. Let x_k be the frequency of allele A_1
in the kth subpopulation, so that the frequencies of genotypes A_1A_1,
A_1A_2, and A_2A_2 in this subpopulation are given by x_k^2, $2x_k(1 - x_k)$, and
$(1 - x_k)^2$, respectively. We also denote by w_k the relative size of the kth
population with $\Sigma\, w_k = 1$. The mean frequencies of A_1A_1, A_1A_2, and
A_2A_2 in the entire population are then given by

$$X_{11} = \sum_{k=1}^{s} w_k x_k^2 = \bar{x}^2 + \sigma^2, \tag{7.12a}$$

$$X_{12} = 2\sum_{k=1}^{s} w_k x_k(1 - x_k) = 2\bar{x}(1 - \bar{x}) - 2\sigma^2, \tag{7.12b}$$

$$X_{22} = \sum_{k=1}^{s} w_k(1 - x_k)^2 = (1 - \bar{x})^2 + \sigma^2, \tag{7.12c}$$

where $\bar{x} \equiv \Sigma \; w_k x_k$ and $\sigma^2 \equiv \Sigma \; w_k(x_k - \bar{x})^2$ are the mean and variance of gene frequency among subpopulations. Comparison of these equations with those in (7.11) shows that σ^2 corresponds to $F\bar{x}(1 - \bar{x})$. Therefore,

$$F = \sigma^2 / [\bar{x}(1 - \bar{x})]. \qquad (7.13)$$

This indicates that if a population is subdivided into many breeding units, the frequency of homozygotes tends to be higher than the Hardy-Weinberg proportion. This property was first noted by Wahlund (1928), and it is called *Wahlund's principle*.

Selection.   Selection may increase or decrease the F value, depending on the type of selection. If heterozygotes have a lower fitness than both homozygotes, then F becomes positive. On the other hand, if heterozygotes have a higher fitness than both homozygotes, F becomes negative. However, negative F does not necessarily mean that there is heterozygote advantage (Lewontin and Cockerham 1959).

Estimation of Fixation Index

We have seen that in the case of two alleles the genotype frequencies in a population can be specified by the gene frequencies and fixation index F. Suppose that n individuals are examined, and $n_{11} \; A_1 A_1$, $n_{12} \; A_1 A_2$, and $n_{22} \; A_2 A_2$ individuals are observed (table 7.2). We now want to estimate the frequency (x_1) of A_1 and F. The maximum likelihood estimates of these quantities have been worked out by Li and Horvitz (1953). They are

$$\hat{x}_1 = (2n_{11} + n_{12})/(2n), \qquad (7.14)$$

$$\hat{F} = \frac{4n_{11}n_{22} - n_{12}^2}{(2n_{11} + n_{12})(2n_{22} + n_{12})}. \qquad (7.15)$$

Table 7.2   Expected and observed numbers
of genotypes for two alleles in a
non-Hardy-Weinberg population.

Genotype	Expected number	Observed number
$A_1 A_1$	$n[(1 - F)x_1^2 + Fx_1]$	n_{11}
$A_1 A_2$	$n[2(1 - F)x_1 x_2]$	n_{12}
$A_2 A_2$	$n[(1 - F)x_2^2 + Fx_2]$	n_{22}
Total	n	n

The deviation of F from 0 can be tested by $\chi^2_{(1)} = n\hat{F}^2$ with one degree of freedom.

EXAMPLE

Tracey et al. (1975) studied the protein polymorphism in various populations of American lobsters. A population off the Maine coast has two electrophoretic alleles, *100* and *98*, at the phosphoglucose isomerase-4 (*Pgi*-4) locus. Examining 60 individuals, Tracey et al. obtained 40 individuals (n_{11}) of genotype *100/100*, 12 individuals (n_{12}) of *100/98*, and 8 individuals (n_{22}) of *98/98*. The sample gene frequencies of *100* and *98* are therefore 0.767 and 0.233, respectively. The fixation index is estimated by equation (7.15) and becomes $\hat{F} = 0.441$. This indicates that there is a high level of heterozygote deficiency at this locus. This $\hat{F}$ is significant at the 0.1 percent level ($\chi^2_{(1)} = 11.7$).

American lobster populations show heterozygote deficiency at most polymorphic loci, though the extent is not as high as that for the *Pgi*-4 locus. Heterozygote deficiency has also been observed in many other marine crustacean organisms (see Zouros and Foltz 1984). Curiously, heterozygote deficiency is more pronounced in the juvenile stage than in the adult stage. However, the reason for the deficiency is unknown, though there are several hypotheses (Zouros and Foltz 1984). These organisms are believed to mate at random, so that the Wahlund principle cannot be used to explain the deficiency. It should be noted that if we consider various groups of organisms these data are exceptional and that most outbreeding organisms show Hardy-Weinberg proportions at enzyme loci.

Multiple Alleles

When there are m alleles at a locus, we generally need $m(m-1)/2$ fixation indices to specify all genotype frequencies in terms of gene frequencies and fixation indices. However, if the deviations from Hardy-Weinberg equilibrium occur solely by inbreeding or random differentiation of gene frequencies among subpopulations, the deviations can be described by a single fixation index. In this case, the frequency (X_{ii}) of homozygote $A_i A_i$ ($i = 1, 2, \cdots, m$) is given by

$$X_{ii} = (1-F)x_i^2 + Fx_i, \tag{7.16}$$

whereas the frequency (X_{ij}) of heterozygote $A_i A_j$ is

$$X_{ij} = 2(1 - F)x_i x_j. \tag{7.17}$$

In the case of subdivided populations, the F in equations (7.16) and (7.17) can be related to the variances and covariances of gene frequencies. Let x_{ki} be the frequency of the ith allele in the kth subpopulation, and assume that in each subpopulation Hardy-Weinberg equilibrium holds. We then have

$$X_{ii} = \Sigma_k w_k x_{ki}^2 = x_i^2 + \sigma_i^2, \tag{7.18}$$

$$X_{ij} = 2\Sigma_k w_k x_{ki} x_{kj} = 2x_i x_j + 2\sigma_{ij}, \tag{7.19}$$

where $x_i = \Sigma_k w_k x_{ki}$, $\sigma_i^2 = \Sigma_k w_k (x_{ki} - x_i)^2$, and $\sigma_{ij} = \Sigma_k w_k (x_{ki} - x_i)(x_{kj} - x_j)$. Comparison of these expressions with (7.16) and (7.17) gives

$$\sigma_i^2 = Fx_i(1 - x_i) \tag{7.20}$$

$$\sigma_{ij} = -Fx_i x_j. \tag{7.21}$$

Therefore, $\sigma_i^2/[x_i(1 - x_i)]$ and $-\sigma_{ij}/x_i x_j$ are expected to be the same for all i and j under random differentiation. Nei and Imaizumi (1966a) and Selander (1975) have used this property to test the random differentiation of allele frequencies.

It is also interesting to note that the correlation between the frequencies of alleles A_i and A_j among subpopulations is given by

$$r_{ij} = -\left[\frac{x_i x_j}{(1 - x_i)(1 - x_j)} \right]^{1/2} \tag{7.22}$$

(Nei 1965). This equation can also be used for testing the hypothesis of random differentiation of allele frequencies.

EXAMPLE

Gene frequencies at protein loci in the brown snail *Helix aspersa* show extensive local differentiation. Selander (1975) and Selander and Whittam (1983) studied the extent of gene frequency differentiation among 140 samples (city blocks) from 40 different cities in California. This snail was introduced to California from France in 1859 and subsequently spread over much of the southwestern United States. The organism occurs primarily in residential gardens on city blocks, and the environment is subdivided by the way human habitations are organized (Selander and

Kaufman 1975). Selander and Whittam tested the hypothesis of random differentiation by comparing the observed correlations of allele frequencies with the expected correlations obtainable from equation (7.22). For each multiallelic locus, they computed both the expected and observed correlations between the common allele and each alternative allele occurring at a frequency greater than 0.05.

The results obtained are presented in figure 7.1(a). The observed correlation is in good agreement with the expected one for all but one locus (*Mdh*-1). These data therefore suggest that the differentiation of gene frequencies in *H. aspersa* is largely due to random genetic drift. The exceptional result for the *Mdh*-1 locus is apparently caused by the gene frequency cline observed at this locus, which in turn may be due to selection or an historical sampling effect (see Selander 1975).

Selander and Whittam (1983) studied the pattern of gene frequency differentiation for another land snail species, *Cepaea memoralis*. This snail is native to Europe and has inhabited river valleys in and around the Spanish Pyrenees at least since the most recent glaciation, and probably for millions of years. The topography determined by the mountain slopes and river valleys of the Pyrenees fragments the total population into a series of linear patches of habitat. For this species, the observed correlations are quite different from the expected for most loci examined [figure

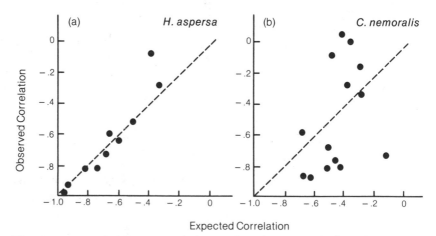

Figure 7.1.   Comparison of observed correlations of allele frequencies with those expected under a model of random differentiation for two species of snails. From Selander and Whittam (1983).

7.1(b)], indicating that the differentiation of gene frequencies is clearly nonrandom. This nonrandom differentiation suggests either that there is strong selection or that the gene frequency differentiation is affected by historical factors. Selander and Whittam's further study of the gene frequency data by principal component analysis has indicated that the total population is subdivided into three major geographical groups which were apparently formed as a result of the expansion and contraction of *Cepaea* populations in and around the Pyrenees in a past period of climatic fluctuation. From this result, Selander and Whittam concluded that the discrepancy between the observed and expected correlations is caused mainly by historical factors rather than by natural selection.

Fixation Indices

As mentioned earlier, Wahlund (1928) showed that the gene frequency differences among subdivided populations cause a deficiency of heterozygotes compared with the case of a single random mating population. Wright (1943, 1951, 1965, 1969) proposed to measure the deviations of genotype frequencies in a subdivided population in terms of three parameters F_{IS}, F_{IT}, and F_{ST}, which are often called *fixation indices* or *F-statistics*. In Wright's terminology, F_{IS} and F_{IT} are the correlations between the two uniting gametes relative to the subpopulation and relative to the total population, respectively, whereas F_{ST} is the correlation between two gametes drawn at random from each subpopulation and measures the degree of genetic differentiation of subpopulations. They are related by the following formula

$$1 - F_{IT} = (1 - F_{IS})(1 - F_{ST}). \qquad (7.23)$$

The fixation indices are useful for understanding the breeding structure of populations or the pattern of selection associated with polymorphic alleles.

Wright's original proof of equation (7.23) depends on the assumption of a pair of neutral alleles or some other simplifying assumptions (Wright 1969). Later, Nei (1977c), and Wright (1978) [see also Cockerham (1969, 1973) for a different formulation] redefined the fixation indices without using the concept of correlation of uniting gametes and showed that (7.23) holds for any situation irrespective of the number of alleles and whether there is selection or not. Particularly, Nei (1977c) showed that all fixation indices can be defined by using the observed and expected heterozygosities for the population under investigation. In the following, we shall use this approach, first considering the case of two alleles and then the case of multiple alleles. The reader who is interested in the details of this subject may refer to Jorde (1980) and Chakraborty and Leimar (1987).

Consider a population divided into s subpopulations in each of which Hardy-Weinberg equilibrium does not necessarily hold. We designate the two alleles by A_1 and A_2. Let x_k and F_{ISk} be the frequency of A_1 and the fixation index in the kth subpopulation, respectively. The frequency of homozygote A_1A_1 in the kth subpopulation (X_k) is then given by $X_k = x_k^2 + F_{ISk}x_k(1-x_k)$. Therefore,

$$F_{ISk} = \frac{X_k - x_k^2}{x_k(1-x_k)}. \tag{7.24}$$

Note that the above formulas hold true whether there is selection or not.

Following Wright's suggestion (see Kirby 1975), we define F_{IS} as a weighted average of F_{ISk}. Namely,

$$F_{IS} = \frac{\sum\limits_{k=1}^{s} w_k x_k(1-x_k)F_{ISk}}{\sum\limits_{k=1}^{s} w_k x_k(1-x_k)}, \tag{7.25}$$

where w_k is the relative size of the kth subpopulation with $\Sigma w_k = 1$. In most instances, $w_k = 1/s$ may be assumed because population size is quite transitory and geneticists are generally interested in gene frequency differences between populations, disregarding the effect of population size. A number of authors (e.g., Neel and Ward 1972) defined F_{IS} as $\Sigma_k w_k F_{ISk}$. However, if we use this definition, equation (7.23) cannot be obtained. To obtain this equation, we must use an average weighted with heterozygosities for all F's, as will be seen in the following. At any rate, if we note $X_k - x_k^2 = F_{ISk}x_k(1-x_k)$, (7.25) reduces to

$$F_{IS} = \frac{\Sigma w_k(X_k - x_k^2)}{\Sigma w_k x_k(1-x_k)}$$

$$= \frac{X - \overline{x^2}}{\bar{x} - \overline{x^2}} = 1 - \frac{\bar{x} - X}{\bar{x} - \overline{x^2}}, \tag{7.26}$$

where $X=\Sigma X_k/s$ is the relative frequency of A_1A_1 in the total popula-
tion, $\bar{x}=\Sigma x_k/s$ is the mean gene frequency, and $\overline{x^2}=\Sigma x_k^2/s$. Here, $w_k=1/s$
is assumed. On the other hand, X may be written as $X=(1-F_{IT})\bar{x}^2+F_{IT}\bar{x}$
from (7.11). Therefore,

$$F_{IT}=\frac{X-\bar{x}^2}{\bar{x}-\bar{x}^2}=1-\frac{\bar{x}-X}{\bar{x}-\bar{x}^2}. \tag{7.27}$$

As mentioned earlier, F_{ST} measures the degree of genetic differentiation
of subpopulations and is defined by $F_{ST}=\sigma_x^2/\bar{x}(1-\bar{x})$, where σ_x^2 is the
variance of gene frequency x among the subpopulations (Wright 1951).
Since $\sigma_x^2=\overline{x^2}-\bar{x}^2$, we can write

$$F_{ST}=1-\frac{\bar{x}-\overline{x^2}}{\bar{x}-\bar{x}^2}. \tag{7.28}$$

It is now clear that the quantities defined by (7.26), (7.27), and (7.28)
satisfy equation (7.23).

When there are m alleles, $m(m-1)/2$ F_{IS} parameters are required for
complete specification of genotype frequencies in a population, as mentioned earlier. If we consider only homozygotes, however, m F_{IS} parameters are sufficient. Thus, the frequency of homozygotes for the ith allele, i.e., A_iA_i, in the kth subpopulation may be written as

$$X_{kii}=x_{ki}^2+F_{ISki}x_{ki}(1-x_{ki}),$$

where the subscript i refers to the ith allele. Therefore,

$$F_{ISki}=\frac{X_{kii}-x_{ki}^2}{x_{ki}(1-x_{ki})}. \tag{7.29}$$

It is clear that the F_{IS}, F_{IT}, and F_{ST} for the ith allele can be defined as
in the case of two alleles. Hence,

$$F_{ISi}=\frac{X_{ii}-\overline{x_i^2}}{\bar{x}_i-\overline{x_i^2}}, \quad F_{ITi}=\frac{X_{ii}-\bar{x}_i^2}{\bar{x}_i-\bar{x}_i^2}, \quad F_{STi}=\frac{\overline{x_i^2}-\bar{x}_i^2}{\bar{x}_i-\bar{x}_i^2}, \tag{7.30}$$

where $X_{ii}=\Sigma_k X_{kii}/s$, $\bar{x}_i=\Sigma_k x_{ki}/s$, and $\overline{x_i^2}=\Sigma_i x_{ki}^2/s$.

As noted earlier, F_{ISki} may vary from allele to allele. It is, however, possible to define a unified fixation index for all alleles in the kth subpopulation. Namely,

$$F_{ISk} = \frac{\Sigma_i x_{ki}(1 - x_{ki})F_{ISki}}{\Sigma_i x_{ki}(1 - x_{ki})}$$

$$= \frac{\Sigma_i(X_{kii} - x_{ki}^2)}{\Sigma_i x_{ki}(1 - x_{ki})} = (h_{Sk} - h_{Ok})/h_{Sk}, \qquad (7.31)$$

where $h_{Sk} = 1 - \Sigma_i x_{ki}^2$ and $h_{Ok} = 1 - \Sigma_i X_{kii}$. The former is the expected heterozygosity under Hardy-Weinberg equilibrium, whereas the latter is the observed heterozygosity. Namely, F_{ISk} is defined as the ratio of the difference between the expected and observed heterozygosities to the expected heterozygosity.

Following (7.25), F_{IS} is now defined as $\Sigma_k h_{Sk} F_{ISk}/\Sigma_k h_{Sk}$ and becomes

$$F_{IS} = (h_S - h_O)/h_S, \qquad (7.32)$$

where h_S and h_O are the averages of h_{Sk} and h_{Ok} over all subpopulations, respectively. That is, $h_S = 1 - \Sigma_k \Sigma_i x_{ki}^2/s$ and $h_O = 1 - \Sigma_k \Sigma_i X_{kii}/s$. It is clear that the fixation index in the total population (F_{IT}) can be defined in the same way as (7.32) by using the gene and genotype frequencies in the total population. It becomes

$$F_{IT} = (h_T - h_O)/h_T, \qquad (7.33)$$

where $h_T = 1 - \Sigma \bar{x}_i^2$. Similarly, we have

$$F_{ST} = (h_T - h_S)/h_T. \qquad (7.34)$$

Obviously, F_{IS}, F_{IT}, and F_{ST} thus defined satisfy equation (23). Note that equations (7.32), (7.33), and (7.34) hold even in the case of two alleles ($m = 2$). It is interesting to see that all F_{IS}, F_{IT}, and F_{ST} can be defined in terms of expected and observed heterozygosities and take a similar mathematical form. Since these are defined in terms of the present gene and genotype frequencies, they can be applied to any situation, whether there is selection or not.

In the above formulation, we considered the case of one-rank hierarchical subdivision of populations. In practice, each subpopulation may be further subdivided, but the above method can easily be extended (Wright 1965; Nei 1977c; Chakraborty 1980).

At this point, it should be indicated that $d_{ST} \equiv h_T - h_S$ in (7.34) can be written as

$$d_{ST} = [\Sigma_k \Sigma_l d_{kl}]/s^2,$$

where $d_{kl} = \Sigma_i (x_{ki} - x_{li})^2/2$ [see equation (8.24)]. Obviously, $d_{kk} = 0$, so that we have $d_{ST} = (s-1)d'_{ST}/s$, where

$$d'_{ST} = \sum_{k \neq l} d_{kl}/[s(s-1)]. \qquad (7.35)$$

Here, the summation is taken over all d_{kl} except d_{kk}'s. [d'_{ST} is equal to $\bar{D}_m$ in equation (8.29).] Therefore, the average allele frequency differentiation between populations is measured by d'_{ST} rather than by d_{ST}. Thus, as a measure of the extent of genetic differentiation of subpopulations, one can also use

$$F'_{ST} = d'_{ST}/h'_T, \qquad (7.36)$$

where $h'_T = h_S + d'_{ST}$. This has an advantage over F_{ST} in that it is independent of s. However, h'_T is no longer the heterozygosity for the total population. In this case, one must redefine F_{IT} as

$$F'_{IT} = (h'_T - h_0)/h'_T, \qquad (7.37)$$

in order to maintain relationship (7.23), whereas F_{IS} remains the same.

The above definitions of F'_{ST}, F'_{IT}, and F_{IS} are equivalent to Cockerham's (1969, 1973) F-statistics, which are based on the analysis of variance of gene frequencies. In the above (Wright's) approach, F-statistics are defined in terms of the gene and genotype frequencies of the subpopulations that actually exist. In Cockerham's approach, the existing subpopulations are considered as a random sample from a population that is composed of infinitely many "replicate subpopulations." The latter approach is certainly traditional in statistics and is thus preferred if there are truly replicate populations as in the case of controlled experiments. In natural populations, however, each subpopulation usually has a unique evolutionary history, so that the concept of "replicate subpopulations" does not apply. Nevertheless, if we use F'_{ST}, F'_{IT}, and F_{IS}, both approaches give essentially the same result.

Estimation of Fixation Indices

In the above formulation, we defined fixation indices in terms of population gene and genotype frequencies. In practice, however, we must estimate these indices from sample gene and genotype frequencies. This problem has been studied by Nei and Chesser (1983). [The reader who is interested in Cockerham's (1969) definition of fixation indices should use Weir and Cockerham's (1984) method.]

We assume that n_k individuals are randomly chosen from the kth subpopulation. Let $\hat{x}_{ki}$ and $\hat{X}_{kij}$ be the frequencies of allele A_i and genotype A_iA_j in the sample from the kth subpopulation, respectively. Since the fixation indices are defined in terms of h_O, h_S, and h_T or h'_T, we must have estimates of these quantities. For the reason mentioned earlier, we assume $w_k = 1/s$. Estimation of h_O is simple, because $\hat{X}_{ii}$ is an unbiased estimate of X_{ii} under our assumption. An unbiased estimate of h_O in the kth subpopulation is $1 - \Sigma_i \hat{X}_{kii}$, and thus h_O may be estimated by

$$\hat{h}_O = 1 - \sum_{k=1}^{s} \sum_{i=1}^{m} \hat{X}_{kii}/s. \qquad (7.38)$$

Derivation of an equation for the unbiased estimate of h_S is more complicated, since $\hat{x}_{ki}^2$ is not an unbiased estimate of x_{ki}^2. It can be shown, however, that an unbiased estimate of h_S is

$$\hat{h}_S = \frac{\bar{n}}{\bar{n}-1}\left[1 - \Sigma_i \overline{\hat{x}_i^2} - \frac{\hat{h}_O}{2\bar{n}}\right], \qquad (7.39)$$

where $\overline{\hat{x}_i^2} = \Sigma_k \hat{x}_{ki}^2/s$, and $\bar{n}$ is the harmonic mean of n_k. Similarly, it can be shown that an unbiased estimate of h_T is

$$\hat{h}_T = 1 - \sum_{i=1}^{m} \bar{x}_i^2 + \frac{\hat{h}_s}{\bar{n}s} - \frac{\hat{h}_o}{2\bar{n}s}. \qquad (7.40)$$

where $\bar{x}_i = \Sigma_k \hat{x}_{ki}/s$. Therefore, the estimates of F_{IS}, F_{IT}, and F_{ST} are

$$\hat{F}_{IS} = 1 - \hat{h}_O/\hat{h}_S, \qquad (7.41)$$

$$\hat{F}_{IT} = 1 - \hat{h}_0/\hat{h}_T, \tag{7.42}$$

$$\hat{F}_{ST} = 1 - \hat{h}_S/\hat{h}_T. \tag{7.43}$$

Similarly, one can estimate F'_{IT} and F'_{ST} by estimating $\hat{h}'_T$. $\hat{h}'_T$ is given by $\hat{h}_S + \hat{d}'_{ST}$, where $\hat{d}'_{ST}$ is $s(\hat{h}_T - \hat{h}_S)/(s - 1)$.

Note that $\hat{h}_0$ is not affected by sample size $\bar{n}$, whereas $\hat{h}_S$ and $\hat{h}_T$ are. Furthermore, the effects of $\bar{n}$ on $\hat{h}_S$ and $\hat{h}_T$ are negligibly small if $\bar{n}$ is sufficiently large, say, $\bar{n} > 50$. Therefore, the above sample size corrections are necessary only when $\bar{n}$ is very small.

$\hat{F}_{ST}$ is of special importance, because this measures the extent of genetic differentiation of subpopulations. The statistical significance of the deviation of $\hat{F}_{ST}$ from 0 can be tested by using the usual χ^2 test of heterogeneity of gene frequencies (Workman and Niswander 1970).

EXAMPLE

Gershowitz et al. (1967) studied the genotype frequencies for the MNSs blood group system in three populations (São Domingos, São Marcos, and Simões Lopes) of the Xavante Indians in South America. Treating MS, Ms, NS, and Ns chromosome segments as multiple alleles, they estimated the gene and genotype frequencies for the three populations. The gene frequencies and the four homozygote frequencies are presented in table 7.3. In the present case, the sample size is so large that the fixation indices can be estimated by equating the sample gene and genotype frequencies to the population frequencies. The F_{ISki} values esti-

Table 7.3   Gene (x_k) and homozygote (X_k) frequencies for the MNSs locus in three populations of the Xavante American Indians and estimates of F_{ISki}. Data from Gershowitz et. al. (1967).

Allele	São Domingos (79)[a]			São Marcos (287)			Simões Lopes (171)		
	x	X	F_{ISki}	x	X	F_{ISki}	x	X	F_{ISki}
MS	.33	.1392	.1370	.37	.1812	.1900	.39	.1637	.0488
Ms	.38	.1646	.0857	.36	.1185	−.0482	.48	.2222	−.0328
NS	.09	.0000	−.0989	.15	.0174	−.0400	.01	.0000	−.0101
Ns	.20	.0380	−.0125	.12	.0035	−.1032	.12	.0176	−.0303
Weighted average			.0578			.0244			.0108

[a]Number of individuals examined.

Table 7.4 Fixation indices for the *MNSs* locus in the population of Xavante American Indians.

Allele	Average gene frequency $(\bar{x})$	X^a	$\bar{x}^2$	$\overline{x^2}$ [b]	F_{ISi}	F_{ITi}	F_{STi}
MS	.3633	.1613	.1320	.1326	.1244	.1266	.0026
Ms	.4067	.1684	.1654	.1681	.0013	.0124	.0112
NS	.0833	.0058	.0069	.0102	−.0602	−.0144	.0432
Ns	.1467	.0197	.0215	.0229	−.0258	−.0144	.0112
Sum		.3552	.3258	.3338			
Weighted average					.0321	.0436	.0119

[a] $\underline{X}$: Average frequency of homozygotes over the three populations with equal weight.
[b] $\overline{x^2}$: Average of x^2 over the three populations with equal weight.

mated by equation (7.29) are given in table 7.3. F_{ISki} is high for allele *MS* but low for *NS* in all three populations. This suggests the possibility that either selection or nonrandom mating or both caused an excess of homozygote *MS/MS* and a deficiency of *NS/NS*. χ^2 tests ($\chi^2 = nF_{ISki}^2$) show that the excess in São Marcos is significant at the 1 percent level, whereas all other excesses and deficiencies are not significant. The fixation indices for the entire population of Xavante Indians are presented in table 7.4. In the computation of these values, $w_k = 1/3$ was assumed. F_{IT} is quite large (0.044), and this is largely due to F_{IS} ($= 0.032$) rather than to F_{ST} ($= 0.012$). The fixation indices for individual alleles, particularly F_{ISi} and F_{ITi}, vary considerably. A χ^2 test shows that the F_{ISi} for allele *MS* is again significantly different from 0. F'_{ST} and F'_{IT} can be computed by (7.36) and (7.37), respectively. They become $F'_{ST} = 0.0177$ and $F'_{IT} = 0.0492$. In the present example, s is only 3, so that the difference between F_{ST} and F'_{ST} or between F_{IT} and F'_{IT} is appreciably large.

Nonrandom Association of Genes

Linkage Disequilibrium

When two or more loci are considered together, the genotype frequencies in a randomly mating population are not necessarily given by the products of gene frequencies. This is because the alleles at different loci are not always randomly combined in chromosomes.

Let us consider two loci each with two alleles, A_1, A_2 and B_1, B_2. There are four different types of chromosomes possible for these loci, i.e., A_1B_1, A_1B_2, A_2B_1, and A_2B_2. Let X_1, X_2, X_3, and X_4 be the frequencies of these chromosomes, respectively. The gene frequencies of A_1, A_2, B_1, and B_2 are then given by $x_1 = X_1 + X_2$, $x_2 = X_3 + X_4$, $y_1 = X_1 + X_3$, and $y_2 = X_2 + X_4$, respectively. The chromosome frequencies are not necessarily given by the products of gene frequencies involved. Namely,

$$X_1 = x_1 y_1 + D, \tag{7.44a}$$

$$X_2 = x_1 y_2 - D, \tag{7.44b}$$

$$X_3 = x_2 y_1 - D, \tag{7.44c}$$

$$X_4 = x_2 y_2 + D, \tag{7.44d}$$

where

$$D = X_1 X_4 - X_2 X_3. \tag{7.45}$$

This D is called *linkage disequilibrium*. When $D = 0$ in a population, this population is said to be in linkage equilibrium. Only in this case can the chromosome frequencies be expressed as the products of gene frequencies.

When there are more than two alleles, linkage disequilibrium may be defined for each pair of alleles from loci A and B. Thus, the linkage disequilibrium for alleles A_i and B_j may be defined as

$$D_{ij} = X_{ij} - x_i x_j, \tag{7.46}$$

where X_{ij} is the frequency of chromosome A_iB_j, and x_i and x_j are the frequencies of alleles A_i and B_j, respectively. In general, $D_{ij} \neq D_{kl}$ except when there are only two alleles at each locus. In usual data analysis, however, multiallelic data are often reduced to two-allele data by considering one particular allele and the sum of all remaining alleles at each locus.

With two loci each with two alleles, there are nine possible genotypes. The frequencies of these genotypes under random mating can be obtained by expanding $(X_1 + X_2 + X_3 + X_4)^2$. They are given in table 7.5. Unlike the case of a single locus, however, the genotype frequencies

Table 7.5 Frequencies and observed numbers of nine possible genotypes for two loci each with two alleles.

		A_1A_1	A_1A_2	A_2A_2	*Sum*
	Expected frequency	X_1^2	$2X_1X_3$	X_3^2	
B_1B_1	Fitness	W_{11}	W_{13}	W_{33}	
	Observed number	n_{11}	n_{12}	n_{13}	$n_1.$
	Expected frequency	$2X_1X_2$	$2(X_1X_4+X_2X_3)^a$	$2X_3X_4$	
B_1B_2	Fitness	W_{12}	$W_{14}=W_{23}$	W_{34}	
	Observed number	n_{21}	n_{22}	n_{23}	$n_2.$
	Expected frequency	X_2^2	$2X_2X_4$	X_4^2	
B_2B_2	Fitness	W_{22}	W_{24}	W_{44}	
	Observed number	n_{31}	n_{32}	n_{33}	$n_3.$
Sum		$n._1$	$n._2$	$n._3$	n

[a]The double heterozygotes are composed of coupling (A_1B_1/A_2B_2) and repulsion (A_1B_2/A_2B_1) genotypes. The frequencies of A_1B_1/A_2B_2 and A_1B_2/A_2B_1 are $2X_1X_4$ and $2X_2X_3$, respectively.

(as well as the chromosome frequencies) do not reach equilibrium values with a single generation of random mating.

In the absence of selection, the chromosome frequencies in the next generation can be obtained in the following way. We note that there are two ways in which chromosome A_1B_1 in generation $t+1$ is produced from the genotypes in generation t. First, it may be derived from genotypes $A_1B_1/--$ without recombination, where notation $-$ refers to an arbitrary allele at the specified locus. The probability of this event is $1-r$, where r is the recombination value between the two loci. Second, the A_1B_1 chromosome may be a product of recombination in genotypes $A_1-/-B_1$. The probability of this event is r. The frequency of genotypes $A_1-/-B_1$ is, of course, x_1y_1. Since the gene frequencies in a large random mating population remain constant in all generations, we have

$$X_1^{(t+1)}=(1-r)X_1^{(t)}+rx_1y_1. \qquad (7.47a)$$

Similarly,

$$X_2^{(t+1)}=(1-r)X_2^{(t)}+rx_1y_2, \qquad (7.47b)$$

$$X_3^{(t+1)}=(1-r)X_3^{(t)}+rx_2y_1, \qquad (7.47c)$$

$$X_4^{(t+1)}=(1-r)X_4^{(t)}+rx_2y_2. \qquad (7.47d)$$

If we note that $x_1y_1 = (X_1 + X_2)(X_1 + X_3) = X_1(1 - X_4) + X_2X_3$, $x_1y_2 = (X_1 + X_2)(X_2 + X_4) = X_2(1 - X_3) + X_1X_4$, etc., the above equations can also be written as

$$X_1^{(t+1)} = X_1^{(t)} - rD^{(t)}, \tag{7.48a}$$

$$X_2^{(t+1)} = X_2^{(t)} + rD^{(t)}, \tag{7.48b}$$

$$X_3^{(t+1)} = X_3^{(t)} + rD^{(t)}, \tag{7.48c}$$

$$X_4^{(t+1)} = X_4^{(t)} - rD^{(t)}, \tag{7.48d}$$

where $D^{(t)}$ is $X_1^{(t)}X_4^{(t)} - X_2^{(t)}X_3^{(t)}$. From (7.44a), we have $X_1^{(t)} = x_1y_1 + D^{(t)}$ and $X_1^{(t+1)} = x_1y_1 + D^{(t+1)}$. Substitution of these into (7.48a) gives

$$D^{(t+1)} = (1 - r)D^{(t)},$$

or

$$D^{(t)} = (1 - r)^t D^{(0)}, \tag{7.49}$$

where $D^{(0)}$ is the initial value of linkage disequilibrium.

Equation (7.49) indicates that linkage disequilibrium declines at a rate of r per generation under random mating and eventually becomes 0 unless $r = 0$. Therefore, we would expect that alleles at different loci are generally combined at random in a randomly mating population.

The above formulation depends on the assumption of no selection. If there is selection, the pattern of the change in D is quite different. Particularly, if there is nonadditive (epistatic) selection operating at two or more loci, linkage disequilibrium may be generated even in a randomly mating population (chapter 12). For this reason, many authors have examined the linkage disequilibrium for polymorphic enzyme loci. In practice, however, there are several other factors that generate non-random association of alleles (chapter 12), so that one must be cautious about the interpretation of linkage disequilibrium data.

Estimation of Linkage Disequilibrium

In haploid organisms or in experimental organisms such as *Drosophila*, chromosome frequencies are directly observable. In this case, the linkage disequilibrium can easily be estimated. Let us consider the case of two alleles at each locus and denote by n_1, n_2, n_3, and n_4 the observed numbers of chromosomes A_1B_1, A_1B_2, A_2B_1, and A_2B_2, respectively, where $n_1+n_2+n_3+n_4=n$. Under the assumption of multinomial sampling, $\hat{X}_1=n_1/n$, $\hat{X}_2=n_2/n$, $\hat{X}_3=n_3/n$, and $\hat{X}_4=n_4/n$ are unbiased estimates of X_1, X_2, X_3, and X_4, respectively, and the maximum likelihood estimates of x_1, y_1, and D are given by

$$\hat{x}_1=(n_1+n_2)/n, \tag{7.50a}$$

$$\hat{y}_1=(n_1+n_3)/n, \tag{7.50b}$$

$$\hat{D}=(n_1n_4-n_2n_3)/n^2. \tag{7.50c}$$

The large-sample variance of $\hat{D}$ is

$$V(\hat{D})=[x_1x_2y_1y_2+(x_1-x_2)(y_1-y_2)D-D^2]/n \tag{7.51}$$

(Hill 1974b; Brown 1975).

When $D=0$, (7.51) reduces to $V(\hat{D})=x_1x_2y_1y_2/n$. Therefore, the null hypothesis of $D=0$ can be tested by

$$\chi^2_{(1)}=n\,\hat{D}^2/[\hat{x}_1\hat{x}_2\hat{y}_1\hat{y}_2], \tag{7.52}$$

with one degree of freedom. This is identical with the usual χ^2 test for the 2×2 contingency table.

In diploid organisms, the maximum likelihood estimation of linkage disequilibrium is more complicated (Bennett 1965), and no explicit solution is available. However, Hill (1974b) developed an iteration method of obtaining a maximum likelihood estimate. Consider two codominant alleles at each locus and denote the observed numbers of genotypes as given in table 7.5. Maximum likelihood estimates of allele frequencies (x_1 and y_1) are then given by

$$\hat{x}_1=(2n_{.1}+n_{.2})/(2n),$$
$$\hat{y}_1=(2n_{1.}+n_{2.})/(2n),$$

where $n_{.1}=n_{11}+n_{21}+n_{31}$, $n_{.2}=n_{12}+n_{22}+n_{32}$, $n_{1.}=n_{11}+n_{12}+n_{13}$, $n_{2.}=n_{21}+n_{22}+n_{23}$, and n is the total number of individuals examined. Since $D=X_1-x_1y_1$ and we have the estimates of x_1 and y_1, an estimate of D is obtainable if we know X_1. Hill showed that X_1 can be estimated by the following iteration formula

$$\hat{X}_1 = [m_{11} + n_{22}\hat{X}_1'(\hat{x}_2 - \hat{y}_1 + \hat{X}_1')/\{\hat{X}_1'(\hat{x}_2 - \hat{y}_1 + \hat{X}_1')$$

$$+ (\hat{x}_1 - \hat{X}_1')(\hat{y}_1 - \hat{X}_1')\}]/(2n), \qquad (7.53)$$

where $m_{11} = 2n_{11} + n_{12} + n_{21}$ and $\hat{X}_1'$ is a trial value of $\hat{X}_1$. This formula is repeatedly used until $\hat{X}_1$ becomes equal to $\hat{X}_1'$. A maximum likelihood estimate of D is then obtained by $\hat{D} = \hat{X}_1 - \hat{x}_1\hat{y}_1$. A suitable starting value for $\hat{X}_1'$ is

$$\hat{X}_1' = \frac{1}{4n}(m_{11} - m_{12} - m_{21} + m_{22}) + \frac{1}{2} - \hat{x}_2\hat{y}_2, \qquad (7.54)$$

where $m_{12} = 2n_{13} + n_{12} + n_{23}$, $m_{21} = 2n_{31} + n_{21} + n_{31}$, and $m_{22} = 2n_{33} + n_{23} + n_{32}$. This is obtained by assuming that the genotype frequency of the double heterozygote class is exactly equal to that computed from the other classes.

The sampling variance of $\hat{D}$ when $D - 0$ again becomes $V(\hat{D}) = x_1x_2y_1y_2/n$. Therefore, the null hypothesis of $D = 0$ can be tested by (7.52). Note that n is the number of individuals studied rather than the number of chromosomes studied in this case.

When there is dominance at one or both loci, the estimation of D becomes much more complicated. The reader who is interested in this problem may refer to Hill (1974b).

So far we have been concerned with the linkage disequilibrium for two loci. If we consider more than two loci, we must consider many different types of higher order linkage disequilibria (Bennett 1954; Hill 1974a). There are several different statistical methods for studying higher order linkage disequilibria (see Mukai et al. 1974; Smouse 1974; Langley et al. 1974; Brown 1975).

EXAMPLE

Mukai et al. (1971) examined the extents of linkage disequilibrium for 6 pairs of polymorphic enzyme loci (all located on the second chromosome) in a North Carolina population of *Drosophila melanogaster*. For the pair of the alcohol dehydrogenase *(Adh)* and the α glycerophosphate dehydrogenase-1 *(α Gpdh*-1) loci, they observed the following numbers of chromosome types (allelic combinations) in the 1968 and 1969 surveys.

Allelic combination	FF	FS	SF	SS	Total	$\hat{D}$
1968	67	15	193	40	315	−0.002
1969	38	13	77	18	146	−0.015

Here, the first and second letters in the allelic combination refer to the allele (F or S) at the Adh and α $Gpdh$-1 loci, respectively. The linkage disequilibrium for the 1968 sample therefore becomes $D = (67 \times 40 - 15 \times 193)/315^2 = -0.002$ from equation (7.50c). The χ^2 for this linkage disequilibrium is 0.05 from equation (7.52). For the 1969 sample, we have $\hat{D} = 0.015$ and $\chi^2 = 0.85$. Therefore, there is no evidence of linkage disequilibrium for this pair of loci.

Mukai et al. (1971) estimated the D values for all six pairs of enzyme loci but found no evidence of linkage disequilibrium except for one pair for the 1968 sample. Similar results have been obtained by several different groups of authors (e.g., Kojima et al. 1970; Charlesworth and Charlesworth 1973; Langley et al. 1974). These results suggest that there is no strong epistatic selection among enzyme loci.

Other Measures of Nonrandom Association

The linkage disequilibrium discussed above is the most commonly used measure of nonrandom association of alleles at different loci. It has nice statistical and mathematical properties. From the biological point of view, however, it has some drawbacks. The most serious one is its dependence on gene frequencies. Thus, if the allele frequencies A_1 and B_1 are both 0.5, D defined in (7.45) takes a value between -0.25 to 0.25. However, if x_1 and y_1 are 0.1 and 0.05, respectively, the minimum and maximum possible values of D are -0.005 and 0.045, respectively. Therefore, we cannot compare the D values for two pairs of loci if the gene frequencies are not the same.

This problem can be solved if we use the ratio of D to its maximum possible absolute value of D for given gene frequencies (Lewontin 1964). Equation (7.45) can be written as

$$D = (x_1 y_1 + \delta)(x_2 y_2 + \delta) - (x_1 y_2 - \delta)(x_2 y_1 - \delta) = \delta,$$

where δ is the deviation of chromosome frequencies from the values for linkage equilibrium. Therefore, the largest positive value δ can take is $x_1 y_2$ or $x_2 y_1$, whichever is smaller, whereas the largest negative value δ can take is either $-x_1 y_1$ or $-x_2 y_2$, whichever is smaller. Therefore, if we compute the ratio (D') of D to its maximum possible value for the positive and negative sides separately, D' takes a value between -1 and $+1$. One can then use D' for comparing linkage disequilibrium values for different pairs of loci or populations.

Another measure of nonrandom association which is often used is

$$R = D/[x_1 y_1 x_2 y_2]^{1/2} \qquad (7.55)$$

(Hill and Robertson 1968). If we give value 0 to alleles A_1 and B_1 and 1 to A_2 and B_2, then the correlation coefficient between the allelic values at loci A and B becomes equal to (7.55). However, its maximum and minimum possible values for given gene frequency values are not necessarily -1 and 1, respectively. Therefore, this measure cannot be used for comparing the extent of nonrandom association for different loci.

In some cases, it is useful to consider one locus (say A) as the primary locus and the other (say B) as the secondary locus. For example, Prakash and Lewontin (1968) showed that gene arrangement (inversion chromosome) ST in $Drosophila\ pseudoobscura$ always carries electromorph (allele detectable by electrophoresis) 1.04 at the Pt-10 locus, whereas gene arrangement SC generally carries electromorph 1.06. A similar but less conspicuous nonrandom association of electromorphs with inversion chromosomes was also observed at other loci. For studying this type of nonrandom association, the following measures are convenient.

Let P be the frequency of an inversion chromosome in a population and $Q(=1-P)$ be the frequency of noninversion chromosomes. We consider a pair of alleles (A_1 and A_2) at a locus in these chromosomes. There are four different possible types of chromosomes, i.e., inversion chromosomes with alleles A_1 (designated as IA_1) and A_2 (IA_2) and noninversion chromosomes with A_1 (NA_1) and A_2 (NA_2), of which the frequencies are denoted by X_1, X_2, X_3, and X_4, respectively. We denote the frequency of A_1 in the group of inversion chromosomes by $x = X_1/P$ and that in the group of noninversion chromosomes by $y = X_3/Q$. The frequencies of A_2 in the inversion and noninversion groups are obviously $1-x$ and $1-y$, respectively. With these definitions, D can be written as $X_1 X_4 - X_2 X_3 = PQ(x - y)$. Therefore, the extent of nonrandom association can also be measured by

$$d = x - y. \qquad (7.56)$$

We call d the coefficient of nonrandom association (Nei and Li 1980). Obviously, d takes a value between -1 and 1 for given values of P and

Q. $d = 1$ or -1 indicates an extreme case where one type of chromosome is always associated with allele A_1 and the other type of chromosome with allele A_2. If P and Q remain constant for all generations, the value (d_t) of d in the tth generation is given by

$$d_t = d_0(1 - r)^t, \tag{7.57}$$

since $D = PQd$.

Another measure of nonrandom association is Yule's coefficient of association, which is given by

$$A_s = (X_1X_4 - X_2X_3)/(X_1X_4 + X_2X_3)$$
$$= (x - y)/(x + y - 2xy). \tag{7.58}$$

This coefficient also takes a value between -1 to 1 and is again independent of P and Q.

EXAMPLE

Langley et al. (1974) published chromosome frequencies for many different pairs of inversions and electromorphs in two populations of Dro-

Table 7.6 Linkage disequilibrium *(D)*, correlation *(R)*, nonrandom associations *(d)*, and coefficients of associations *(A_s)* for the pairs of electromorphs and inversions in the same arms of chromosomes II and III in *Drosophila melanogaster*. *x* and *y* are the frequencies of a given allele in the inversion and noninversion chromosomes. B = Brownsville, Texas; K = Katsunuma, Japan. Data from Langley et al. (1974).

Locus and inversion		D	R	d	A_s	x	y
$\alpha Gpd/In(2L)t$	B	0.020*	0.17	0.14	1.00	1.00	0.86
	K	0.012	0.08	0.08	0.32	0.90	0.82
$Adh/In(2L)t$	B	0.047**	0.28	0.32	1.00	1.00	0.68
	K	0.111**	0.59	0.75	1.00	1.00	0.25
$Amy/In(2R)NS$	B	0.022*	0.17	0.11	0.79	0.98	0.87
	K	0.009	0.10	0.08	0.63	0.97	0.89
$Est-6/In(3L)P$	B	0.084**	0.39	0.45	0.83	0.90	0.45
	K	0.016**	0.17	0.25	0.58	0.41	0.16

*Significant at the 5 percent level.   **Significant at the 1 percent level. The significance levels are the same for all D, R, d, and A_s.

sophila melanogaster. Table 7.6 gives the values of D, R, d, and A_s for four pairs of enzyme locus and inversion in the same arms of chromosomes II and III. D is generally very low compared with the other association measures, but six of the eight D's are statistically significant. (The significance test was done by using the usual 2×2 χ^2 test.) By contrast, A_s generally has a very high value. In the case of $Adh - In(2L)t$, it is 1.0 for both Brownsville and Katsunuma. That is, one allele (A_1) at the Adh locus is always associated with inversion $In(2L)t$. This complete association cannot be revealed by D, R, or d.

Significant nonrandom association is often observed between electromorphs and inversion chromosomes (e.g., Prakash and Lewontin 1968; Kojima et al. 1970; Mukai et al. 1971; Pinsker and Sperlich 1981; Prevosti et al. 1983). This is because the recombination value between two different gene arrangements is extremely small, and thus the genes in different gene arrangements are almost completely isolated (see Nei and Li 1975, 1980; Ishii and Charlesworth 1977). In some cases, however, nonrandom association between electromorphs and inversions may be caused by natural selection (see Prevosti et al. 1983).

CHAPTER EIGHT

GENETIC VARIATION WITHIN SPECIES

Measurement of Genetic Variation

One of the main objectives of population genetics is to describe the amount of genetic variation in populations and study the mechanism of maintenance of variation. Genetic variation may be measured at various levels, but in this chapter we will be concerned mainly with allelic variation at structural loci, particularly variation identified by electrophoresis.

The genome of higher organisms probably contains 4,000 to 50,000 structural loci. Therefore, to determine the exact amount of genetic variability of a population, we must study all these loci. In practice, this is virtually impossible, so that only a small proportion of genes are sampled to estimate the total amount of genetic variability. Since the extent of genetic polymorphism varies from locus to locus, we must use a random sample of loci. In electrophoretic studies, the choice of a locus generally depends on the availability of a proper staining technique for the protein encoded. This staining technique has nothing to do with the extent of polymorphism, so that the loci chosen by this criterion are generally considered to be a random sample (Hubby and Lewontin 1966). Of course, this view is not firmly established, and some authors (e.g., Leigh Brown and Langley 1979) have questioned it. In the following, we assume that the loci chosen for studying genetic variability are a random sample from the genome.

Proportion of Polymorphic Loci

When a large number of loci are examined for studying genic variation of a population, the amount of variation is often measured by the proportion of polymorphic loci and average heterozygosity per locus. Usually a locus is called *polymorphic* when the frequency of the most common allele (x_c) is equal to or less than 0.99. This definition is obviously arbitrary, and there is no reason why the distinction between polymorphic and monomorphic loci should not be made at $x_c = 0.95$ or 0.995 or at any other similar value. Furthermore, when sample size (n) is smaller than 50, the criterion of $x_c = 0.99$ is not really reasonable because in this case an allele of which the population frequency is less than $1/(2n)$ may not be represented in the sample, even if this allele has a frequency of 0.01 or more. Another problem is that when the number of loci examined (r) is small, the proportion of polymorphic loci is subject to a large sampling error and becomes useless. Because of the above problems, the proportion of polymorphic loci is not a good measure of genic variation. Nevertheless, as long as a large number of loci and a large number of individuals per locus are studied, it gives one important aspect of genic variation within populations.

Gene Diversity (Heterozygosity)

Definition and estimation. A more appropriate measure of genic variation is *average heterozygosity* or *gene diversity*. This measure does not depend on the arbitrariness of the definition of polymorphism; it can be defined unambiguously in terms of gene frequencies. Consider a randomly mating population, and let x_i be the *population frequency* of the ith allele at a locus. The *heterozygosity* for this locus is then defined as

$$h = 1 - \sum_{i=1}^{m} x_i^2, \qquad (8.1)$$

where m is the number of alleles. Average heterozygosity (H) is the average of this quantity over all loci.

It is clear that H is the average proportion of heterozygotes per locus in a randomly mating population. It is also equal to the expected proportion of heterozygous loci in a randomly chosen individual. Therefore, it has a clear-cut biological meaning. In haploid or polyploid organisms, the concept of heterozygosity is not applicable. Even in diploid organisms, H is not equal to the average proportion of heterozygotes, if genotype frequencies deviate from Hardy-Weinberg proportions. However, we can consider the probability that two randomly chosen genes from a population are different. This probability is equivalent to the heterozygosity in a randomly mating diploid population and can be defined in terms of gene frequencies in the same way as that of heterozygosity. Nei

(1973a) has called this quantity the *gene diversity* and indicated that this word can be used for any organism, whether it is haploid, diploid, or polyploid. It can also be used for any type of reproductive system, whether it is random mating, selfing, or asexual reproduction. In this case, gene diversity is simply a measure of genetic variability but still has a nice statistical property. In practice, however, the word heterozygosity is widely used in the literature, so that I shall use both words, gene diversity and heterozygosity, in this book, depending on the situation. Furthermore, Nei has called $1-h$ *gene identity*. In some circumstances, this quantity is more useful than gene diversity.

Let us now study the sampling property of gene diversity in a diploid population. We denote by X_{ij} the population frequency of genotype A_iA_j at a locus, and assume that n individuals are randomly sampled and their genotypes are determined. In this and the following sections, we consider only codominant alleles. Let $\hat{X}_{ij}$ be the frequency of A_iA_j in the sample. Since our sampling is multinomial, $\hat{X}_{ij}$ is an unbiased estimate of X_{ij}. Therefore, the estimate $(\hat{x}_i)$ of allele frequency x_i is given by

$$\hat{x}_i = \hat{X}_{ii} + \sum_{j\neq i}\hat{X}_{ij}/2, \tag{8.2}$$

whereas the sample gene diversity is

$$h_1 = 1 - \sum \hat{x}_i^2. \tag{8.3}$$

However, this is not an unbiased estimate of h, as shown in the preceding chapter. Under the assumption of multinomial sampling, the expection of $1-\sum\hat{x}_i^2$ is given by $(1-\sum x_i^2)(1-1/2n)$, where n is the number of individuals sampled. Therefore, an unbiased estimate of h is given by

$$\hat{h} = 2n(1-\sum \hat{x}_i^2)/(2n-1) \tag{8.4}$$

(Nei and Roychoudhury 1974a). It can be seen from (8.4) that the bias in h_1 is negligibly small when $n\geq 50$. It should also be noted that $\hat{h}$ in (8.4) has a smaller sampling variance than $\hat{h}_s$ in (7.39). Therefore, it is recommended that in a randomly mating population equation (8.4) be used unless the deviations from Hardy-Weinberg proportions are significant.

On the other hand, in selfing populations, h_0 in equation (7.39) is virtually 0 so that we have

$$\hat{h} = n(1 - \Sigma \hat{x}_i^2)/(n-1).$$ (8.5)

It is interesting to see that $2n$ in (8.4) is replaced by n in (8.5).

The average gene diversity (H) is estimated by sampling r loci from the genome. Namely,

$$\hat{H} = \sum_{j=1}^{r} \hat{h}_j/r,$$ (8.6)

where $\hat{h}_j$ is the value of $\hat{h}$ for the jth locus. Here, n may vary from locus to locus. The sampling variance of $\hat{H}$ may be obtained by

$$V(\hat{H}) - V(\hat{h})/r,$$ (8.7)

where $V(\hat{h})$ is the variance of $\hat{h}$ and is given by

$$V(\hat{h}) = \sum_{j=1}^{r} (\hat{h}_j - \hat{H})^2/(r-1).$$ (8.8)

As will be seen later, the distribution of h among loci is usually L-shaped. Therefore, an exact test of significance for $\hat{H}$ is difficult to conduct. However, if r is large, the distribution of $\hat{H}$ will be approximately normal because of the central limit theorem. Thus, the ordinary statistical test may be made by using the variance obtained by (8.7) (see Archie 1985).

Sampling variance. In the estimation of H, there are two sampling processes, i.e., sampling of loci from the genome and sampling of genes. Let us now evaluate the variances associated with these sampling processes. First, we note that the observed heterozygosity for the jth locus can be written as

$$\hat{h}_j = h_j + s_j,$$ (8.9)

where h_j is the population heterozygosity $(1 - \Sigma x_i^2)$, and s_j is the sampling error with mean 0 and variance $V_{sj}(h)$. Since h_j and s_j are independent, the variance of $\hat{h}_j$ over all loci (the entire genome) is

$$V(\hat{h}) = V(h) + V_s(h), \tag{8.10}$$

where $V(h)$ is the variance of h and $V_s(h)$ is the mean of $V_{sj}(h)$ over all loci. Here, we have assumed that there are linkage equilibria among different loci and genes are sampled independently at each locus. Obviously, $V(h)$ is associated with the sampling of loci from the genome, and $V_s(h)$ with the sampling of individuals at each locus. Nei and Roychoudhury (1974a) and Nei (1978a) have called $V(h)$ and $V_s(h)$ the *interlocus* and *intralocus variances,* respectively. The latter variance can be reduced by increasing the number of individuals sampled, but the former cannot. The sampling variance of $\hat{H}$ due to the interlocus variation of h can only be reduced by increasing the number of loci. This variance is determined by mutation, selection, and genetic drift, and it is generally much larger than the intralocus variance.

In a randomly mating population, the variance $V_s(h)$ can be determined relatively easily. We first consider $V_s(h)$ for a particular locus and denote it by $V_{s1}(h)$. This is the variance of h among samples for a given set of population gene frequencies at a locus. From (8.4), we note

$$V_{s1}(h) = \left(\frac{2n}{2n-1}\right)^2 V(\Sigma \ \hat{x}_i^2), \tag{8.11}$$

where $V(\Sigma \ \hat{x}_i^2)$ is the variance of $\Sigma \ \hat{x}_i^2$. Therefore, $V_{s1}(h)$ can be determined by evaluating $V(\Sigma \ \hat{x}_i^2)$. Nei and Roychoudhury (1974a) have worked out a formula for $V(\Sigma \ \hat{x}_i^2)$. Using their formula, we have

$$V_{s1}(h) = \frac{2}{2n(2n-1)}\{2(2n-2)[\Sigma \ x_i^3 - (\Sigma \ x_i^2)^2] + \Sigma \ x_i^2 - (\Sigma \ x_i^2)^2\}. \tag{8.12}$$

An estimate $[\hat{V}_{s1}(h)]$ of $V_{s1}(h)$ can be obtained by replacing $\Sigma \ x_i^2$ and $\Sigma \ x_i^3$ by $\Sigma \ \hat{x}_i^2$ and $\Sigma \ \hat{x}_i^3$, respectively.

Note that $V_{s1}(h)$ is 0 when the locus is monomorphic, as it should be. Note also that when sample size is large, (8.12) becomes

$$V_{s1}(h) = 2[\Sigma \ x_i^3 - (\Sigma \ x_i^2)^2]/n, \tag{8.13}$$

approximately.

At any rate, if we have the estimates of $V_{s1}(h)$ for r loci, $V_s(h)$ may be estimated by

$$\hat{V}_s(h) = \sum_{j=1}^{r} \hat{V}_{sj}(h)/r, \qquad (8.14)$$

where $\hat{V}_{sj}(h)$ is the value of $\hat{V}_{s1}(h)$ for the jth locus. The estimate $[\hat{V}(h)]$ of $V(h)$ can then be obtained by subtracting $\hat{V}_s(h)$ from $V(\hat{h})$ in (8.10). The relationship between $\hat{H}$ and $\hat{V}(h)$ has been used for testing the neutral mutation hypothesis (see chapter 13).

EXAMPLE

Smith and Coss (1984) studied the allele frequencies for 37 protein loci in two subspecies (*douglasii* and *beecheyi*) of *Spermophilus beecheyi* (ground squirrels) in California and another species of ground squirrels, *S. parryii*, in Alaska. The results obtained are presented in table 8.1. The estimates of single-locus heterozygosities and their standard errors can be computed by (8.4) and (8.12), respectively. For example, in the case of the transferrin (*Tf*) locus for *S. b. beecheyi*, $\Sigma \hat{x}_i^2 = 0.8136$, $\Sigma \hat{x}_i^3 = 0.7204$, and $n = 53$. Therefore, $\hat{h} = (106 \times 0.1864)/105 = 0.1881$, and $\hat{V}_{s1}(h) = 0.00221209$ from (8.12). Thus, the standard error due to sampling is 0.047. For all other polymophric loci, these quantities were computed, and the results obtained are presented in table 8.2.

The estimate of average heterozygosity ($\hat{H}$) is the average of single-locus heterozygosities over all loci, including monomorphic loci. This becomes 0.0449 for *S. b. douglasii*, 0.0442 for *S. b. beecheyi*, and 0.0155 for *S. parryii*. Apparently, *S. parryii* has a lower average heterozygosity than *S. beecheyi*. Table 8.2 also includes the proportion of polymorphic loci ($\hat{P}$). *S. parryii* again shows a lower $\hat{P}$ value than *S. beecheyii*, but the difference is not statistically significant.

The estimate of intralocus variance $[V_s(h)]$ is obtained by (8.14). This becomes 0.00034 for *douglasii*, 0.00022 for *beecheyi*, and 0.00103 for *parryii*. On the other hand, the estimates of the corresponding interlocus variance $[\hat{V}(h)]$ for these three taxa are 0.01790, 0.01474, and 0.00252, respectively. Therefore, the interlocus variance is much larger than the intralocus variance except in *parryii*. The standard errors of $\hat{H}$'s obtained by (8.7) are given in table 8.2.

Table 8.1   Gene frequencies for 11 protein loci in two subspecies *(douglasii* and *beecheyi)* of *Spermophilus beecheyi* and *S. parryii* (ground squirrels). Data from Smith and Coss (1984).[a]

| | S. beecheyi | | |
Locus and alleles	douglasii	beecheyi	S. parryii
Number of individuals *(n)*	36	53	8
1. Tf A	0.000	0.000	1.000
Tf B	0.000	0.896	0.000
Tf C	1.000	0.104	0.000
2. GPI 1	0.000	0.000	1.000
GPI 2	0.972	1.000	0.000
GPI 3	0.028	0.000	0.000
3. PGMI 1	0.889	1.000	0.938
PGMI 2	0.111	0.000	0.062
4. PGMII 1	0.097	0.453	1.000
PGMII 2	0.903	0.547	0.000
5. Hb S'	0.000	0.000	1.000
Hb S	0.264	0.000	0.000
Hb F	0.319	0.849	0.000
Hb SF	0.403	0.151	0.000
Hb M	0.014	0.000	0.000
6. Al 1	0.014	0.000	0.062
Al 2	0.972	1.000	0.938
Al 3	0.014	0.000	0.000
7. AAT 1	0.000	0.991	0.000
AAT 2	0.319	0.009	0.000
AAT 3	0.681	0.000	1.000
8. IDH 1	0.000	0.642	1.000
IDH 2	1.000	0.358	0.000
IDH 3	0.000	0.000	0.000
9. Pep C 1	0.000	0.019	0.188
Pep C 2	0.972	0.887	0.812
Pep C 3	0.028	0.094	0.000
10. Cat 1	1.000	1.000	0.000
Cat 2	0.000	0.000	1.000
11. Dial 1	0.000	1.000	0.000
Dial 2	1.000	0.000	1.000

[a] In addition to these 11 loci, there were 9 loci in which *S. b. douglasii* and *S. b. beecheyi* had the same allele fixed and *S. parryii* had another allele fixed. Furthermore, the three populations were fixed for the same allele at 17 other loci.

Table 8.2   Estimates of heterozygosities for nine polymorphic loci for two subspecies *(douglasii* and *beecheyi)* of *Spermophilus beecheyi* and *S. parryii* (ground squirrels). Locus numbers refer to those in table 8.1. $\hat{H}$ = average heterozygosity. $\hat{P}$ = proportion of polymorphic loci.

Locus	S. beecheyi		S. parryii
	douglasii	*beecheyi*	*S. parryii*
1	0.0000 ± 0.0000	0.1881 ± 0.0470	0.0000 ± 0.0000
2	0.0552 ± 0.0367	0.0000 ± 0.0000	0.0000 ± 0.0000
3	0.2001 ± 0.0577	0.0000 ± 0.0000	0.1241 ± 0.1062
4	0.1776 ± 0.0563	0.5003 ± 0.0113	0.0000 ± 0.0000
5	0.6753 ± 0.0186	0.2588 ± 0.0487	0.0000 ± 0.0000
6	0.0556 ± 0.0373	0.0000 ± 0.0000	0.1241 ± 0.1062
7	0.4406 ± 0.0407	0.0180 ± 0.0180	0.0000 ± 0.0000
8	0.0000 ± 0.0000	0.4641 ± 0.0272	0.0000 ± 0.0000
9	0.0552 ± 0.0367	0.2060 ± 0.0496	0.3257 ± 0.1250
$\hat{H}$[a]	0.0449 ± 0.0222	0.0442 ± 0.0201	0.0155 ± 0.0098
$\hat{P}$	$0.19 \quad \pm 0.06$	$0.14 \quad \pm 0.06$	$0.08 \quad \pm 0.04$

[a]$\hat{H}$ and $\hat{P}$ were computed by including 28 monomorphic loci (see table 8.1). The standard error of $\hat{h}$ at each locus was obtained by $[V_{s1}(h)]^{1/2}$, whereas the standard error of $\hat{H}$ was obtained from the total variance in (8.7).

Test of Significance

We have discussed several types of variances related to heterozygosity. Let us now consider the utility of these variances. As mentioned earlier, $V_{s1}(h)$ is the variance generated at the time of gene frequency survey for a particular locus. Therefore, if one is interested in testing the difference in single-locus heterozygosity between two populations, this variance should be used. For example, in table 8.2 the $\hat{h}$ value of locus 4 is 0.1776 ± 0.0563 in *douglasii* and 0.5003 ± 0.0133 in *beecheyi*. Therefore, the difference in $\hat{h}$ is $d = 0.3227$, whereas the standard error of this difference is $s_d = (0.0563^2 + 0.0133^2)^{1/2} = 0.0666$. Thus, the normal deviate is $t = 0.3227/0.0666 = 4.8$, indicating that the difference is statistically significant.

$V(\hat{H})$ is the total variance of $\hat{H}$ when r loci are randomly sampled. Therefore, if the $\hat{H}$ values of two independent populations are to be compared, this variance can be used. For example, there seems to be little correlation between the allele frequencies of *douglasii* and *parryii* if we exclude monomorphic loci (table 8.1). Therefore, we may use $V(\hat{H})$'s to test the difference (0.0294) in $\hat{H}$ between these two species. The standard error of this difference is $s_d = 0.0243$. Therefore, the difference is not statistically significant.

When the two populations to be compared are closely related with each other, the above test is not valid because the heterozygosities of the two populations are historically correlated. Li and Nei (1975) have shown that when genetic differentiation of populations occurs by mutation and genetic drift, the correlation (r) of heterozygosities of the two populations that diverged t generations ago is given by

$$r = e^{-(4v + 1/N)t}, \tag{8.15}$$

where v and N are the mutation rate and effective population size, respectively. The mutation rate for electrophoretically detectable alleles seems to be about 10^{-7} per year (chapter 2). Therefore, it takes a long time for the correlation to become close to 0 when N is large. Note that the correlation of heterozygosities is also generated by variation in the mutation rate among different loci (Chakraborty et al. 1978).

When single-locus heterozygosities are correlated, there is another way of testing the difference in $\hat{H}$. It is the usual t-test for paired observations. For example, the difference in $\hat{H}$ between *douglasii* and *parryii* is based on the differences in single-locus heterozygosity for seven polymorphic loci. The mean difference for these seven polymorphic loci is $\bar{d} = 0.155$, whereas its standard error (s_d) is 0.316. Therefore, $t_{(6)} = \bar{d}/s_d = 0.49$, which is again not significant.

Number of Individuals to Be Studied

As mentioned earlier, the intralocus variance is caused by the limited number of individuals sampled at each locus and can be reduced by increasing n. In contrast, the interlocus variance can be reduced only by increasing the number of loci examined (r). We can then ask the question: how many loci and how many individuals per locus should be studied to estimate average heterozygosity? This problem has been studied by Nei and Roychoudhury (1974a), Nei (1978a), and Gorman and Renzi (1979). Their general conclusion is that in electrophoretic surveys a large number of loci should be examined even if the number of individuals per locus is small. Gorman and Renzi (1979) stated that even a few individuals are sufficient for estimating H if the number of loci examined is large.

The reason for this is that in almost all natural populations the variation in single-locus heterozygosity among loci is so great that average heterozygosity cannot be estimated adequately unless the number of loci studied is large. Figure 8.1 shows the observed distributions of single-locus heterozygosity for three different organisms together with the theoretical distribution for neutral alleles. It is clear that in all cases the distribution is L-shaped and 60 to 70 percent of loci show no genetic variability (monomorphism). In other words, the interlocus variance is very large. Therefore, unless a large number of loci is studied, the estimate of average heterozygosity is not reliable.

However, some warning should be given against using an extremely small number of individuals. The argument mentioned above is based on the assumption that a large number of loci can be studied relatively easily. In practice, the number of loci studied is often limited because of technical difficulties, and in most electrophoretic studies less than 30

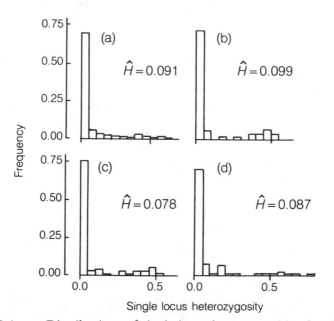

Figure 8.1.   Distributions of single-locus heterozygosities for *(b)* man, *(c)* the house mouse *(Mus musculus)*, and *(d)* the *Drosophila mulleri* species group. The theoretical distribution *(a)* was obtained by computer simulation with the infinite-allele model. $\hat{H}$ stands for average heterozygosity. From Nei et al. (1976b).

loci have been examined. In this case, a large number of individuals per locus still help to reduce the standard error of average heterozygosity. This is particularly so when average heterozygosity is high. If the number of loci examined is about 25, it is recommended that at least 20 or 30 individuals be examined for each locus. Note also that if one is interested in some other properties, such as allele frequency distributions, it is essential to examine a large number of individuals.

Number of Alleles

A number of authors (e.g., Nevo 1978) have used the number of alleles per locus (n_a) as a measure of genetic variability. This quantity seems to be appealing to experimental geneticists, because it is a clear-cut quantity and is expected to be large when the extent of polymorphism is high. However, it has a serious drawback as a general measure of genetic variability. The drawback is that the number of observed alleles is highly dependent on sample size, and thus comparison of this quantity between different samples is not very meaningful unless sample size is more or less the same. The sample size dependence occurs because there are many low-frequency alleles in natural populations, and the number of observed alleles increases with increasing sample size. Gene diversity is also affected by sample size to some extent, but the effect is very small, since low-frequency alleles hardly contribute to this quantity.

To obtain a rough idea about the relationship between sample size and the number of alleles, let us consider neutral alleles maintained by mutation and genetic drift. Ewens (1972) has shown that in an equilibrium population the expected number of neutral alleles in a sample of n diploid individuals is given by

$$E(k) = \frac{M}{M} + \frac{M}{M+1} + \frac{M}{M+2} + \cdots + \frac{M}{M+2n-1}, \qquad (8.16)$$

where $M = 4Nv$. Here, v and N are the mutation rate and effective population size, respectively. The relationships among $E(k)$, n, and M are given in table 8.3. It is clear that when M is small, the expected number of alleles is not affected by sample size very much, but for a large value of M the expected number increases substantially with increasing sample size.

However, the number of alleles in a sample is useful for estimating

Table 8.3   Expected number of different alleles *(k)* in a sample
of *n* individuals *(2n* genes) for various values of $M = 4Nv$ under
the assumption of neutral mutations. From Ewens (1972).

Value of n	Value of M = 4Nv										
	0.01	0.2	0.4	0.6	0.8	1.0	1.2	1.5	2.0	5.0	10.0
10	1.04	1.65	2.22	2.72	3.18	3.60	3.98	4.51	5.29	8.5	11.3
20	1.04	1.79	2.50	3.14	3.73	4.28	4.79	5.52	6.61	11.5	16.5
40	1.05	1.93	2.78	3.55	4.28	4.97	5.62	6.54	7.96	14.7	22.4
60	1.05	2.01	2.94	3.80	4.60	5.37	6.10	7.14	8.75	16.6	26.1
80	1.06	2.07	3.05	3.97	4.83	5.66	6.44	7.57	9.32	18.0	28.8
100	1.06	2.12	3.14	4.10	5.01	5.88	6.71	7.90	9.77	19.1	30.9
150	1.06	2.20	3.31	4.35	5.34	6.28	7.19	8.51	10.57	21.1	34.8
200	1.07	2.26	3.42	4.52	5.57	6.57	7.54	8.94	11.14	22.5	37.6
250	1.07	2.30	3.51	4.65	5.74	6.79	7.81	9.27	11.59	23.6	39.8

the population parameter *M,* which plays an important role in the theory of neutral mutations (see chapter 13). Under the assumption of neutral mutations and population equilibrium, *M* can be estimated by using
equation (8.16) backwards. The estimate obtained in this way is known
to be the maximum likelihood estimate. In practice, however, this estimate is often inflated by the existence of deleterious alleles of which
the contribution to genetic variability is small. Furthermore, we generally do not know whether or not the population is in equilibrium.
Therefore, some caution should be exercised in this approach.

Kimura and Crow (1964) introduced the concept of effective number
of alleles. This number is defined as the reciprocal of homozygosity, i.e.,

$$n_e = 1/\Sigma \; x_i^2. \tag{8.17}$$

This number is equal to the actual number of alleles *(n_a)* only when all
alleles have the same frequency, i.e., $x_i = 1/k$. Otherwise, n_e is smaller
than n_a. Statistically this quantity does not behave nicely. In theoretical
studies, n_e is sometimes defined as $1/E(\Sigma \; x_i^2)$, where the expectation is
taken over the allele frequency distribution in the population (chapter 13).

Analysis of Gene Diversity in Subdivided Populations

In the preceding section, we discussed the extent of gene diversity
within populations. Natural populations are, however, often subdivided

into a number of subpopulations, and in this case it is useful to study the gene diversities within and between populations. The analysis of gene diversity in the total population into its components can be made by using Nei's (1973a) method. In this method, gene diversity is not related to genotype frequencies except in randomly mating populations. In other words, we disregard the distribution of genotype frequencies within populations and consider the decomposition of genomic variation into the interpopulational and intrapopulational variation. A similar study can be done by using the information measure (Lewontin 1972), but the biological meaning of this measure is not clear.

Theory

The following theory is intended to be applied to the average gene diversity for a large number of loci, but for simplicity we consider a single locus. The results obtained are directly applicable to the average gene diversity. For this reason, we shall use the notations for the average gene diversity and identity rather than those for a single locus.

Consider a population which is divided into s subpopulations. Let x_{ki} be the frequency of the ith allele in the kth subpopulation. The gene identity ($1 -$ gene diversity) in this subpopulation is given by

$$J_k = \sum_i x_{ki}^2, \tag{8.18}$$

whereas the gene identity in the total population is

$$J_T = \sum_i \bar{x}_i^2, \tag{8.19}$$

where $\bar{x}_i = \sum_k w_k x_{ki}$, in which w_k is the weight for the kth subpopulation with $\sum w_k = 1$. In many instances, $w_k = 1/s$ may be used for the reason mentioned in chapter 7. J_T can then be written as

$$J_T = \left(\sum_k \sum_i x_{ki}^2 + \sum_{k \neq l} \sum_i x_{ki} x_{li} \right)/s^2$$

$$= \left(\sum_k J_k + \sum_{k \neq l} J_{kl} \right)/s^2, \tag{8.20}$$

where $J_{kl}=\sum_i x_{ki}x_{li}$ is the gene identity between the kth and lth sub-populations.

Let us now define the gene diversity between the kth and lth sub-populations as

$$D_{kl}=H_{kl}-(H_k+H_l)/2$$
$$=(J_k+J_l)/2-J_{kl}, \tag{8.21}$$

where $H_k=1-J_k$ and $H_{kl}=1-J_{kl}$. This quantity is identical with the minimum genetic distance between two populations, which will be discussed in chapter 9. Note that D_{kl} is $\sum_i(x_{ki}-x_{li})^2/2$, so that it is non-negative. If we use (8.21) and note that $D_{kk}=0$, J_T reduces to

$$J_T=\sum_k J_k/s-\sum_k\sum_l D_{kl}/s^2$$
$$=J_S-D_{ST}, \tag{8.22}$$

where

$$J_S\equiv\sum_k J_k/s=\sum \overline{x_i^2} \tag{8.23}$$

is the average gene identity within subpopulations, and

$$D_{ST}=\sum_k\sum_l D_{kl}/s^2 \tag{8.24}$$

is the average gene diversity between subpopulations, including the comparisons of subpopulations with themselves. The average gene diversity within subpopulations (H_S) is defined as

$$H_S=1-J_S. \tag{8.25}$$

Therefore, the gene diversity in the total population ($H_T=1-J_T$) is

$$H_T=H_S+D_{ST}. \tag{8.26}$$

Thus, the gene diversity in the total population can be decomposed into the gene diversities within and between subpopulations. As mentioned earlier, the above formula holds true for the average gene diversity for any number of loci. In fact, in order to have a general picture of gene differentiation among subpopulations, a large number of loci should be used, including both polymorphic and monomorphic loci.

The relative magnitude of gene differentiation among subpopulations may be measured by

$$G_{ST} = D_{ST}/H_T. \tag{8.27}$$

This varies from 0 to 1 and will be called the *coefficient of gene differentiation*.

Equation (8.26) can easily be extended to the case where each subpopulation is further subdivided into a number of colonies. In this case, H_S may be decomposed into the gene diversities within and between colonies (H_C and D_{CS}, respectively). Therefore,

$$H_T = H_C + D_{CS} + D_{ST}. \tag{8.28}$$

This sort of analysis can be continued to any degree of hierarchical subdivision. The relative degree of gene differentiation attributable to colonies within subpopulations can be measured by $G_{CS(T)} = D_{CS}/H_T$. It can also be shown that $(1 - G_{CS})(1 - G_{ST})H_T = H_C$, where $G_{CS} = D_{CS}/H_S$. For a further discussion of this problem, see Chakraborty et al. (1982) and Chakraborty and Leimar (1987).

Although G_{ST} (or G_{CS}) is a good measure of the relative degree of gene differentiation among subpopulations, it is highly dependent on the value of H_T. When this is small, G_{ST} may be large even if the absolute gene differentiation is small. Note also that D_{ST} includes comparisons of subpopulations with themselves, i.e., the cases of $D_{kk} = 0$. To measure the extent of absolute gene differentiation, however, we had better exclude these cases. Nei (1973a), therefore, proposed that the following quantity be used as a measure of absolute gene differentiation.

$$\bar{D}_m \equiv \sum_{k \neq l} D_{kl}/[s(s-1)]$$
$$= sD_{ST}/(s-1). \tag{8.29}$$

This measure is the average minimum genetic distance between subpopulations and is independent of the gene diversity within subpopulations (see chapter 9). It can, therefore, be used for comparing the degrees of gene differentiation in different organisms. $\bar{D}_m$ may also be used to compute the interpopulational gene diversity relative to the intrapopulational gene diversity. That is,

$$R_{ST} = \bar{D}_m / H_S. \qquad (8.30)$$

In this connection, it should be noted that G_{ST} defined in (8.27) depends on the number of subpopulations (s) used even if H_S and the extent of absolute gene differentiation ($\bar{D}_m$) remain the same. One can remove this dependence by replacing D_{ST} by $\bar{D}_m$ and redefining the total gene diversity as $H'_T = H_S + \bar{D}_m$, as in the case of fixation indices (chapter 7). We then have $G'_{ST} = \bar{D}_m / H'_T$, which is independent of s. In this case, however, the simple concept of decomposition of the total gene diversity no longer holds.

Estimation of Gene Diversities

In the foregoing section, we considered the analysis of gene diversity in terms of population gene frequencies. As long as the number of individuals studied for each subpopulation is about 50 or larger, the theory developed applies to sample gene frequencies as well. However, if sample size is smaller than this, certain biases may arise in the estimates of gene diversities, as in the case of single random mating populations discussed earlier. Unbiased estimates of H_S, H_T, and D_{ST} may be obtained by (7.39), (7.40), and $\hat{D}_{ST} = \hat{H}_T - \hat{H}_S$, respectively. When the Hardy-Weinberg equilibrium applies in each subpopulation, $H_S = \hat{H}_0$ may be assumed. In this case,

$$\hat{H}_S = 2n(1 - \overline{\Sigma\hat{x}_i^2})/(2n - 1), \qquad (8.31)$$

$$\hat{H}_T = 1 - \Sigma\bar{x}_i^2 + \hat{H}_S/(2ns). \qquad (8.32)$$

These estimators generally have smaller sampling variances than those in (7.39) and (7.40).

Gene diversity analysis has been applied to many organisms. Some of the results obtained are presented in table 8.4. Both H_T and H_S vary considerably with organism. The proportion of genetic variability attributable to population differentiation (G_{ST}) also varies. In most organisms, G_{ST} is less than 0.1, but in kangaroo rats *(D. ordii)* it is as high as

Table 8.4   Analysis of gene diversity and degree of gene differentiation among local populations of various organisms.

Population	No. of Loci	H_T	H_S	G_{ST}	$\bar{D}_m$
Man—3 major races[a]	62	0.148	0.135	0.088	0.019
Yanomama Indians—37 villages[b]	15	0.039	0.036	0.069	0.003
House mouse—4 populations[c]	40	0.097	0.086	0.119	0.015
Dipodomys ordii—9 populations[d]	18	0.037	0.012	0.674	0.028
Brown trout—38 populations[e]	35	0.040	0.025	0.292	0.015
Drosophila equinoxialis—5 populations[f]	27	0.201	0.179	0.109	0.026
Horsehoe crab—4 populations[g]	25	0.066	0.061	0.072	0.006
Lycopodium lucidulum—4 populations[h]	13	0.071	0.051	0.284	0.027
Escherichia coli—Iowa, Sweden, Tonga[i]	12	0.518	0.499	0.036	0.028

[a]Nei and Roychoudhury (1982).   [b]Weitkamp et al. (1972); Weitkamp and Neel (1972).   [c]Selander et al. (1969).   [d]Johnson and Selander (1971).   [e]Ryman (1983).   [f]Ayala et al. (1974).   [g]Selander et al. (1970).   [h]Levin and Crepet (1973).   [i]Whittam et al. (1983).

0.674. This high value of G_{ST} is caused by a small value of H_T. D_{ST} or $\bar{D}_m$ is not particularly high compared with that of other organisms. This burrowing rodent is known to be sedentary and does not migrate very much. An opposite case is that of E. coli where H_T is very high but G_{ST} is very low. Since the samples for this species were collected from North America, Europe, and the Pacific, various E. coli strains seem to be distributed worldwide. Nevertheless, the absolute measure of differentiation ($\bar{D}_m$) indicates that the extent of gene differentiation is as great as that of kangaroo rats.

Extent of Protein Polymorphism

In the last two decades, the extent of protein polymorphism has been studied for numerous organisms ranging from microorganisms to mammals by using electrophoresis. In most of these studies, the extent was measured by average gene diversity or heterozygosity. In early days, the estimate of heterozygosity was based on a small number of loci, so that its reliability was low. In recent years, however, most authors are examining a fairly large number of loci (20 loci or more).

Average heterozygosity or gene diversity varies from organism to or-

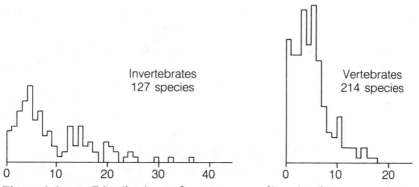

Figure 8.2.   Distributions of average gene diversity (heterozygosity) for species of invertebrates and vertebrates. Only species in which 20 or more loci were examined are included. From Nei and Graur (1984).

ganism. In general, vertebrates tend to show a lower heterozygosity than invertebrates (figure 8.2). If we consider only those species in which 20 or more loci are studied, $\hat{H}$ is generally lower than 0.1 in vertebrates and rarely exceeds 0.15. In invertebrates, a large fraction of species again show an average heterozygosity lower than 0.1, but there are many species showing a value between 0.1 and 0.4. In plants, the number of loci studied is generally very small, so that the estimates are not very reliable. However, if we consider only those species in which 20 or more loci are studied, the average heterozygosity is generally lower than 0.15 except in *Oenothera*, where permanent heterozygosity is enforced by chromosomal translocations (Levin 1975; Nevo 1978; Hamrick et al. 1979; Nevo et al. 1984). The highest level of gene diversity so far observed is that of bacteria ($\hat{H} = 0.48$ based on 20 loci in *Escherichia coli*, Selander and Levin 1980; $\hat{H} = 0.49$ based on 29 loci in *Klebsiella oxytoca*, Howard et al. 1985).

As data on average heterozygosity accumulated, many authors examined the relationship between heterozygosity and environmental factors such as temperature, humidity, availability of resource, multitude of niches, etc., in the hope of identifying the environmental determinants of genetic variability (e.g., Bryant 1974; Powell 1975; Valentine 1976; Nevo 1978; Hamrick et al. 1979; Nelson and Hedgecock 1980; Nevo et al. 1984). Many of these authors have found some correlations between heterozygosity and environmental factors at least in limited groups of species. However, critical reviews of these works have shown that many of the correlations identified are spurious and do not indicate real causal relationships (e.g., Schnell and Selander 1981; Nei and Graur 1984). Some other authors have attempted to relate heterozygosity to general properties of proteins such as substrate specificity, regulatory and nonregulatory enzymes, etc. (e.g., Gillespie and Langley 1974; Johnson 1974). Significant correlations have been obtained in some organisms, but they do not appear to hold for all groups of organisms (Selander 1976).

However, there are at least three factors which are clearly related to heterozygosity level. They are (1) quartenary structure of proteins, (2) molecular weight of protein subunit, and (3) species population size. Zouros (1976) noticed that in several species of animals and plants monomeric enzymes consisting of one polypeptide tend to show a higher heterozygosity than multimeric enzymes. Similar observations were subsequently reported by Harris et al. (1977), Ward (1977), and Koehn and Eanes (1978). The results obtained from humans are presented in table 8.5. Both the proportion of polymorphic loci and average heterozygosity are higher for monomers than for dimers, which in turn show higher values of P and H than trimers and tetramers. The reason for multimeric enzymes showing a lower degree of polymorphism seems to be that they have a higher degree of functional constraint because different polypeptides have to form a single functional protein (Koehn and Eanes 1978; Kimura 1983a).

A significant correlation between heterozygosity and protein subunit molecular weight was first noted by Koehn and Eanes (1977) for 11 enzymes from *Drosophila*. Later, this was confirmed by Ward (1978),

Table 8.5   Proportions of polymorphic loci *(P)* and average heterozygosity *(H)* for monomeric, dimeric, trimeric, and tetrameric enzymes in man. From Harris et al. (1977).

Subunit structure	Number of loci	P	H
Monomeric	27	0.56	0.096
Dimeric	37	0.35	0.071
Trimeric	4	0.25	0.015
Tetrameric	19	0.21	0.050
Total	87	0.38	0.073

Nei et al. (1978), and Turner et al. (1979) for many other enzymes from various organisms. Table 8.6 shows the correlations for six different groups of organisms. There are two types of correlations, i.e., the correlations between single-locus heterozygosity and molecular weight (r_k) and the correlation between average heterozygosity over species and molecular weight (r_H). Since single-locus heterozygosity is subject to a large stochastic error, r_k is lower than r_H. Yet it is statistically significant in most groups of organisms studied.

This positive correlation between heterozygosity and molecular weight is expected to occur under the neutral mutation theory, since the expected heterozygosity in an equilibrium population is given by $E(h) = 4Nv/(1 + 4Nv)$, where N and v are the effective population size and the mutation rate, respectively [equation (13.42) in chapter 13]. Therefore, if a larger molecule has a higher mutation rate than a smaller one, it will have a higher expected heterozygosity. The fact that r_H is not close to 1 probably indicates that the correlation between molecular weight and the mutation rate is not perfect (Nei et al. 1978).

As soon as electrophoretic study became popular in the early 1970s, a number of authors noted that small, isolated populations often have a lower heterozygosity than large populations. For example, Selander et al. (1971) showed that the Santa Rosa Island (off the Gulf Coast of the Florida panhandle) population of *Peromyscus polionotus*, of which the size is known to be of the order of 12,000, has a heterozygosity of 0.018, compared with the heterozygosity of 0.086 in the Florida population.

Table 8.6 Correlation coefficients between molecular weight and heterozygosity in six different groups of organisms. From Nei et al. (1978).

Organism	No. of proteins	No. of species	Correlation	
			r_k	r_H
Primates	23	9	0.208**	0.347
Rodents	14	32	0.198**	0.653*
Reptiles	12	56	0.188**	0.649*
Salamanders	13	12	0.136	0.369
Fishes	11	64	0.117**	0.366
Drosophila	11	29	0.394**	0.722*

*Significant at the 5 percent level.   **Significant at the 1 percent level.

r_k: Correlation between single-locus heterozygosity and molecular weight.

r_H: Correlation between average heterozygosity over species and molecular weight.

The cave populations (200 ~ 500 individuals) of the characid fish *Astyanax mexicanus* in Mexico also have a very low heterozygosity compared with the nearby surface population (Avise and Selander 1972). One of the most extreme examples of low heterozygosity is that of the cheetah, the fastest running land animal. The population size of this species has been estimated to be from 1,500 to 25,000. O'Brien et al. (1985b) examined 52 electrophoretic loci for 55 individuals but found no variability. They also showed that unrelated individuals of this species can accept skin grafts from each other, suggesting that even major histocompatibility complex (MHC) loci, which are highly polymorphic in most mammals, are virtually monomorphic. It is interesting to see that this highly evolved animal species has little genetic variability at the protein level.

Recently, Nei (1983) and Nei and Graur (1984) examined the relationship between average heterozygosity and population size for 77 different species. Although the estimates of population sizes were very rough, they still found a significant correlation (figure 8.3) (see also Soulé 1976).

Mechanism of Maintenance of Protein Polymorphism

Although the extent of protein polymorphism varies greatly with organism, most natural populations seem to be highly polymorphic. Before electrophoresis was introduced in evolutionary studies, a relatively small number of polymorphic characters were known, and whenever a new polymorphism was discovered, population geneticists tended to believe that it was maintained by some kind of balancing selection (see chapter 12). The discovery of a vast amount of protein variability within and between species has changed this view, and now the possibility that even neutral mutations can produce a large amount of variation is widely accepted. Neutral mutations may spread through the population by genetic drift, and if they occur with a sufficiently high rate, many such mutations may coexist in the population (see chapter 13). Indeed, as mentioned earlier, Kimura (1968a) proposed the hypothesis that protein polymorphism as well as amino acid substitution in evolution are mainly due to random fixation of neutral or nearly neutral mutations (see chapter 14 for the detailed property of this hypothesis). Although this hypothesis is supported by a substantial amount of data, there is still disagreement about the proportion of neutral mutations contributing to protein polymorphism.

This disagreement is caused by three major problems. (1) We still do

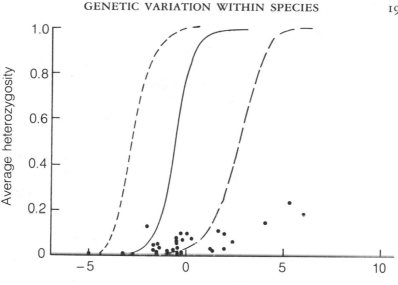

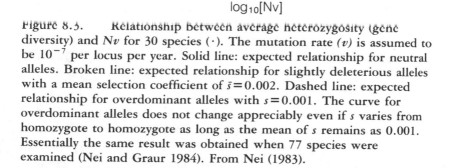

Figure 8.3. Relationship between average heterozygosity (gene diversity) and Nv for 30 species ($\cdot$). The mutation rate (v) is assumed to be 10^{-7} per locus per year. Solid line: expected relationship for neutral alleles. Broken line: expected relationship for slightly deleterious alleles with a mean selection coefficient of $\bar{s} = 0.002$. Dashed line: expected relationship for overdominant alleles with $s = 0.001$. The curve for overdominant alleles does not change appreciably even if s varies from homozygote to homozygote as long as the mean of s remains as 0.001. Essentially the same result was obtained when 77 species were examined (Nei and Graur 1984). From Nei (1983).

not know the actual rate of mutation contributing to protein polymorphism. (2) Even a small amount of selection (say 0.5 percent selective advantage) affects the frequency of mutant genes substantially in a large population, and it is difficult to detect this small amount of selection experimentally except in microorganisms (Dykhuizen and Hartl 1980). (3) Even if one can detect natural selection for a pair of alleles, it is often difficult to know whether it is solely due to the pair of alleles under investigation or to the alleles at other closely linked loci.

Statistical Studies

Because of these difficulties, many authors have turned to statistical tests of the "null hypothesis" of neutral mutations in the hope of detecting weak selection. In these studies, theoretical predictions about such quantities as heterozygosity, number of alleles per locus, distribution of allele frequencies, etc., are derived under the assumption of neutral mutations, and these predictions or predicted relationships among various quantities are compared with observed data to test the "null hypothesis" of neutral mutations (Kimura and Ohta 1971b; Ewens 1972; Yamazaki and Maruyama 1972, 1974; Latter 1975; Nei 1975, 1980; Nei et al. 1976b; Ohta 1976; Fuerst et al. 1977; Eanes and Koehn 1977; Watterson 1978; Chakraborty et al. 1978, 1980; Skibinski and Ward 1981, 1982; Nei and Graur 1984; and others). Although there are some disagreements among authors, these studies have indicated that the general pattern of protein polymorphism can be explained by the neutral theory, particularly when the bottleneck effect and slightly deleterious mutations are taken into account (Nei 1983; Nei and Graur 1984).

Of course, the agreement of polymorphism data with predictions from the neutral theory does not mean that this theory is valid. The same set of data may be explained equally well by other alternative hypotheses. This problem has been studied by a number of authors. Chakraborty et al. (1977) examined the distributions of allele frequencies, single-locus heterozygosity, single-locus genetic distance, etc., for sequentially advantageous mutations and showed that these distributions are similar to those for neutral mutations. Nei (1980), however, showed that in this case an extremely high rate of gene substitution—about 100 times the observed value—is required to maintain the observed level of average heterozygosity. Nei (1980) and Maruyama and Nei (1981) examined the pattern of protein polymorphism expected for overdominant selection with mutation (see chapter 12). Their conclusion is that overdominant selection is so powerful to maintain polymorphism that an absurdly low rate of gene substitution is required to explain the observed level of polymorphism. Most other types of balancing selection seem to have the same property as that of overdominant selection. The view that overdominant selection plays a minor role in the maintenance of protein polymorphism is also supported by the observation that haploid species *E. coli* and *K. oxytoca* have the highest level of heterozygosity so far observed. A high level of heterozygosity has also been observed in such haploid organisms as *Neurospora* (Spieth 1975) and mosses (Krzakowa and Szweykowski 1979; Yamazaki 1981). Gillespie (1979) stated that his SAS-CFF model, a special type of fluctuating selection, can explain protein polymorphism, but his conclusion does not seem to be justified when finite populations are considered (Takahata 1981).

In this connection, it is important to note that the observed heterozygosity in natural populations is usually much lower than the expected heterozygosity for neutral alleles when the current population size is considered. Nei and Graur (1984) computed the expected heterozygosity for each of 77 species, considering the mutation rate per locus per year (see chapter 3), generation time, and the current population size, and compared it with the observed value (figure 8.3). Although their computation is very rough, the observed value is much lower than the expected in almost all species. Note that the gap between the observed and expected values is not narrowed substantially by decreasing the population size estimates by a factor of 10. We can, therefore, conclude that any selection that would increase genetic variability compared with the case of neutral genes (diversity-enhancing selection) cannot explain observed data. If there is any selection operating for protein polymorphism, it must be diversity-reducing selection.

Nei and Graur (1984) interpreted their results as being due to the reduction in population size that apparently occurred in the last glaciation period. As mentioned in chapter 2, the last glaciation (Wisconsin-Würm period) ended about 10,000 years ago, and in this glaciation period the population size was apparently much smaller than the current size for many species. As shown by Nei et al. (1975), the effect of population size reduction (bottleneck effect) on average heterozygosity is expected to last for hundreds of thousands of years after the recovery of population size (figure 8.4). Therefore, we would expect that the observed heterozygosity is generally lower than the expected heterozygosity for neutral mutations obtained from the current population size.

Of course, there are several types of selection that would reduce the level of genetic variability compared with the case of neutral mutations. One of them is selection against slightly deleterious mutations with very small selection coefficients. If the selection coefficient is of the order of 10^{-5}, these mutations would behave just like neutral alleles in relatively small populations; they may become polymorphic or even fixed in the population (chapter 13). In large populations, however, there would be a mutation-selection balance between the type (best) allele and other deleterious alleles, and this balance would give an upper limit for heterozygosity. This hypothesis of slightly deleterious mutations, originally proposed by Ohta (1973b, 1974), is thus capable of explaining the ap-

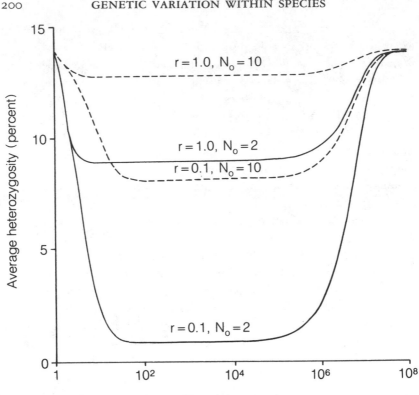

Figure 8.4.   Changes in average heterozygosity when population size goes through a bottleneck. The solid lines refer to the case where the bottleneck size (N_0) is 2, whereas the broken lines refer to the case of $N_0 = 10$. The population is assumed to grow logistically after the bottleneck, and r stands for the intrinsic rate of growth. The original and final heterozygosity is 0.138. From Nei et al. (1975).

parent upper limit of average heterozygosity observed for protein loci.

However, there are some problems with this hypothesis. First, since the mutation-selection balance is supposed to obtain between the type allele and other slightly deleterious mutations in large populations, gene substitution is expected to occur very slowly or even stop in these populations. In practice, however, gene substitution seems to have proceeded roughly at a constant rate, whether the population size is large or small. Second, if this hypothesis is correct, the distribution of allele frequencies in large populations is expected to be bell-shaped (Li 1978).

In all species so far examined, however, the actual distribution is U-shaped (figure 13.2) and is in accordance with the neutral expectation (Chakraborty et al. 1980). The mathematical theory of slightly deleterious mutations has recently been extended to the case where the selection coefficients of newly arisen mutations are continuously distributed from 0 to 1 (Ohta 1977; Li 1978; Kimura 1979). This extension mitigates the deficiency of Ohta's original hypothesis, but even this modified form does not seem to be able to explain observed data without recourse to the bottleneck effect (Nei and Graur 1984).

The second class of diversity-reducing selection is the model of fluctuating selection as conceived by Nei and Yokoyama (1976). If we use the usual Wrightian model of natural selection, fluctuating selection in different generations may lead to a stable equilibrium in large populations (see Karlin and Lieberman 1974; Gillespie 1978, for reviews). However, if selection occurs through competition among different genotypes, fluctuating selection does not lead to stable polymorphism but results in a reduced amount of genetic variability in finite populations (Nei and Yokoyama 1976). Recently, Takahata (1981) and Gillespie (1985) showed that even in the Wrightian model of selection the amount of genetic variability is reduced under fluctuating selection unless the population size is very large.

In my opinion, fluctuating selection becomes important only when the selection coefficient (s) is very small. In the hypothesis of slightly deleterious mutations, a selection intensity of the order of the mutation rate is considered. Namely, s is of the order of 10^{-5}. If the selection coefficient is as small as this, it would be very difficult to maintain a constant value of s in all generations. In this case, I believe, the selection coefficient fluctuates from generation to generation to a considerable extent, and thus the effect of variation of s must be considered. In Nei and Yokoyama's formulation, alleles are expected to behave just like neutral alleles when the mean of s is equal to zero, and the rate of gene substitution is not affected by population size. The only effect of this type of fluctuating selection is to increase the amount of genetic drift per generation and thus to reduce genetic variability. At any rate, if we combine this effect with the bottleneck effect, most of the discrepancies between the observed gene diversity and the neutral expectation for equilibrium populations can be explained (Nei and Graur 1984).

It should be noted, however, that the above discussion refers to the behavior of the majority of polymorphic mutations and does not preclude the possibility that a small proportion of mutations are advantageous or even overdominant. Furthermore, the power of the statistical tests used is not very high, and the above conclusion is partially derived from additional qualitative arguments. Therefore, it is still possible that a substantial proportion of polymorphic alleles are subject to some kind of selection.

Heterozygosity and Growth Rate

A simple way of studying selection operating at protein loci is to examine the relative fitnesses of different genotypes. At individual loci, this is not an easy job because the differences in fitness between genotypes are usually very small. However, if fitness is determined cumulatively by many loci, one might detect selection by examining many loci. A number of authors have used this approach and found evidence of selection associated with protein polymorphisms. The general approach taken by these authors is to determine the number of heterozygous loci for each individual and relate it to growth rate or developmental stability of the individual. The general pattern emerging from these studies is that individuals heterozygous for more loci tend to show a higher growth rate or stronger developmental homeostasis (stability) than those heterozygous for fewer loci (e.g., Schaal and Levin 1976; Singh and Zouros 1978; Mitton 1978; Koehn and Shumway 1982; Ledig et al. 1983; Leary et al. 1984), though there are many exceptions [see Mitton and Grant (1984) and Zouros and Foltz (1986) for references.]

One example of this type of study is presented in figure 8.5, where the mean and variance of log-transformed weights of one-year-old American oysters are given for each heterozygosity class. Although the variance of weights is quite large, the mean clearly increases as the number of heterozygous loci increases. Furthermore, the variance which may be used as a measure of developmental stability decreases as the number of heterozygous loci increases. [When a quantitative character is controlled by a large number of loci, a group of individuals showing a character value close to the mean is expected to have a large variance compared with the groups showing extreme values, simply because the former would contain many different genotypes (Chakraborty and Ryman 1983). Since the group showing a character close to the mean is also expected to be highly heterozygous, the number of heterozygous loci will be positively correlated with the variance even in the absence of developmental ho-

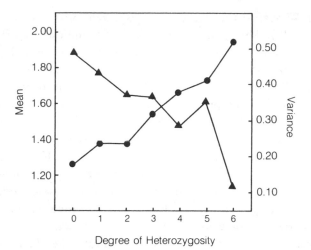

Figure 8.5.   Means (circles) and variances (triangles) of log-transformed weights of oysters for seven different heterozygosity classes. From Zouros et al. (1980).

meostasis. Therefore, this correlation alone cannot be taken as evidence of developmental homeostasis; see Leary et al. (1984).]

These observations have been interpreted as being due either to over-dominant selection at the protein loci or to selection at closely linked protein loci (associative overdominance). This brings us back to the old problem of genetic basis of heterosis which has been debated from the 1910s (see Gowen 1952). It is also the problem of classical theory versus balance theory of which the controversy preceded the recent debate over the neutral mutation hypothesis (Dobzhansky 1955; Morton et al. 1956; Lewontin 1974; Wallace 1981; Kimura 1983a). Unfortunately, these problems are not completely solved and new data do not seem to give any better answers to them (Zouros and Foltz 1986).

Some authors (e.g., Mitton and Grant 1984) prefer the hypothesis of overdominance. In my view, however, there are several problems with this hypothesis. First, if overdominant selection operates for many pro-tein loci, we would expect that average heterozygosity is much higher than the one expected for neutral mutations (figure 8.3). In practice, the observed heterozygosity is even lower than the neutral expectation, as discussed earlier. Second, the observed distribution of allele frequencies is U-shaped in all species so far examined and does not show the bell-shaped or W-shaped patterns that are expected for overdominant alleles (see chapter 13). Third, enzyme kinetic studies indicate that the enzyme activity for heterozygotes is usually intermediate between the activities for two homozygotes or equal to the activity for one homozygote (Kacser and Burns 1981). It is, therefore, difficult to explain overdominance at the enzyme level (Gillespie and Langley 1974). Fourth, in marine mollusks, where many of these observations were made, there is homozygote excess in the early stage of development, and this excess gradually declines as individuals grow. However, the cause of the homozygote excess is unknown. At the adult stage, Hardy-Weinberg proportions are usually observed (e.g., Tracey et al. 1975). Fifth, extensive studies on polygenes controlling viability in *Drosophila* have produced conflicting results about the heterozygous effects of mutations. Some experiments have shown overdominant effects (e.g., Wallace 1959; Crenshaw 1965; Wallace 1981), whereas others (e.g., Hiraizumi and Crow 1960; Mukai 1964; Simmons and Crow 1977) have suggested mildly deleterious effects in heterozygous condition.

In my view, data on the correlation between the number of heterozygous loci and growth rate can be explained much more easily by the hypothesis of associative overdominance, which is originally due to Jones (1917). In a large equilibrium population, the linkage disequilibrium between neutral and selective loci is certainly expected to be 0 (chapter 12), but this is not the case in small or moderately large populations. In finite populations, linkage disequilibrium is generated by genetic drift (Hill and Robertson 1968; Sved 1968), and the linkage disequilibrium causes associative overdominance. Ohta (1971) has shown that when many deleterious genes are closely linked to a protein locus with two neutral alleles, heterozygotes at this locus tend to show a higher fitness than homozygotes.

This can be seen by referring to figure 8.6, where A and a stand for the two neutral alleles at a protein locus and $+$ and $-$ represent the wild-type and deleterious alleles at closely linked loci. Because of the different arrangement of $+$ and $-$ alleles at the neighboring loci, there are two types of chromosomes for each of alleles A and a (A_1, A_2 and a_1, a_2, respectively). We assume that allele $+$ is completely dominant over $-$ and the homozygote for any $-$ allele has a fitness of $1-s$ (see chapter 12). Thus, homozygotes for all chromosome types will have a fitness of $1-s$, whereas the heterozygotes have a fitness of 1. However, if we consider alleles A and a disregarding the chromosome types, het-

Chromosome type	Allele	Frequency
A_1	$+ + A + -$	x_1
A_2	$- + A + +$	x_2
a_1	$+ + a - +$	y_1
a_2	$+ - a + +$	y_2

Figure 8.6.   An example showing associative overdominance for neutral alleles A and a.

erozygote Aa will have a higher fitness than homozygotes AA and aa. Let x_1 and x_2 be the frequencies of A_1 and A_2 among A-carrying chromosomes and y_1 and y_2 be the frequencies of a_1 and a_2 among a-carrying chromosomes, respectively. Under random mating, the fitnesses of AA, Aa, and aa are then given by $1 - s(x_1^2 + x_2^2)$, 1, and $1 - s(y_1^2 + y_2^2)$, respectively. Therefore, although there is no selection at the A locus itself, the neighboring loci generate an overdominant effect for this locus.

Of course, this type of overdominance is not stable. If recombination occurs between the A locus and neighboring loci, the associative overdominance may disappear. On the other hand, mutation and random genetic drift may produce linkage disequilibria, leading to new associative overdominance. Furthermore, the actual number of neighboring loci that are closely linked would be substantial, so that the dynamics of generation of associative overdominance would be more complicated (see Ohta 1971).

At this point, one might wonder whether or not associative overdominance is effectively the same as genuine overdominance in the maintenance of the polymorphism at the A locus. Sved's (1972) computer simulation shows that associative overdominance produces some stability of neutral polymorphism. However, when long-term evolution is considered by taking into account mutation and genetic drift, associative overdominance has very little effect on the average heterozygosity and the rate of gene substitution at neutral loci (Ohta 1973a; P. Pamilo and M. Nei, unpublished). In this model, therefore, we can explain both polymorphism and gene substitution at protein loci. In the overdominance hypothesis, however, it is not easy to explain protein polymorphism and long-term evolution simultaneously (Nei 1980). Of course, this argument does not preclude the existence of overdominant selection at a small proportion of loci.

Selection at Individual Loci

Selection at individual loci should eventually be understood at the molecular and physiological level, as in the case of sickle cell anemia polymorphism (Clarke 1975). In recent years, considerable effort has been made in this direction (Koehn et al. 1983). One of the most interesting results so far obtained is the difference in adaptive characters between two polymorphic alleles (*Ldh-B^a* and *Ldh-B^b*) at the lactate dehydrogenase locus of the teleost fish *Fundulus heteroclitus*. Allele *Ldh-B^b* has a high frequency in the northern part of the Atlantic coast of the United States, whereas allele *Ldh-B^a* has a high frequency in the southern part. Enzyme kinetic analyses have shown that the lactate dehydrogenase-B (Ldh-B^bB^b) produced by genotype *Ldh-B^bB^b* exhibits greater reaction velocities than that (Ldh-B^aB^a) produced by genotype *Ldh-B^aB^a* at low temperature (10°C), but at high temperature (25°C) there is no significant difference (Place and Powers 1979). This observation correlates with the clinal distribution of these alleles along the Atlantic coast. Furthermore, DiMichele and Powers (1982) have shown that when the fish were swum to exhaustion at 10°C, genotype *Ldh-B^bB^b* was able to sustain a swimming speed 20 percent higher than that of *Ldh-B^aB^a*. By contrast, when the fish were swum at 25°C, there was no significant difference between the two genotypes. These results suggest that the allelic difference at this locus is directly involved in the adaptation of this organism to its environmental condition.

There are many other reports indicating that polymorphic alleles at enzymic loci are directly associated with adaptive characters such as relative rates of lipid versus protein synthesis in *Drosophila melanogaster* (Cavener and Clegg 1981), the osmoregulation of the blue mussel, *Mytilus edulis* (Koehn et al. 1983), and the differential survival and flight activity of butterflies (Watt et al. 1983). The antigenic differences in some surface proteins such as the influenza virus hemagglutinin are also directly related to adaptive difference (Webster et al. 1982). The reader who is interested in this problem may refer to Koehn (1985), Watt (1985), Zera et al. (1985), and Hochachka and Somero (1984).

It should be noted that although these studies have received a great deal of attention in recent years, there are many experiments in which no significant difference in fitness was observed between different alleles or electromorphs (e.g., Langridge 1974; Yoshimaru and Mukai 1979; Mukai et al. 1980; Hartl and Dykhuizen 1985; Laurie-Ahlberg et al.

1985). It seems that, as concluded from statistical studies of gene frequency data, a majority of polymorphic alleles at protein loci are more or less neutral, but there are polymorphisms that are definitely related to adaptation. To understand the mechanisms of adaptive evolution, we must eventually study the effects of allelic differences on phenotypic characters at the molecular level. We shall discuss this problem again in chapters 10 and 14.

□ CHAPTER NINE □
GENETIC DISTANCE BETWEEN POPULATIONS

Genetic distance is the extent of gene differences (genomic difference) between populations or species that is measured by some numerical quantity. Thus, the number of nucleotide substitutions per nucleotide site or the number of gene substitutions per locus is a measure of genetic distance. Historically, however, it usually refers to the gene differences as measured by a function of gene frequencies.

The concept of genetic distance was first used by Sanghvi (1953) for an evolutionary study, but a similar concept of measuring population differences goes back to Czekanowski (1909) and Pearson (1926). The methods invented by these authors were primarily for classification of populations in terms of quantitative characters. Later, Fisher (1936) and Mahalanobis (1936) improved these methods from the statistical point of view. Mahalanobis' D^2-statistic is still widely used for quantitative characters. This was later extended to the case of discrete characters by various authors. The evolutionary tree of human races in terms of gene frequency data was first constructed by Cavalli-Sforza and Edwards (1964, 1967). They used a genetic distance based on an angular transformation of gene frequencies at the suggestion of R. A. Fisher.

Since Cavalli-Sforza and Edwards (1964) published their work, many geneticists were interested in this problem, and various measures of genetic distance were proposed. Cavalli-Sforza and Edwards (1967) attempted to relate their distance measure to the evolutionary changes of gene frequencies. Latter (1972) proposed several measures that are closely related to Wright's (1951) fixation index. Nei (1971, 1972) proposed a genetic distance measure with which one can estimate the number of gene or codon substitutions per locus between two populations. Using this distance measure, he attempted to relate the population dynamics of gene frequencies to codon substitution in genes. In the following, I shall discuss the statistical and genetic properties of several typical distance measures. They will be classified into two groups, i.e., measures for population classification and measures for evolutionary study.

Distance Measures for Population Classification

Czekanowski's Mean Difference and Its Variations

Czekanowski's (1909) distance was originally proposed for quantitative characters, but it can be used for gene frequency data as well. Let x_i and y_i be the frequencies of the ith allele at a locus in populations X and Y, respectively. Czekanowski's mean difference is then given by

$$C_Z = \frac{1}{m}\sum_{i=1}^{m}|x_i - y_i|, \tag{9.1}$$

where m is the number of alleles. A version of this measure, i.e.,

$$C_M = \sum_{i=1}^{m}|x_i \quad y_i|, \tag{9.2}$$

is called the "Manhattan metric" and often used by numerical taxonomists (Sneath and Sokal 1973). When there are many loci, the average of C_Z or C_M over the loci is used.

C_Z and C_M are primarily used for population classification, particularly when the populations studied are closely related. In this case, they have an advantage over some other measures, since they satisfy the so-called triangle inequality. Consider populations $1, 2, \cdots, s$, and let D_{ij} be the distance between populations i and j. In most distance measures, $D_{ii}=0$ and $D_{ij}=D_{ji}$. In some distance measures such as C_Z and C_M, however, the triangle inequality, i.e.,

$$D_{ij} \leq D_{ik} + D_{kj},$$

also holds. A distance measure satisfying this principle is called a *metric*. A metric is required if one wants to represent populations as points in an Euclidean space and measure the distance between populations in terms of the geometric distance between the corresponding points in the space. In classical numerical taxonomy, this is the general practice so that metrics rather than nonmetrics are used for population classification. Metrics are also used for principal components analysis. In evolutionary studies or in the construction of phylogenetic trees, however, metricity is not a requirement (see chapter 11).

From the evolutionary point of view, C_Z and C_M have some undesirable properties. First, the maximum value of C_Z depends on the number of alleles. When the two populations are fixed for different alleles, C_Z becomes 1, as expected. However, when each of populations X and Y has five alleles with various frequencies but there is no shared allele, C_Z becomes 0.2. In both cases, the two populations have no shared alleles, but the numerical values are quite different. From this point of view, the Manhattan metric C_M is better than C_Z. C_M is 0 when the two populations are identical but always 2 when they have no shared alleles. One can adjust C_M so that it takes a value between 0 and 1. That is,

$$C_P = \frac{1}{2} \sum_{i=1}^{m} |x_i - y_i|. \tag{9.2a}$$

This measure has been used by Prevosti et al. (1975) and Thorpe (1979). Without the knowledge of population genetics, C_M or C_P might be appealing, since they are functions of gene frequency differences between two populations. Note, however, that the difference in gene frequency is not proportional to the time since divergence between two populations. In general, the gene frequency change per generation is larger when the frequency is close to 0.5 than when it is close to 0 or 1, whatever the cause is. For this reason, C_Z, C_M, and C_P do not seem to be suitable for evolutionary studies. It should also be noted that the sampling properties of these quantities have not been studied.

Pearson's Coefficient of Racial Likeness

This measure was also originally proposed for quantitative characters. Let $\bar{x}_i$ and u_i be the mean and variance of the ith character in population X, respectively, and $\bar{y}_i$ and v_i be the corresponding values in population Y. Pearson's (1926) coefficient of racial likeness (CRL) is then given by

$$CRL = \frac{1}{m} \sum_{i=1}^{m} \frac{(\bar{x}_i - \bar{y}_i)^2}{u_i/n_X + v_i/n_Y} - \frac{2}{m}, \tag{9.3}$$

where n_X and n_Y are the number of individuals examined in populations X and Y, respectively, and m is the number of characters used. The last term $2/m$ is the correction for sampling error and is usually neglected. A measure similar to this was earlier used by F. Heinke (see Sneath and Sokal 1973) in his study of the classification of herring. It is given by

$$C_H = \frac{1}{m} \sum_{i=1}^{m} (\bar{x}_i - \bar{y}_i)^2. \tag{9.4}$$

This is similar to Nei's (1973b) minimum genetic distance, which will be discussed later.

Rogers' Distance

Suppose that there are m alleles at a locus, and let x_i and y_i be the frequencies of the ith allele in populations X and Y, respectively. Each allele frequency may take a value between 0 and 1. Therefore, it is possible to represent populations X and Y in an m-dimensional space. The distance between the two populations in the space is then given by

$$D_P = \left[\sum_{i=1}^{m} (x_i - y_i)^2 \right]^{1/2}. \tag{9.5}$$

This distance is often used in numerical taxonomy, because it permits a simple geometric interpretation. In the case of gene frequency data, D_P takes a value between 0 and $\sqrt{2}$, the latter value being obtained when the two populations are fixed for different alleles. This property is not very desirable. Rogers (1972) rectified this property and proposed the following measure, which takes a value between 0 and 1.

$$D_R = \left[\frac{1}{2} \sum_{i=1}^{m} (x_i - y_i)^2 \right]^{1/2}. \tag{9.6}$$

When gene frequency data for many loci are available, the average of this value is used.

This distance satisfies the triangle inequality, so that it can be used for population classification. It has also been used for phylogeny construction in combination with Farris' (1972) method. However, it is proportional neither to evolutionary time nor to the number of gene substitutions. D_R has another deficiency similar to that of Czekanowski's distance. That is, when the two populations are both polymorphic but share no common alleles, D_R is given by $[(\Sigma x_i^2 + \Sigma y_i^2)/2]^{1/2}$. This value can be much smaller than 1 even if the populations have entirely different sets of alleles. For example, when there are five nonshared alleles in each population and all allele frequencies are equal, we have $D_R = 0.45$.

Mahalanobis' D^2 Statistic and Its Variations

Mahalanobis' distance (1936) was proposed as an extension of Pearson's coefficient of racial likeness to the case where the characters used are correlated. It was again developed for quantitative characters, but its modified forms have been used for gene frequency data. Consider m quantitative characters, and let $\bar{x}_i$ and $\bar{y}_i$ be the means of the ith character in populations X and Y, respectively. For simplicity, we assume that the variances (v_{ii}) and covariances (v_{ij}) of the characters are the same for the two populations, and the variance-covariance matrix is given by

$$
V = \begin{bmatrix}
v_{11} & v_{12} & \cdots & v_{1m} \\
v_{21} & v_{21} & \cdots & v_{2m} \\
\cdot & \cdot & & \cdot \\
\cdot & \cdot & & \cdot \\
\cdot & \cdot & & \cdot \\
v_{m1} & v_{m2} & \cdots & v_{mm}
\end{bmatrix}.
\tag{9.7}
$$

Mahalanobis' D^2 is then defined in the following way.

$$
D^2 = \sum_{i=1}^{m} \sum_{j=1}^{m} v^{ij}(\bar{x}_i - \bar{y}_i)(\bar{x}_j - \bar{y}_j),
\tag{9.8}
$$

where v^{ij} is the element of the ith row and jth column of the inverse matrix of V in (9.7). It can be shown that D in (9.8) is the geometric distance between the two populations represented in an m-dimensional space. When the covariances, $v_{ij}(i \neq j)$, are all 0, (9.8) reduces to

$$
D^2 = \sum_{i=1}^{m} (\bar{x}_i - \bar{y}_i)^2 / v_{ii}.
\tag{9.9}
$$

If we divide this D^2 by m, it becomes equivalent to Pearson's (1926) coefficient of racial likeness. Mahalanobis' D^2 statistic is the most commonly used distance measure for quantitative characters, and the statistical properties of this measure have been studied extensively (Rao 1952).

A number of authors have modified D^2 for gene frequency data. Considering the first $m - 1$ alleles at a locus, Steinberg et al. (1967) assumed $v_{ii} = z_i(1 - z_i)$ and $v_{ij} = -z_i z_j (i \neq j)$, where $z_i = (x_i + y_i)/2$ and V is a $(m-1) \times (m-1)$ square matrix. Here, x_i and y_i again denote the frequencies of the ith allele in populations X and Y, respectively. They then showed that D^2 can be written as

$$X^2 = \sum_{i=1}^{m} \frac{(x_i - y_i)^2}{z_i}. \tag{9.10}$$

This is identical with the distance measure first used by Sanghvi (1953), so it is often called *Sanghvi's distance*. Note that X^2 is closely related to the χ^2 for testing the allele frequency differences between the two populations. That is, when the sample size (n) is the same for the two populations, the χ^2 is given by $2nX^2$. Note also that X^2 in (9.10) takes a value of 0 when the two populations are identical and a value of 4 when the two populations do not have any shared allele.

Balakrishnan and Sanghvi (1968) introduced another definition of matrix V. They first noted that the sampling variance of gene frequency x_i in population X is $v_{xii} = x_i(1 - x_i)/n_X$ and the covariance between gene frequencies x_i and x_j is $v_{xij} = -x_i x_j/n_X$, where n_X is the sample size for population X. Similarly, the variances (v_{Yii}) and covariances (v_{Yij}) for population Y can be written in terms of the gene frequencies $(y_i$'s) and sample size (n_Y). They then defined the elements of V in (9.7) by

$$v_{ij} = (n_X^2 v_{Xij} + n_Y^2 v_{Yij})/(n_X + n_Y).$$

Here, the mth allele A_m is excluded. With these definitions, D^2 in (9.8) becomes

$$B^2 = \sum_{i=1}^{m-1} \sum_{j=1}^{m-1} v^{ij}(x_i - y_i)(x_j - y_j). \tag{9.11}$$

When the number of populations studied is s rather than 2, v_{ij} is defined as

$$v_{ij} = \sum_{k=1}^{s} n_k^2 v_{kij} / \sum_{k=1}^{s} n_k,$$

where k refers to the kth population. In the above definition of X^2 or B^2, we considered only one locus. When there are data for many loci, the mean or sum of these quantities are used.

Although X^2 and B^2 are both derived from the D^2 statistic, they give quite different numerical values. It is worth noting that X^2 depends only on the gene frequencies of the populations compared, whereas B^2 depends on the gene frequencies of all the populations studied. Thus, the B^2 distance for a particular pair of populations varies according to whether the other populations are included or not. Furthermore, B^2 is highly dependent on sample size. Kurczynski (1970) has given an example in which B^2 varies twentyfold when sample size alone changes. These seem to be serious drawbacks of the B^2 distance. It should also be mentioned that the absolute values of X^2 and B^2 have no biological meaning, and only their relative magnitudes are important. This is because they were developed for the purpose of population classification. Kurczynski (1970) proposed another version of the D^2 statistic, removing the sample size dependence of the B^2 statistic, but his distance measure is also for classification of closely related populations.

Bhattacharyya's Angular Transformation and Its Modifications

Representing two populations on the surface of a multidimensional hypersphere, Bhattacharyya (1946) measured the extent of differentiation of populations in terms of the angle (θ) between the two lines projecting from the origin to the two populations $(X$ and $Y)$ on the hypersphere (figure 9.1). When there are m alleles, we consider an m dimensional

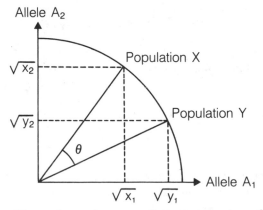

Figure 9.1.   Bhattacharyya's geometric representation of populations X and Y for the case of two alleles.

hypersphere with radius 1 and let each axis represent the square root of allele frequency, i.e., $\xi_i = \sqrt{(x_i)}$ and $\eta_i = \sqrt{(y_i)}$. Therefore, $\Sigma \xi_i^2 = \Sigma \eta_i^2 = 1$. When there are only two alleles, populations X and Y can be represented on a circle, as shown in figure 9.1. Elementary geometry shows that in the case of m alleles the angle θ is given by

$$\cos \theta = \sum_{i=1}^{m} \xi_i \eta_i = \sum_{i=1}^{m} \sqrt{x_i y_i}. \tag{9.12}$$

Bhattacharyya (1946) proposed that the square of θ, i.e.,

$$\theta^2 = [\text{arc } \cos(\Sigma \sqrt{x_i y_i})]^2, \tag{9.13}$$

be used as a measure of population divergence. This θ^2 is closely related to X^2 in (9.10). That is, when θ is small, $\cos \theta = 1 - \theta^2/2$ approximately (the first two terms of Taylor's series). Therefore, θ^2 is $2(1 - \cos \theta)$ approximately. On the other hand, $\cos \theta$ can be written as

$$\cos \theta = \Sigma \sqrt{x_i y_i} = \frac{1}{2} \Sigma \left[(x_i + y_i)^2 - (x_i - y_i)^2\right]^{1/2}$$

$$= \frac{1}{2} \Sigma (x_i + y_i) \left[1 - \frac{(x_i - y_i)^2}{(x_i + y_i)^2}\right]^{1/2}$$

$$= \frac{1}{2} \Sigma (x_i + y_i) \left[1 - \frac{(x_i - y_i)^2}{2(x_i + y_i)^2} + \frac{(x_i - y_i)^4}{4(x_i + y_i)^4} \cdots \right]$$

$$\approx 1 - \frac{1}{4} \Sigma \frac{(x_i - y_i)^2}{(x_i + y_i)}.$$

Therefore,

$$\theta^2 = \frac{1}{2} \Sigma \frac{(x_i - y_i)^2}{(x_i + y_i)} \tag{9.14}$$

approximately. This is 1/4 of X^2 in (9.10) and takes a value between 0 and 1.

Cavalli-Sforza and Edwards (1967) also used angular transformation. They proposed that the genetic distance between two populations be measured by the chord length between points X and Y on the m-dimensional hypersphere. Geometry shows that this chord length is given by $[2(1 - \cos \theta)]^{1/2}$. In practice, they used

$$d_C = (2/\pi)[2(1 - \cos \theta)]^{1/2} \qquad (9.15)$$

as a distance measure, because $\theta = \pi/2$ corresponds to the case of complete gene substitution. However, the division of the chord length by $\pi/2$ does not have much meaning. If one wants to make the distance take a value between 0 and 1, $d = (1 - \cos \theta)^{1/2}$ should be used. The utility of d_C for the study of molecular evolution has been studied by Latter (1972). He has shown that d_C is nonlinearly related to evolutionary time.

In a computer simulation, Nei et al. (1983a) noted that the following distance measure is quite efficient in recovering the true topology of an evolutionary tree when it is estimated from gene frequency data.

$$D_A = \sum_{k=1}^{r} (1 - \sum_{i=1}^{m_k} \sqrt{x_{ik}y_{ik}})/r, \qquad (9.16)$$

where m_k and r are the number of alleles at the kth locus and the number of loci examined, respectively. This measure takes a value between 0 and 1, the latter value being obtained when the two populations share no common alleles. Since the maximum of D_A is 1, D_A is nonlinearly related to the number of gene substitutions. When D_A is small, however, it increases roughly linearly with evolutionary time (Nei et al. 1983a).

Distance Measures for the Study of Evolution

Wright's F_{ST} and Its Modifications

The mathematical model of genetic differentiation of populations was first studied by Wright (1931, 1943, 1951). He considered the case where a foundation stock splits into a large (theoretically infinite) number of populations at an evolutionary time (t generations ago) and thereafter no migration occurs among the populations. He further assumed that the effective size of each population is N and random mating occurs within populations. In this case, the gene frequency in each population fluctuates every generation because of random genetic drift even in the absence of mutation and selection. Consider a pair of alleles, A_1 and A_2, and let x be the frequency of A_1 in a population. The mean and variance of x over all populations are then given by $\bar{x} = p$ and σ_x^2

$= p(1-p)[1-\exp(-t/2N)]$, where p is the initial gene frequency (see chapter 13). Therefore, F_{ST} defined by equation (7.15) becomes

$$F_{ST} = 1 - e^{-t/2N}. \qquad (9.17)$$

This measure of population differentiation takes a value between 0 and 1.

Although the above property of F_{ST} is interesting mathematically, F_{ST} is not a measure of genetic distance because it is defined for a large number of populations; genetic distance should be defined for a pair of populations. Of course, one can redefine F_{ST} considering only two populations. That is,

$$F_{ST1} = (x_1 - y_1)^2/[2z_1(1-z_1)], \qquad (9.18)$$

where x_1 and y_1 are the frequency of A_1 in populations X and Y, respectively, and $z_1 = (x_1 + y_1)/2$. However, the range of this F_{ST1} is no longer between 0 and 1 but between 0 and 2. Furthermore, the expectation of F_{ST1} is not equal to F_{ST} in (9.17), as will be shown later.

A better estimate of F_{ST} for the case of two populations has been derived by Latter (1972). In the presence of multiple alleles at many loci, it is given by

$$\phi^* = \frac{\sum_k^r \sum_i^{mk}(x_{ik} - y_{ik})^2/r}{2[1 - \sum_k^r \sum_i^{mk} x_{ik} y_{ik}/r]}, \qquad (9.19)$$

where m_k and r are the same as those of equation (9.16). This takes a value between 0 and 2, but the expectation of ϕ^* is equal to (9.17), as will be shown later. Therefore, it can be used for estimating t if N is known and the effect of mutation is negligible.

Cavalli-Sforza (1969) attempted to relate Bhattacharyya's angular transformation to F_{ST} in (9.17). As mentioned earlier, when θ is small, θ^2 is given by (9.14) approximately. Furthermore, when there are only two alleles, θ^2 can be expressed as

$$\theta^2 = (x_1 - y_1)^2/[4z_1(1-z_1)]. \qquad (9.20)$$

Comparison of this formula with (9.18) indicates that $2\theta^2 = F_{ST1}$ approximately. Therefore, if we note $\theta^2 = 2(1 - \cos \theta)$ approximately when θ is small, F_{ST1} is $4(1 - \cos \theta)$ approximately. Assuming that this relationship holds even for the case of multiple alleles, Cavalli-Sforza (1969) proposed that

$$f_\theta = 4(1 - \cos \theta)/(m - 1) \qquad (9.21)$$

be used as a measure of genetic distance. He also assumed that the expectation of f_θ is given by $1 - e^{-t/2N}$ when N is large. Therefore, according to his theory, the time (t) since divergence between two populations can be estimated by $-2N\log_e (1 - f_\theta)$.

In practice, there are several problems with f_θ. First, the expectation of f_θ is a complicated function of N and t even in the absence of mutation and selection and depends on the initial gene frequency, which is generally unknown (Heuch 1975). Second, while F_{ST} in (9.17) takes a value between 0 and 1, f_θ varies from 0 to 4. Third, when there are multiple alleles, the expectation of f_θ is considerably different from (9.17) (Nei 1976). Furthermore, when a long evolutionary time is considered, the effect of mutation cannot be neglected. Nevertheless, computer simulations by Latter (1973a) and Nei (1976) have shown that the relationship between f_θ and t is approximately linear when the evolutionary time considered is short.

Genetic Distances for Measuring the Number of Gene Substitutions

In chapters 4 and 5, we discussed various methods of estimating the number of amino acid or nucleotide substitutions between two homologous sequences. In these chapters, we considered distantly related organisms or genes, so that the effect of polymorphism within populations was neglected. However, when the genetic differences between closely related species or populations are to be studied, this effect is no longer negligible, because the number of codon differences per codon site is very small in these cases. It is, therefore, necessary to examine a large number of loci for many individuals and compute the average number of codon differences between and within populations.

In practice, of course, it is difficult to sequence amino acids or nucleotides for many genes. Fortunately, however, it is possible to estimate the number of codon differences per locus from polymorphism data. For this purpose, data obtained from electrophoretic studies are most appropriate, but other data such as blood group data can also be used with certain qualifications. The basic principle of this method is that any allelic difference in electrophoretic mobility or immunological reaction is caused by at least one codon difference at the gene level and thus the average number of codon differences per locus can be estimated from allele frequency data statistically. Since this number is a direct measure of gene differences, a statistic measuring this number is a good measure of genetic distance between populations. Nei (1972, 1973b) proposed three different statistics for measuring this number, considering various uncertainties that arise in actual situations. Of course, the estimate of the number of codon differences obtained by these methods refers to those codon differences that are detectable by the biochemical technique used. For example, electrophoresis detects only about $1/4 \sim 1/3$ of amino acid substitutions in proteins (chapter 3). Therefore, to estimate the actual number of codon differences from these data, we must multiply the genetic distances obtained by 3 or 4.

Minimum Genetic Distance

The simplest measure of genetic distance proposed by Nei is called the *minimum genetic distance* (D_m) and is intended to measure the minimum number of codon differences per locus. We again denote by x_i and y_i the frequencies of allele A_i in populations X and Y, respectively. Suppose that we choose one allele at random from each of the two populations and compare them. The alleles chosen will then be identical with probability $j_{XY} = \Sigma x_i y_i$ and different with probability $1 - \Sigma x_i y_i$. When the alleles are different, there is at least one codon difference between them. Therefore, $d_{XY} = 1 - \Sigma x_i y_i$ gives the minimum number of codon differences. However, when there is polymorphism, two alleles randomly drawn from the same population are not always the same. Therefore, it is necessary to subtract the intrapopulational codon differences from d_{XY} to obtain the net minimum codon differences. The minimum number of codon differences between two alleles randomly drawn from population X is given by $d_X = 1 - \Sigma x_i^2$, since the probability of drawing the same alleles is $j_X = \Sigma x_i^2$. Similarly, the minimum number of codon differences between two alleles drawn from population Y is $d_Y = 1 - j_Y$, where $j_Y = \Sigma y_i^2$. Therefore, the number of *net minimum codon differences* between the two populations is given by

$$d = d_{XY} - (d_X + d_Y)/2$$

$$= \Sigma(x_i - y_i)^2/2. \qquad (9.22)$$

In practice, d varies considerably with locus, and to estimate the net minimum codon differences between two populations, we must take the average of d over all loci. We call this average the *minimum genetic distance*. It is given by

$$D_m = D_{XY(m)} - [D_{X(m)} + D_{Y(m)}]/2, \qquad (9.23)$$

where $D_{XY(m)} = 1 - J_{XY}$, $D_{X(m)} = 1 - J_X$, and $D_{Y(m)} = 1 - J_Y$. Here, J_{XY}, J_X, and J_Y are the averages of j_{XY}, j_X, and j_Y over all loci in the genome.

It is interesting to note that (9.22) is the square of Rogers' distance in (9.6). However, when data from many loci are used, D_m is not the square of D_R, since D_R is the average of single-locus D_R values. While D_m is easy to understand, it has the same disadvantage as that of D_R. That is, even if the two populations have entirely different sets of alleles, it may be considerably smaller than 1 when polymorphism exists within populations. In other words, D_m can be a serious underestimate of codon differences when D_m is large. Thus, the use of this measure should be limited to the case of comparison of intraspecific populations. However, the mathematical properties of D_m or d have been well studied (Li and Nei 1975), and they can be used for the study of maintenance of genetic polymorphism (Chakraborty et al. 1978).

Standard Genetic Distance

The drawback of D_m is that $D_{X(m)}$, $D_{Y(m)}$, and $D_{XY(m)}$ are the proportions of different genes between two randomly chosen genomes, so that D_m may be a gross underestimate of the number of net codon substitutions when $D_{XY(m)}$ is large. As will be shown later, when individual codon changes occur independently, the mean number of net codon substitutions is given by

$$D = -\log_e I, \qquad (9.24)$$

where

$$I = J_{XY}/(J_X J_Y)^{1/2}. \qquad (9.25)$$

I have called D the *standard genetic distance*. Note that D can be written as $D = D_{XY} - (D_X + D_Y)/2$, where $D_{XY} = -\log_e J_{XY}$, $D_X = -\log_e J_X$, and $D_Y = -\log_e J_Y$.

I in (9.25) is 1 when the two populations have identical gene frequencies over all loci and is 0 when they share no alleles. Because of this property, I itself has been used for measuring the extent of genetic similarity between populations. It is called the *normalized identity of genes* or simply *genetic identity*. While I takes a value between 0 and 1, D varies from 0 to ∞.

Distance Measures for the Case of Unequal Rates of Gene Substitution

Equation (9.24) is based on the assumption that the rate of gene substitution is the same for all loci. In reality, this assumption does not hold (chapter 3). One distance measure that takes care of unequal rates of gene substitution is the *maximum genetic distance* defined by

$$D' = -\log_e I', \tag{9.26}$$

where $I' = J'_{XY}/(J'_X J'_Y)^{1/2}$. Here, J'_{XY}, J'_X, and J'_Y are the geometric means of j_{XY}, j_X, and j_Y over loci, respectively. In practice, however, D' is subject to a large sampling error, and this error is expected to inflate the estimate of the mean number of net codon differences. If any one of the j_{XY}'s is small, D' can be a gross overestimate. For this reason, I have called D' the maximum genetic distance. However, when local races within a species are compared, the differences among D_m, D, and D' are generally very small, and all of them give similar conclusions about the genetic differentiation of populations (Nei and Roychoudhury 1974b). In most practical cases, $D_m < D < D'$, but this relationship does not necessarily hold when the values of these quantities are extremely small and are estimated from sample gene frequencies.

Nei et al. (1976a) developed a somewhat different formula, assuming that the rate of codon substitution per locus varies among loci following the gamma distribution with coefficient of variation 1. It is given by

$$D_v = (1 - I)/I. \tag{9.27}$$

The rationale for this formula will be discussed later. This distance measure seems to be superior to D', since it is less sensitive to sampling error.

Estimation of Genetic Distance

Theoretically, the genetic distance between two populations is defined in terms of population allele frequencies for all loci in the genome. In practice, however, it is virtually impossible to examine all genes for all loci in the populations. Therefore, we must estimate the genetic distance by sampling a certain number of individuals from the populations and examining a certain number of loci. Let us now consider how to estimate genetic distance from actual data and how to compute the variance of the estimate of genetic distance, following Nei and Roychoudhury (1974a) and Nei (1978a). In this section, we shall be concerned mainly with Nei's distance measures, since the sampling theory of other distance measures is not well developed.

Unbiased Estimates

Let $\hat{x}_i$ be the frequency of the ith allele (A_i) in a sample of $2n_X$ genes from population X and $\hat{y}_i$ be the frequency of the same allele in a sample of $2n_Y$ genes from population Y. Many authors have used these allele frequencies directly in (9.23), (9.24), and other equations to estimate genetic distance. However, when the sample size is small, this method gives a biased estimate.

Let us now evaluate the amount of this bias for D_m. If we use sample allele frequencies, D_m in (9.23) becomes

$$D_{ms} = (1 - J_{XY1}) - \frac{1}{2}[(1 - J_{X1}) + (1 - J_{Y1})], \qquad (9.28)$$

where J_{X1}, J_{Y1}, and J_{XY1} are the means of $\Sigma \hat{x}_i^2$, $\Sigma \hat{y}_i^2$, and $\Sigma \hat{x}_i \hat{y}_i$, respectively, over all loci examined. In a randomly mating population, the expectations of $\Sigma \hat{x}_i^2$ and $\Sigma \hat{y}_i^2$ under the assumption of multinomial sampling are $\Sigma x_i^2 + (1 - \Sigma x_i^2)/2n_x$ and $\Sigma y_i^2 + (1 - \Sigma y_i^2)/2n_Y$, respectively, whereas the expectation of $\Sigma \hat{x}_i \hat{y}_i$ is $\Sigma x_i y_i$. Here, x_i and y_i are the population gene frequencies of A_i in X and Y, respectively. Therefore, the expectation of D_{ms} is

$$E(D_{ms}) = (1 - J_{XY}) - \frac{1}{2}\left[(1 - J_X)\left(1 - \frac{1}{2n_X}\right) + (1 - J_Y)\left(1 - \frac{1}{2n_Y}\right)\right]$$

$$= D_m + \frac{1}{2}[(1 - J_X)/2n_X + (1 - J_Y)/2n_Y]. \tag{9.29}$$

Thus, D_{ms} tends to give an overestimate of D_m when n_X and n_Y are small. The bias in D_{ms} can be easily corrected, and an unbiased estimate of D_m is given by

$$\hat{D}_m = (1 - \hat{J}_{XY}) - \frac{1}{2}[(1 - \hat{J}_X) + (1 - \hat{J}_Y)],$$

$$= \frac{1}{2}(\hat{J}_X + \hat{J}_Y) - \hat{J}_{XY}, \tag{9.30}$$

where $\hat{J}_X$, $\hat{J}_Y$, and $\hat{J}_{XY}$ are the means of $\hat{j}_X \equiv (2n_X \Sigma \hat{x}_i^2 - 1)/(2n_X - 1)$, $\hat{j}_Y \equiv (2n_Y \Sigma \hat{y}_i^2 - 1)/(2n_Y - 1)$, and $\hat{j}_{XY} \equiv \Sigma \hat{x}_i \hat{y}_i$ over all loci examined, respectively (chapter 8). We note that $\hat{D}_m$ can also be written as

$$\hat{D}_m = \sum_{k=1}^{r} \hat{d}_k/r, \tag{9.31}$$

where r is the number of loci examined, and $\hat{d}_k$ is the value of

$$\hat{d} = (\hat{j}_X + \hat{j}_Y)/2 - \hat{j}_{XY} \tag{9.32}$$

for the kth locus.

The expectation of the sample value (D_s) of standard genetic distance D is more complicated. However, if the number of loci studied is large, it is approximately given by

$$E(D_s) \simeq D + \frac{1 - J_X}{4n_X J_X} + \frac{1 - J_Y}{4n_Y J_Y} \tag{9.33}$$

(Nei 1978a). Therefore, D_s gives an overestimate of D when n_X and n_Y are small. This bias is again easily corrected, and an unbiased estimate of D is given by

$$\hat{D} = -\log_e \hat{I}, \tag{9.34}$$

where

$$\hat{I} = \hat{J}_{XY}/(\hat{J}_X \hat{J}_Y)^{1/2}. \tag{9.35}$$

An unbiased estimate of D_ν in (9.27) is also given by $\hat{D}_\nu = (1 - \hat{I})/\hat{I}$.

When D_m, D, and D_ν are small, their estimates can be negative because of sampling error. These negative values can be avoided by increasing sample sizes. If sample sizes cannot be increased, one may assume that the genetic distance is 0 when the estimate is negative.

The above method of sample size correction can be used for any distance measure as long as the distance measure is a function of second moments of allele frequencies. For example, an unbiased estimate of Latter's $\phi*$ is given by

$$\hat{\phi}* = [(\hat{J}_X + \hat{J}_Y)/2 - \hat{J}_{XY}]/(1 - \hat{J}_{XY}) \tag{9.36}$$

[see Reynolds et al. (1983) for a similar correction]. When a distance measure is a function of $\Sigma|x_i - y_i|$, evaluation of the bias is more complicated. However, sampling error is again expected to give an overestimate of the distance.

Sampling Variance

Let us now consider the variance of the estimate of genetic distance under the assumption of linkage equilibrium among all loci. The variance of the estimate of minimum genetic distance can be computed easily, since it is the average of single-locus distances (d). That is,

$$V(\hat{D}_m) = V(\hat{d})/r, \tag{9.37}$$

where

$$V(\hat{d}) = \sum_{k=1}^{r} (d_k - \hat{D}_m)^2/(r - 1) \tag{9.38}$$

It is not easy to get the exact variance of standard genetic distance, but the asymptotic variance when sample size is large can be obtained easily by the so-called delta method (Nei 1978a,b). It becomes

$$V(\hat{D}) = \frac{V(\hat{J}_X)}{4\hat{J}_X^2} + \frac{V(\hat{J}_Y)}{4\hat{J}_Y^2} + \frac{V(\hat{J}_{XY})}{\hat{J}_{XY}^2} + \frac{Cov(\hat{J}_X,\hat{J}_Y)}{2\hat{J}_X\hat{J}_Y}$$

$$- \frac{Cov(\hat{J}_X,\hat{J}_{XY})}{\hat{J}_X\hat{J}_{XY}} - \frac{Cov(\hat{J}_Y,\hat{J}_{XY})}{\hat{J}_Y\hat{J}_{XY}}. \tag{9.39}$$

In the above equation, $V(\hat{J}_X)$ and $V(\hat{J}_Y)$ are nothing but the variance of average heterozygosity discussed in chapter 8, since $\hat{H} = 1 - \hat{J}$. $V(\hat{J}_{XY})$, $Cov(\hat{J}_X,\hat{J}_Y)$, etc. are also easily obtainable, since $\hat{J}_X$, $\hat{J}_Y$, and $\hat{J}_{XY}$ are the averages of single-locus gene identities. For example,

$$Cov(\hat{J}_X,\hat{J}_Y) = \sum_{k=1}^{r} (\hat{j}_{Xk} - \hat{J}_X)(\hat{j}_{Yk} - \hat{J}_Y)/[r(r-1)]. \tag{9.40}$$

Therefore, $V(\hat{D})$ can be computed. This method seems to give a quite accurate value even if the number of loci used is as small as 20 (Chakraborty 1985).

The variances of $\hat{I}$ and D_v can also be obtained by the delta method. They become

$$V(\hat{I}) = \frac{\hat{J}_{XY}^2}{4\hat{J}_X^3\hat{J}_Y}V(\hat{J}_X) + \frac{\hat{J}_{XY}^2}{4\hat{J}_X\hat{J}_Y^3}V(\hat{J}_Y) + \frac{1}{\hat{J}_X\hat{J}_Y}V(\hat{J}_{XY})$$

$$+ \frac{\hat{J}_{XY}^2}{2\hat{J}_X^2\hat{J}_Y^2}Cov(\hat{J}_X,\hat{J}_Y) - \frac{\hat{J}_{XY}}{\hat{J}_X^2\hat{J}_Y}Cov(\hat{J}_X,\hat{J}_{XY})$$

$$- \frac{\hat{J}_{XY}}{\hat{J}_X\hat{J}_Y^2}Cov(\hat{J}_Y,\hat{J}_{XY}), \tag{9.41}$$

$$V(\hat{D}_v) = \frac{\hat{J}_Y}{4\hat{J}_X\hat{J}_{XY}^2}V(\hat{J}_X) + \frac{\hat{J}_X}{4\hat{J}_Y\hat{J}_{XY}^2}V(\hat{J}_Y) + \frac{\hat{J}_X\hat{J}_Y}{\hat{J}_{XY}^4}V(\hat{J}_{XY})$$

$$+ \frac{1}{2\hat{J}_{XY}^2}Cov(\hat{J}_X,\hat{J}_Y) - \frac{\hat{J}_Y}{\hat{J}_{XY}^3}Cov(\hat{J}_X,\hat{J}_{XY})$$

$$- \frac{\hat{J}_X}{\hat{J}_{XY}^3}Cov(\hat{J}_Y,\hat{J}_{XY}). \tag{9.42}$$

Theoretically, computation of the above variances is straightforward, but in practice it is very tedious, particularly when there are many polymorphic loci. It is therefore necessary to use a computer unless the number of loci used is very small.

However, when $\hat{I}$ is lower than about 0.85 and average heterozygosity is low ($\hat{H} < 0.2$), the variances of $\hat{I}$, $\hat{D}$, and $\hat{D}_v$ are approximately given by

$$V(\hat{I}) = \hat{I}(1 - \hat{I})/r, \tag{9.43}$$

$$V(\hat{D}) = (1 - \hat{I})/(\hat{I}r), \tag{9.44}$$

$$V(\hat{D}_v) = (1 - \hat{I})/(\hat{I}^3 r), \tag{9.45}$$

(Nei 1971, 1978b). The rationale for this is that the single-locus identity $\hat{I}_j = j_{XY}/(\hat{j}_X \hat{j}_Y)^{1/2}$ usually takes a value close to 1 or 0 (figure 9.2), so that we can assume a binomial distribution for $\hat{I}_j$. When all populations are monomorphic or asexual haploid populations as in the case of bacteria (e.g., Howard et al. 1985), $J_X = J_Y = 1$, $J_{XY} = I$, and all variances and covariances except $V(J_{XY})$ are 0 in (9.39), (9.41), and (9.42). Therefore, these equations become identical with (9.43), (9.44), and (9.45), respectively.

In the previous section, we discussed several other distance measures. Many of them are given by the average of single-locus distances over all

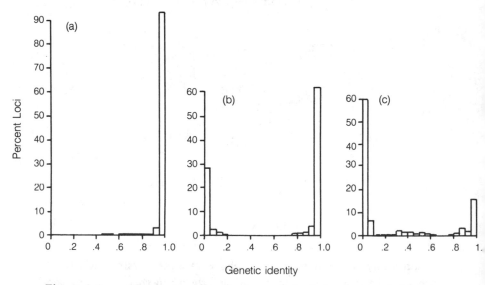

Figure 9.2.   Frequency distributions of single-locus genetic identity (I_j) for protein loci in *Drosophila*. (a) Local populations. (b) Sibling species. (c) Nonsibling species. From Ayala et al. (1974).

loci. In this case, the variance of the genetic distance is given by the variance of single-locus distances among loci divided by the number of loci, as in the case of minimum genetic distance.

There are two sampling processes involved in the estimation of genetic distance, i.e., sampling of loci from the genome and sampling of individuals (genes) from the populations. Therefore, there are two corresponding sampling variances, i.e., the interlocus variance and intralocus variance. Decomposition of the total variance into these two components can be made by the same method as that for average heterozygosity (Nei and Roychoudhury 1974a; Nei 1978a).

Test of Significance

There are two different types of significance tests concerning the estimates of genetic distances. One is the test of the null hypothesis $D_{ij}=0$, and the other is the test of the null hypothesis $D_{ij}=D_{kl}$, where D_{ij} refers to the genetic distance between populations i and j. A simple test of the first null hypothesis is to use the χ^2 test for gene frequency differences at each locus. This test can be done by using the following χ^2.

$$\chi^2 = 2n_X n_Y \sum_{i=1}^{m} \frac{(\hat{x}_i - \hat{y}_i)^2}{\hat{x}_i n_X + \hat{y}_i n_Y},$$

where n_X and n_Y are the numbers of individuals examined in populations X and Y, respectively. The number of degrees of freedom for this χ^2 is $m-1$. If this χ^2 is significant for a locus, any estimate of genetic distance is significantly different from 0. Note that when $n_X=n_Y$, this χ^2 is equal to $2n$ times the X^2 distance given in (9.10). In many cases, however, this test is unnecessary, because visual inspection of allele frequency data usually indicates the significant differences between populations.

The test of the hypothesis $D_{ij}=D_{kl}$ is a little more complicated. When D_{ij} and D_{kl} are completely unrelated to each other, the total variances of the estimates of D_{ij} and D_{kl} can be used for testing the hypothesis. For example, if one is interested in comparing the genetic distance between two species of a mammalian genus with the distance between two species of *Drosophila*, the total variance may be used. (This total variance can also be used when an estimate of genetic distance based on a particular set of loci is compared with an estimate based on a different set of loci.)

However, if one is interested in comparing D_{ij} with D_{kl} for different pairs of closely related species or populations, the total variance is not appropriate, because in this case D_{ij} and D_{kl} are historically correlated as in the case of testing the difference in average heterozygosity (Li and Nei 1975). It is possible to estimate this correlation from gene frequency data (Mueller and Ayala 1982; Nei et al. 1985), but the actual procedure is quite complicated. A simpler method is to compute single-locus minimum distances or single-locus genetic identities and compare them locus by locus. For example, if one is interested in minimum distances, the difference (δ) between single-locus distances d_{ij} and d_{kl} for each locus is computed. The difference between D_{ij} and D_{kl} is equal to the mean of δ, so that it can be tested by the ordinary t test. Although d_{ij} is not normally distributed, this test is known to be quite robust (Chakraborty 1985).

Standard genetic distance is not equal to the mean of single-locus distances, so that this approach cannot be used directly. However, if the difference between the minimum distances for the same set of data is significant, the difference between the standard distances must also be significant. Therefore, the test of standard distances can be replaced by the test of minimum distances. When distance values are large, however, this method is not very powerful because in the presence of polymorphism minimum distances are not one even if the two populations compared have no shared alleles. In this case, it is better to use single-locus genetic identities.

EXAMPLE

The minimum and standard genetic distances for two subspecies (*douglasii* and *beecheyi*) of *Spermophilus beecheyi* and *S. beecheyi* and *S. parryi* can be computed from the gene frequency data given in table 8.1. The $\hat{J}_X$ values for the three populations can be obtained from the $\hat{H}$ values in table 8.2, since $\hat{J}_X = 1 - \hat{H}$. They are 0.9551, 0.9558, and 0.9845 for *S. b. douglasii*, *S. b. beecheyi*, and *S. parryi*, respectively. On the other hand, $\hat{J}_{XY}$'s are the averages of $\Sigma \hat{x}_i \hat{y}_i$ over all loci, and the values obtained are given in table 9.1. Using these $\hat{J}_X$'s and $\hat{J}_{XY}$'s, we can estimate D_m, D, and D_v by using equations (9.30), (9.34), and (9.27), respectively. The results obtained are presented in table 9.1. With both $\hat{D}_m$ and $\hat{D}$, the distance between subspecies is considerably smaller than that between species, as expected. When the genetic distance is small,

Table 9.1   Estimates of D_m, D, and D_v and their standard errors for two subspecies *(douglasii* and *beecheyi)* of *Spermophilus beecheyi* and *S. parryii* (ground squirrels). $\hat{J}_X$'s for *S. b. douglasii*, *S. b. beecheyi*, and *S. parryii* are 0.955, 0.956, and 0.984, respectively. The figures in parentheses are the standard errors obtained by approximate formulas.

Species compared	$\hat{J}_{XY}$	$\hat{D}_m$	Genetic distance $\hat{D}$	$\hat{D}_v$
douglasii vs. *beecheyi*	0.866	0.090 ± 0.040	0.099 ± 0.047 (0.053)	0.104 ± 0.052 (0.058)
douglasii vs. *S. parryii*	0.576	0.394 ± 0.079	0.520 ± 0.137 (0.136)	0.683 ± 0.230 (0.229)
beecheyi vs. *S. parryii*	0.560	0.411 ± 0.080	0.551 ± 0.141 (0.141)	0.734 ± 0.245 (0.244)

the difference between $\hat{D}_m$ and $\hat{D}$ is relatively small, but as the distance increases the difference gradually increases. Note also that D_v is nearly equal to $\hat{D}$ when $\hat{D}$ is 0.099, but is considerably larger than $\hat{D}$ when $\hat{D}$ is as large as 0.6.

The standard errors of $\hat{D}_m$, $\hat{D}$, and $\hat{D}_v$ in table 9.1 were computed by using equations (9.37), (9.39), and (9.42), respectively. Despite the relatively large number of loci used (37 loci), the standard errors are 1/5 to 1/2 of the estimate. Table 9.1 also includes the standard errors of $\hat{D}$ and $\hat{D}_v$ obtained by approximate formulas (9.44) and (9.45). They are very close to the more exact values obtained by the delta method even when I is close to 0.9. This is because average heterozygosities are very low in the present species.

The genetic distance between *S. b. douglasii* and *S. parryii* is nearly the same as that between *S. b. beecheyi* and *S. parryii*, but both of them are significantly higher than the distance between the two subspecies *douglasii* and *beecheyi*. This is true even if we use the total standard errors given in table 9.1. For example, in the comparison of the minimum distance, $D_{m(db)}$, for *douglasii* vs. *beecheyi* and the distance, $D_{m(bp)}$, for *beecheyi* vs. *parryii* the difference is significant at the 1 percent level, t being 3.1. As mentioned earlier, the difference between $D_{m(db)}$ and $D_{m(bp)}$ can also be tested by the t-test of paired differences. This test gives $t_{(36)} = 23.4$, which is significant at the 0.1 percent level.

Theoretical Relationship Between Genetic Distance and Evolutionary Time

In chapter 4, we have seen that the rate of amino acid (codon) substitution in proteins (genes) is approximately constant per year. Therefore, if a distance measure is proportional to the number of codon substitutions, it will also be proportional to evolutionary time. The proportionality with evolutionary time is a useful property, since it can be used for estimating the time since divergence between two populations or species. An evolutionary tree constructed by using this type of distance is also expected to give good estimates of branch lengths. Unfortunately, most of the distance measures proposed for population classification are not linearly related with evolutionary time, except when a short evolutionary time is considered. Even the distance measures proposed for evolutionary studies are not necessarily linear. In this section, let us examine the theoretical relationships of some of these distance measures with evolutionary time.

F_{ST} and $\phi*$

As mentioned earlier, Wright's F_{ST} is given by $1 - e^{-t/2N}$ when there are no selection and no mutation. Therefore, t can be written as

$$t = -2N \log_e(1 - F_{ST}).$$ (9.46)

However, when there are only two populations, this relationship does not apply, and the expectation of F_{ST1} in (9.18) is approximately given by

$$E(F_{ST1}) \simeq \frac{E(x-y)^2}{2E[z(1-z)]},$$

where subscript 1 for x, y, and z is dropped for brevity. $E(x-y)^2$ is $E(x^2) + E(y^2) - 2E(xy) = 2p(1-p)(1 - e^{-t/2N})$, since x and y are independent of each other. Here, p is the frequency of allele A_1 in the common ancestral population, and the expectation is taken over all generations (see chapter 13). On the other hand, $E[(x+y)\{2-(x+y)\}/4] = E(x+y)/2 - E(x+y)^2/4 = p(1-p)(1 + e^{-t/2N})/2$. Therefore, $E(F_{ST1})$ is

$$E(F_{ST1}) = \frac{2(1 - e^{-t/2N})}{1 + e^{-t/2N}},$$ (9.47)

approximately. Thus, the relationship between F_{ST1} and t is a little more complicated than that in equation (9.46). However, when $t/2N \ll 1$,

$$E(F_{ST1}) \simeq t/(2N).$$ (9.48)

Therefore, if the evolutionary time is very short, F_{ST1} is expected to increase linearly.

Unlike F_{ST1}, Latter's $\phi*$ in (9.19) can be related to t by the same equation as (9.46) if we neglect the effect of mutation. The expectation of the numerator of (9.19) for a single locus is

$$\Sigma_i [E(x_i^2) + E(y_i^2) - 2E(x_i y_i)]$$
$$= \Sigma_i [2p_i^2 + 2p_i(1 - p_i)(1 - e^{-t/2N}) - 2p_i^2]$$
$$= 2(1 - \Sigma_i p_i^2)(1 - e^{-t/2N}),$$

whereas the expectation of the denominator is $2(1 - \Sigma p_i^2)$. Therefore, the expectation of $\phi*$ is

$$E(\phi*) = 1 - e^{-t/2N}.$$ (9.49)

Thus, t may be estimated by $t* = -2N\log_e (1 - \phi*)$, if we know N.

In the presence of mutation, however, the expectation of $\phi*$ becomes

$$E(\phi*) = \frac{J^{(\infty)} + (J_0 - J^{(\infty)})e^{-[2v + (1/2N)]t} - J_0 e^{-2vt}}{1 - J_0 e^{-2vt}},$$ (9.50)

where v is the mutation rate per locus per generation, J_0 is the initial homozygosity, and $J^{(\infty)} \equiv 1/(1 + 4Nv)$ is the equilibrium value of expected homozygosity (see chapter 13). When $J_0 = J^{(\infty)}$ as is often assumed, (9.50) becomes

$$E(\phi*) = \frac{J^{(\infty)}(1 - e^{-2vt})}{1 - J^{(\infty)}e^{-2vt}}.$$ (9.51)

Therefore, $t*$ is not proportional to t if we consider a long evolutionary time.

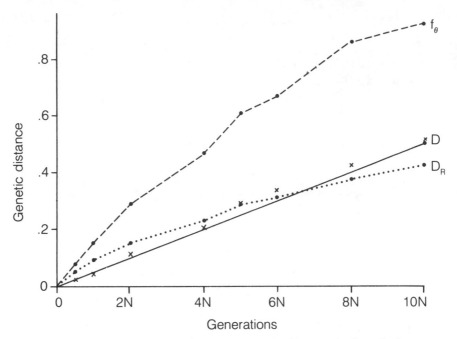

Figure 9.3.   Relationships between genetic distance and evolutionary time (in units of N generations) for distance measures f_θ, D_R, and D. These results were obtained by computer simulation. The straight line for D is a theoretical line, whereas crossmarks represent simulation results. From Nei (1976).

D_R, f_θ, and D_A

These three quantities involve square root terms, so that it is difficult to derive their relationships with t analytically. However, these relationships can be studied by using computer simulation. Results of one such simulation, in which 200 "loci" were used, are presented in figure 9.3. It is clear that D_R and f_θ show a curvilinear relationship except in the early generations. D_A increases more slowly and maintains the initial linear relationship for a longer period of time (Nei et al. 1983a).

D, D_m, and D_v

Suppose that populations X and Y are separated at a particular generation and thereafter no migration occurs between them. We assume that the effective sizes of the populations are both N and random mating occurs within each population. We also assume that new mutations are all different from the extant alleles in the populations and that the effects of mutation, selection, and genetic drift are balanced throughout the evolutionary time within each population. In this case, the value of j_X may vary from generation to generation because of the effect of stochastic factors, but the expected value remains the same. For example, $E(j_X)$ is equal to $1/(1+4Nv)$ in the absence of selection. If there is selection, it takes a more complicated form (chapter 13), but the steady-state value of $E(j_X)$ is equal to J_X, the average of j_X over all loci, if all loci are subject to the same mutation and selection pressures. Similarly, $E(j_Y)$ is equal to J_Y and remains constant throughout the evolutionary process.

The expectation of j_{XY} or J_{XY}, however, gradually decreases with increasing t because of different mutations accumulating in the two populations. Let α be the rate of gene substitution per locus *per year*. As we have seen, α is approximately constant for each protein. For simplicity, we assume that the rate is the same for all loci. We note that if gene substitution is caused by random fixation of neutral mutations, α is equal to the mutation rate per locus per year (v_y). If it is caused by genic selection, α is given by $4Nsv_y$, where s is the average selective advantage of mutant alleles over the original allele (chapter 13). At any rate, J_{XY} in the tth year after separation of the two populations is given by

$$J_{XY}^{(t)}=J_{XY}^{(0)} (1-\alpha)^{2t}\simeq J_{XY}^{(0)} e^{-2\alpha t}, \qquad (9.52)$$

since α is very small compared with 1. This is analogous to the formula for the proportion of identical amino acids between homologous polypeptides in equation (4.4).

Noting $J_X^{(t)}=J_X^{(0)}$ and $J_Y^{(t)}=J_Y^{(0)}$, the I value in the tth year is then given by

$$I=I_0 e^{-2\alpha t}, \qquad (9.53)$$

where I_0 is the initial value of I, i.e., $I_0=J_{XY}^{(0)}/(J_X^{(0)} J_Y^{(0)})^{1/2}$. I_0 is expected to be close to 1 in most cases, since no appreciable gene differentiation occurs as long as there is migration between the two populations. Therefore, we have

$$D \equiv -\log_e I = 2\alpha t, \qquad (9.54)$$

approximately. It is clear from this equation that D measures the accumulated number of gene (codon) substitutions ($2\alpha t$) per locus between the two populations.

In chapter 3, we estimated that the average rate of codon substitution that is detectable by electrophoresis is 10^{-7} per locus per year. Therefore, if we assume $\alpha = 10^{-7}$, we can estimate t if the value of D is known. Namely,

$$t = 5 \times 10^6 D. \qquad (9.55)$$

This formula is useful for estimating the time since divergence between two populations or species. As will be discussed later, however, there are a number of factors that disturb the above relationship, and one must be aware of these factors. Note also that our estimate of α is very crude, and some authors have used different α values (see table 9.4 below).

The relationship between the minimum distance and evolutionary time can easily be derived, since we know the expectations of j_X, j_Y, and j_{XY}. D_m can be written as $(J_X + J_Y)/2 - J_{XY}$. Therefore,

$$D_m = J_X(1 - e^{-2\alpha t}). \qquad (9.56)$$

Thus, D_m increases almost linearly when t is small, but the rate of increase gradually declines. The expected maximum value of D_m is equal to average gene identity $J_X(=J_Y)$. Therefore, it is higher when the extent of polymorphism is low than when this is high.

In the derivation of (9.54), we assumed that α is the same for all loci. This assumption is certainly wrong. As shown in figure 3.5, the rate of gene substitution varies greatly from locus to locus, and the distribution of α roughly follows the gamma distribution with coefficient of variation 1. Studies of protein polymorphism by Nei et al. (1976b), Fuerst et al. (1977), Chakraborty et al. (1978), and Zouros (1979) support this conclusion. Let us therefore assume that α has the following gamma distribution,

$$f(\alpha) = \frac{b^a}{\Gamma(a)} e^{-b\alpha} \alpha^{a-1}, \qquad (9.57)$$

where $a = \bar{\alpha}^2/V(\alpha)$ and $b = \bar{\alpha}/V(\alpha)$, $\bar{\alpha}$ and $V(\alpha)$ being the mean and variance of α, respectively (see Rao 1952). Here, $\Gamma(a)$ is the gamma function defined by

$$\Gamma(a) = \int_0^\infty e^{-t} t^{a-1} dt.$$

The expected genetic identity is then given by

$$I_A = E(\Sigma j_{XY})/[E(\Sigma j_X)E(\Sigma j_Y)]^{1/2}$$

$$= \Sigma F(j_k)e^{-2\alpha_k t}/\Sigma E(j_k)$$

$$\simeq \int_0^\infty f(\alpha)e^{-2\alpha t} d\alpha = \left[\frac{a}{a + 2\bar{\alpha}t}\right]^a, \qquad (9.58)$$

where $E(j_k)$ is the expected homozygosity at the kth locus, and Σ stands for the summation for all loci in the genome. Equation (9.58) is expected to give an overestimate of the true value of genetic identity, since it is based on the assumption of no correlation between $E(j_k)$ and $\exp(-2\alpha_k t)$ though in practice there should be a positive correlation. In the case of neutral alleles, the effect of this assumption on I_A can be evaluated, but unless $2\bar{\alpha}t$ is very large, the effect seems to be quite minor (Griffiths 1980a).

When the coefficient of variation $(a^{-1/2})$ is 1,

$$I_A = 1/(1 + 2\bar{\alpha}t). \qquad (9.59)$$

Therefore, the mean number of gene substitutions per locus $(2\bar{\alpha}t)$ can be estimated by D_v in (9.27). Namely,

$$D_v \equiv 2\bar{\alpha}t = (1 - I_A)/I_A. \qquad (9.60)$$

Mathematically, $D_v > D$, but the difference between (9.54) and (9.60) is small when t is relatively small (see table 9.2). Note that because of the assumption we have made above, equation (9.58) is expected to give an underestimate of $2\bar{\alpha}t$ when $2\bar{\alpha}t$ is large. With this reservation, we can estimate t from D_v by using equation (9.55).

It should be mentioned that formulas (9.54) and (9.60) are valid only when a large number of loci is studied since each event of gene substitution is subject to large stochastic error. Nei and Tateno (1975) studied the distribution of single-locus gene identity (I_j) under the assumption of neutral mutations by using computer simulation. The results obtained indicate that when $2\bar{\alpha}t$ is small, I_j shows an inverse J-shaped distribution, whereas it shows a U-shaped distribution when $2\bar{\alpha}t$ is moderately large (see also figure 9.2). Therefore, to obtain a reliable estimate of I, a large number of loci must be studied. This is true even if gene substitution is mediated by natural selection (Chakraborty et al. 1977). Mathematical formulas for the stochastic variance of genetic distance under the assumption of neutral mutations have been obtained by Li and Nei (1975).

Strictly speaking, equation (9.54) is not appropriate for electrophoretic data, even if α is the same for all loci. This is because at the electrophoretic level the effect of back mutations becomes important as t increases. This problem can be studied by using Ohta and Kimura's (1973) stepwise model of neutral mutations (chapter 13), though some authors (Ramshaw et al. 1979; Fuerst and Ferrell 1980; McCommas 1983) have questioned the appropriateness of this model for electrophoretic data. Nei and Chakraborty (1973), Li (1976a), and Chakraborty and Nei (1977) have studied the expected genetic identity under the stepwise mutation model. The exact formula for the genetic identity for electrophoretic data (I_E) is rather complicated (Li 1976a), but for practical purposes we can use the following equation.

$$I_E = e^{-2vt} \sum_{r=0}^{\infty} (vt)^{2r}/(r!)^2, \tag{9.61}$$

where v is the mutation rate per generation (Nei 1978b). We note that $\alpha = v$ in this case because we are dealing with neutral mutations.

When v varies from locus to locus following the gamma distribution with coefficient of variation 1, the average value of I_E is given by

$$I_{EA} = \frac{1}{1 + 2\bar{v}t}\left[1 + \sum_{r=1}^{\infty} \frac{(2r)!}{(r!)^2}\left(\frac{\bar{v}t}{1 + 2\bar{v}t}\right)^{2r}\right], \tag{9.62}$$

approximately, where $\bar{v}$ is the mean of v over all loci.

Table 9.2 shows the values of genetic identity for the four different models, i.e., equations (9.53), (9.59), (9.61), and (9.62). In this table, calendar year rather than generation is used as a unit of time with $\bar{\alpha} = \bar{v} = 10^{-7}$. The relationship between genetic distance $D = -\log_e I$ and

Table 9.2   Evolutionary time and genetic identity under the infinite-allele model (I, I_A) and the stepwise mutation model (I_E, I_{EA}). I, I_A, I_E, and I_{EA} were obtained by formulas (9.53), (9.59), (9.61), and (9.62), respectively. In this computation, the rate of gene substitution $(\alpha = v)$ was assumed to be 10^{-7} per year. The accumulated number of codon substitutions may be estimated by $2\bar{\alpha}t$, if the observed value of I is given.

Time $(10^3\ yrs)$	I	I_A	I_E	I_{EA}	Time (MY)	I	I_A	I_E	I_{EA}
10	.998	.998	.998	.998	1	.819	.833	.827	.845
50	.990	.990	.990	.990	2	.670	.714	.696	.745
100	.980	.980	.980	.980	3	.547	.625	.599	.674
200	.961	.961	.961	.962	4	.449	.556	.524	.620
300	.942	.943	.943	.945	5	.368	.500	.466	.577
400	.923	.926	.925	.928	6	.301	.455	.420	.542
500	.905	.909	.907	.913	7	.247	.417	.383	.513
600	.887	.893	.890	.898	8	.202	.385	.353	.488
700	.869	.877	.874	.884	9	.165	.357	.329	.466
800	.852	.862	.858	.870	10	.135	.333	.309	.447
900	.835	.847	.842	.857	20	.018	.200	.207	.333

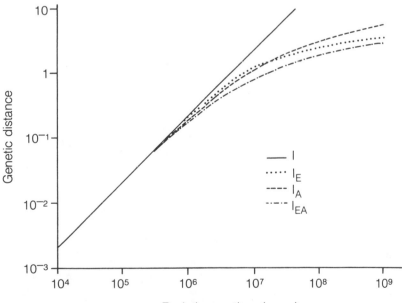

Figure 9.4.   Relationships between genetic distance $(D = -\log_e I)$ and evolutionary time for four different genetic models.

evolutionary time is also given in figure 9.4 for the four different genetic models. It is clear that genetic identity is virtually the same for all four models for the first one million years. Therefore, if the observed value of I is larger than about 0.82, formula (9.53) may be used for estimating divergence time. However, if the divergence time increases further, the difference between the models becomes substantial. In this case, formula (9.53) should not be used, since the assumption of the same mutation rate for all loci is certainly incorrect. The equation for I_E is also expected to give an underestimate for the same reason. In general, equation (9.59) or (9.62) seem to be more appropriate. At any rate, the numerical values in table 9.2 or figure 9.4 can be used for obtaining a rough estimate of divergence time if the genetic identity value is available.

Complete Isolation: Short-Term Evolution

The above theory does not apply to nonprotein loci such as those for blood groups, since the relationship between codon substitution and phenotypic change at these loci may not be as simple as that for protein loci. However, if we consider a very short period of evolutionary time ($t \ll 2N$), D, D_m, and D_v are all approximately linearly related to evolutionary time. In this case, we can neglect the effect of mutation. In the absence of selection, the values of J_X, J_Y, and J_{XY} in generation t [$J_X(t)$, $J_Y(t)$, and $J_{XY}(t)$, respectively] can be written as

$$J_X(t) = J_Y(t) = 1 - [1 - J(0)]\left(1 - \frac{1}{2N}\right)^t$$

$$\simeq J(0) + [1 - J(0)]t/(2N),$$

$$J_{XY}(t) = J_{XY}(0) = J_X(0) = J_Y(0) = J(0),$$

where $t \ll 2N$ is assumed (chapter 13). Therefore, we have

$$D_m = [1 - J(0)]t/(2N), \tag{9.63a}$$

$$D = D_v = [(1 - J(0))/J(0)]t/(2N), \tag{9.63b}$$

approximately. Thus, as long as $t \ll 2N$, our distance measures can be used even for nonprotein loci.

Bottleneck Effect

In our mathematical formulation, we assumed that the average heterozygosities of the two populations in question have remained constant throughout the entire evolutionary process. This assumption, however,

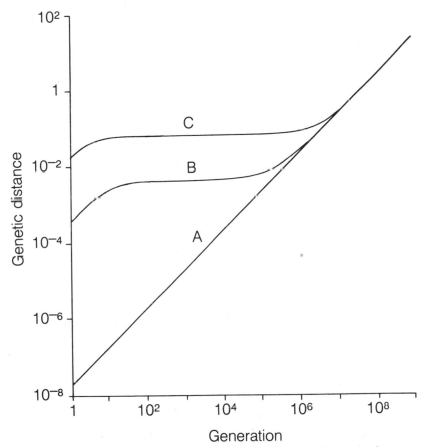

Figure 9.5.   Bottleneck effects on genetic distance. Computations have been made under the assumption that one isolated population (or species) is established through a bottleneck of size N_0 and thereafter population size increases to the level of the parental population following the sigmoid curve. The genetic distance in the ordinate represents the distance between this population and its parental population which has undergone independent evolution. The infinite-allele model is used. A: no bottlenecks. B: $N_0 = 100$. C: $N_0 = 10$. From Chakraborty and Nei (1977).

may not always be satisfied. In fact, there are many cases in which one or both of the populations have gone through bottlenecks. The bottleneck effect on genetic distance has been studied by Chakraborty and Nei (1977). They have shown that genetic distance increases rapidly in the presence of bottlenecks and that the rate of increase is higher when the bottleneck size is small than when this is large. However, if the population size returns to the original level, the bottleneck effect gradually disappears (figure 9.5).

Under certain circumstances, it is possible to make a correction for the bottleneck effect. In the case where only one of the two populations has gone through a bottleneck, the following genetic identity may be computed.

$$I = J_{XY}/J_X,   \qquad (9.64)$$

where J_X is the mean homozygosity (gene identity) for the population whose size has remained constant. If we use this I in (9.54) or (9.60), then D or D_v is linearly related to evolutionary time under the infinite-allele model (Chakraborty and Nei 1974). In the case where both populations have gone through bottlenecks, a similar correction can be made if there is a third population the size of which is known to have remained more or less the same as that of the foundation stock of the two populations under investigation. In this case, I may be computed by replacing J_X in (9.64) by the mean homozygosity for the third population.

Effects of Migration

In the early stage of population differentiation, gene migration usually occurs between populations. Migration retards gene differentiation considerably, and even a small amount of migration is sufficient to prevent any appreciable gene differentiation unless there is strong differential selection. The effect of migration on genetic distance has been studied by Nei and Feldman (1972), Chakraborty and Nei (1974), Slatkin and Maruyama (1975), and Li (1976b) under the assumption of no selection. Their main conclusions are as follows. (1) If there is migration of genes with a rate of m per generation with $v \ll m \ll 1$, the genetic identity (I) eventually reaches a steady-state value, which is given by

$$I = m/(m + v), \tag{9.65}$$

approximately. Here, v is the mutation rate per locus per generation. (2) The approach to the steady state value is generally very slow; the number of generations required is of the order of the reciprocal of the mutation rate. Equation (9.65) indicates that the genetic distance between populations cannot be large unless the migration rate is very small.

If we know v, (9.65) can be used for estimating the maximum amount of migration between two populations. That is, we have $m = vI/(1 - I)$. This formula has been used by a number of authors (e.g., Chakraborty and Nei 1974; Nei 1975; Larson et al. 1984).

Interracial and Interspecific Gene Differences

In the past fifteen years, numerous authors have studied the genetic distances for various groups of organisms. From these studies, some general features have emerged. Let us now discuss the magnitude of interracial and interspecific genetic distances for some typical cases. In this section, we shall use Nei's standard genetic distance unless otherwise mentioned.

Table 9.3 shows estimates of genetic distance D between taxa of various ranks. At each level, various groups of organisms are included. The examples chosen are from those studies with a large number of loci examined. The genetic distance between local races varies from 0.00 to 0.05 and is generally much lower than that for interspecific comparisons. This is true for all kinds of organisms except microorganisms. The smaller genetic distances for local races than for different species are, of

Table 9.3 Estimates of standard genetic distance between taxa of various ranks.

Taxa	No. of taxa	No. of loci	D	Source
A. Local races				
Man	3	62	.011– .029	Nei and Roychoudhury (1982)
Mice (*M. musculus*)	4	41	.010– .024	Selander et al. (1969)
Lizards (*A. carolinensis*)	3	23	.001– .017	Webster et al. (1972)
Fish (*Catostomos*)	4	33	.000– .003	Buth and Crabtree (1982)
Horseshoe crabs	4	25	.001– .013	Selander et al. (1970)
Drosophila willistoni group	—	31	.008– .049	Ayala et al. (1974)

Table 9.3 (continued)

Taxa	No. of taxa	No. of loci	D	Source
B. Subspecies				
Red deer	4	34	.016	Gyllensten et al. (1983)
Mice (M. musculus)	2	41	.194	Selander et al. (1969)
Pocket gophers[a]	10	31	.004– .262	Nevo et al. (1974)
Ground squirrel	2	37	.103	Smith and Coss (1984)
Lizards (A. carolinensis)	4	23	.335– .351	Webster et al. (1972)
Drosophila willistoni group	—	31	.228 ± .026	Ayala et al. (1974)
Plants (peppers)	4	26	.02 – .07	McLeod et al. (1982)
C. Species				
Macaques	6	30	.02 – .10	Kawamoto et al. (1982)
Ground squirrels	2	37	.56	Smith and Coss (1984)
Gophers	2	27	.12	Patton et al. (1972)
Birds (Catharus)	4	27	.01 – .028	Avise et al. (1980a)
Galapagos finches	6	27	.004– .065	Yang and Patton (1981)
Lizards (Anolis)	4	23	1.32 –1.75	Webster et al. (1972)
Lizards (Crotaphytus)	4	27	.12 – .27	Montanucci et al. (1975)
Amphisbaenian (Bipes)	3	22	.61 –1.01	Kim et al. (1976)
Salamanders (Plethodon)	26	29	.18 –3.00	Highton and Larson (1979)
Teleosts (Xiphophorus)	6	42	.36 – .52	Morizot and Siciliano (1982)
Teleosts (Hypentelium)	3	38	.09 – .33	Buth (1979)
Drosophila				
Sibling species				
willistoni group	—	31	.54 ± .05	Ayala et al. (1974)
pseudoobscura vs.				
persimilis	2	24	0.05	Prakash (1969)
Nonsibling species				
obscura group	5	68	.29 – .99	Cabrera et al. (1983)
willistoni group	5	31	1.21 ± .06	Ayala et al. (1974)
Hawaiian species	8	31	.33 –2.82	Ayala (1975)
Plants (peppers)	10	26	.05 – .79	McLeod et al. (1982)
D. Genera				
Insectivora	3	24	.42 –1.10	Patton (unpublished)
Birds (Parulidae)	12	26	.05 – .69	Avise et al. (1980b)
Galapagos finches	5	27	.04 – .14	Yang and Patton (1981)
Fish (Sciaenidae)	5	16	1.1 –2.8(∞)	Shaw (1970)
Fish (Pleuronectidae)	3	31	.47 –1.3	Ward and Galleguillos (1978)
E. Families				
Man—Chimpanzee	2	42	.62	King and Wilson (1975)

[a] The populations studied have different chromosome numbers, so that they are classified as distinct subspecies.

course, expected because usually there is no reproductive isolation among local races and even if they are isolated the time since isolation is much shorter than that for different species.

Different strains of a microbial species sometimes show large genetic distance values, as in the case of *Escherichia coli* (Whittam et al. 1983) and fungus *Mucor racemosus* (Stout and Shaw 1974). These organisms reproduce asexually most of the time, and thus different strains are sexually isolated. Some of these strains seem to have been isolated for a long time, despite their morphological similarities. In these organisms, the distinction between race and species is not always clear when morphological and physiological characters alone are used for classification. For this reason, some authors (e.g., Stout and Shaw 1974) proposed that when two strains share the same allele at about 10 percent of loci or less, they should be ranked as different species.

When two taxa are morphologically quite distinct, though not as much as the usual difference between different species, they are often given different subspecific names. Therefore, the genetic distance between different subspecies within species is expected to be larger than that for local races. This is indeed the case in most groups of organisms, as seen from table 9.3. The intersubspecific distance is usually about 0.05 or larger. There are, however, many exceptions, and subspecific distances are as small as interracial distances in some cases. These exceptions are, of course, expected to occur because the definition of subspecies is quite arbitrary.

In this connection, it is interesting to note that the genetic distances between the three major races of man, Caucasoids, Negroids, and Mongoloids, are 0.01 to 0.03 (Nei and Roychoudhury 1982). Previously, Coon (1965) suggested that these major races should be classified as different subspecies. However, the extent of genetic differentiation among the major races is of the same order of magnitude as that for races in many other species. A similar situation has been observed in the European red deer *(Cervus elaphus)*. In northern Europe, there are four different subspecies (the British, Norwegian, Swedish, and continental red deer) which are distinguishable by morphological characters. Gyllensten et al. (1983) have shown that the average genetic distance among these subspecies was 0.0164. It might be more appropriate to classify them as local races.

The genetic distances between different species are generally larger than those between subspecies, except in birds. If we exclude birds and some mammalian groups, the interspecific genetic distance is about 0.05 or larger and can be as large as 3 or more. In most cases, however, there are some alleles shared between a pair of species within a genus (congeneric species), so that $D<\infty$. Surveying the literature, Thorpe (1982) studied the distributions of I for conspecific populations, congeneric species, and confamilial genera, excluding birds. The results obtained are presented in figure 9.6. In the case of congeneric species, I varies from about 0.98 to 0.02. This corresponds to the variation of D from 0.02 to 3.9. However, in the majority of cases, I is 0.8 ($D=0.22$) to 0.2 ($D=1.6$).

The status of specific differences is usually recognized by examining morphological and physiological differences as well as the extent of reproductive isolation. In practice, however, the recognition of species differences is not always easy. In this case, information on genetic distances

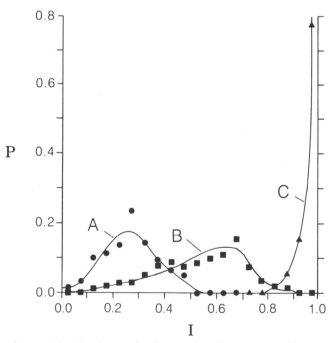

Figure 9.6.   Distributions of I for confamilial genera (A), congeneric species (B), and conspecific populations (C), excluding data from birds. From Thorpe (1982). Copyright by Annual Reviews, Inc.

is of some help. Thorpe (1982) has suggested that when the species status of a population is unresolvable by other criteria, information on genetic distance should be used. Examining the distribution of I for congeneric species, he has concluded that if allopatric populations of dubious status have genetic identities below 0.85 ($D = 0.16$), it is improbable that they should be considered conspecific, while nominate species with I values above 0.85 should be considered doubtful if there is no other evidence of their specific status. This seems to be a reasonable suggestion.

Recently, Sage (1981) examined the genetic distances between various subspecies (e.g., *musculus, domesticus, castaneus*) of the mouse *Mus musculus*. Seeing that the genetic distances are of the order of 0.2, he suggested that these subspecies should be raised to the rank of species. This suggestion was later supported by a study of the extent of genetic differentiation of mitochondrial DNA (Ferris et al. 1983).

The criterion of $I = 0.85$, however, does not apply to bird species. As is clear from table 9.3, most congeneric species in birds show rather small genetic distances. For example, six species of *Geospiza* (finches) in the Galapagos Islands show a genetic distance ranging from 0.004 to 0.065. This low value of genetic distance is quite typical in bird families so far studied (Avise and Aquadro 1982). This suggests that in birds speciation occurs rather rapidly without much genetic change. This is possible if reproductive isolation is established by a small number of gene substitutions and the genes controlling reproductive isolation are subject to strong selection (see chapter 14). Ethological isolation by means of visual appearance and songs is apparently important in birds, and this might have speeded up the evolution of reproductive isolation. Wyles et al. (1983) and Wilson (1985) proposed the hypothesis that the apparent rapid evolution of morphological characters in birds is due to behavioral innovation and social transmission unique in birds (and higher primates).

Genetic distances between confamilial genera have been studied in a relatively small number of groups of organisms. On the average, intergeneric distances are larger than interspecific distances, except in birds. In birds, the distances are again considerably smaller than those for other groups of organisms. This is compatible with Wyles et al.'s (1983) view that most families and genera in birds have evolved more recently than in other groups of organisms.

At the morphological level, there is a great difference between man and the chimpanzee. This difference has led primate taxonomists to classify man and the chimpanzee into different families. However, the genetic distance between these two species is quite small and of the same order as that of different species in *Drosophila* (King and Wilson 1975). Essentially the same conclusion was obtained by Bruce and Ayala (1979) and Nozawa et al. (1982). This contrast between genetic distance and morphological difference has led King and Wilson (1975) to propose the hypothesis that morphological evolution occurs mainly by mutation of regulatory genes.

Empirical Relationship Between Genetic Distance and Evolutionary Time

I have previously discussed the theoretical relationship between genetic distance and the time since divergence between two populations. This relationship can be used for estimating divergence times from genetic distance data. In practice, however, the underlying assumptions are not always satisfied in natural populations, so that some caution is necessary in the application of the theory.

In recent years, a number of authors have estimated evolutionary times from genetic distance data and compared them with the estimates from other sources such as fossil records, separation of lands and seas, island formation, etc. In these studies, most authors have used equation (9.54) or $t = kD$, where k is a proportionality constant and given by $k = (2\alpha)^{-1}$. Table 9.4 shows the results of these studies. It should be emphasized

Table 9.4   Estimates of divergence time from genetic distance and other sources. Proportionality constant k for $t = kD$ is also given.

| | | Time estimate (years) from | | | |
| | | Distance* ($\times 10^5$) | Other sources ($\times 10^5$) | | k** |
Organism	D			Loci	($\times 10^6$)
Mammals					
Negroid & Mongoloid[a]	0.031	1.5 ± 0.5	1–2	62	5
Man & chimpanzee[b]	0.62	40 ± 10	50	44	(5)
Two macaque spp.[c]	0.11	5.4	4–5	28	5
Pocket gophers[d]	0.08	1.2 ± 0.8	1–2	31	1.5
Woodrats[e]	0.18	1.5 ± 0.9	2–4	20	0.8
		9 ± 4.5			5

Organism	D	Time estimate (years) from		Loci	k^{**} ($\times 10^6$)
		Distance* ($\times 10^5$)	Other sources ($\times 10^5$)		
Deer mice spp.[f]	0.15	1.5 ± 0.7 7.5 ± 3	1–5	28	1.0 (5)
Ground squirrels[g] (spp.)	0.56	50 ± 10 40 ± 10	50	37	6.7 (5)
Ground squirrels[g] (subsp.)	0.10	6.9 ± 0.3 5.5 ± 1	7	37	6.7 (5)
Birds					
Galapagos finches[h]	0.12	6 + 3	5–40	27	5
Reptiles					
Bipes spp.[i]	0.62	31 ± 10 40 ± 10	40	22	5 (5)
Lizards[j]	0.28	50 ± 25 16 ± 8	c. 50	22	18 (5)
Fishes					
Cave and surface fishes[k]	0.14	7 ± 4.6	3–20	17	5
Minnows[l]	0.053	2.7	1–20	24	5
Panamanian fishes[m]	0.32	58 19 ± 6	20–50	28	18 (5)
Panamanian fishes[n]	0.24	35 12	2–5	31.4	18 (5)
Echinoids					
Panamanian sea urchins[o]	0.03–0.64 0.39	— 24	20–50	18	 (5)

*When the standard error of D was not given in the original papers, it was computed by using equation (9.44).

**The k value in parentheses was applied to the D_v's obtained from equation (9.60).

Data Source: [a]Nei and Roychoudhury (1982).   [b]King and Wilson (1975).   [c]Nozawa et al. (1977).   [d]Nevo et al. (1974).   [e]Zimmerman and Nejtek (1977).   [f]Gill (1976).   [g]Smith and Coss (1984).   [h]Yang and Patton (1981).   [i]Kim et al. (1976).   [j]Adest (1977).   [k]Chakraborty and Nei (1974).   [l]Avise et al. (1975).   [m]Gorman and Kim (1977).   [n]Vawter et al. (1980), average of ten pairs of species.   [o]Lessios (1979), average of four pairs of species.

that the estimates of evolutionary times obtained from other sources in this table are not as certain as they might look; the numerical values are given only to obtain a rough idea. Despite this problem, table 9.4 shows that the agreement between the estimate obtained from genetic distance and that from other sources is reasonably good in most studies, particularly if we consider the large standard error of genetic distance estimates. However, there is one problem; the proportionality constant is not always the same, and there is a twentyfold difference between the largest and smallest k values. Therefore, the data in this table seem to suggest that there is no molecular clock that is universal for all organisms. Before rushing to this conclusion, however, let us examine each set of data carefully, taking into account other factors that might affect genetic distance estimates.

Proteins Used

The first factor to be considered is the kind of proteins used for electrophoresis. As discussed by Nei (1975), the rate of gene substitution per locus per year [α in equation (9.54)] may be expressed as

$$\alpha = nc\lambda, \tag{9.66}$$

where n = the average number of amino acids per polypeptide, c = proportion of amino acid substitutions that are detectable by electrophoresis, and λ = the average rate of amino acid substitution per year. The value of $\alpha = 10^{-7}$ in equation (9.55) is obtained by using $n = 400$, $c = 0.25$, and $\lambda = 1 \times 10^{-9}$ (see chapter 3). In practice, however, different investigators use different sets of proteins, though large-scale electrophoretic surveys usually contain many commonly used proteins (Avise and Aquadro 1982). For example, the n value (515) for the proteins used by Nevo et al. (1974) was somewhat larger than that (400) of Nei (1975). They also assumed that $\lambda = 2.1 \times 10^{-9}$. Mainly because of these differences, Nevo et al.'s $k = 1.5 \times 10^6$ was about three times smaller than Nei's.

Detectability of Protein Differences by Electrophoresis

The detectability of amino acid differences in proteins by electrophoresis depends on various biological and biochemical conditions of electrophoresis, such as tissue used, pH and type of the gel, etc., and these conditions are not always the same for all organisms studied or for all laboratories. Differences in these conditions are expected to cause some differences in the detectability of protein differences (c). Although Avise and Aquadro (1982) suggested that the technique of electrophoresis used is virtually the same for most laboratories, there is evidence that this is not the case. For example, King and Wilson (1975) reported a genetic distance of $D = 0.62$ between man and the chimpanzee, whereas Bruce and Ayala (1979) and Nozawa et al. (1982) obtained $D = 0.39$ and 0.45, respectively, for the same pair of species. These differences are partly due to the differences in the proteins used, but even if the same 15 protein loci that are common to the three studies are used, there are substantial differences, the D values for the data of King and Wilson, Bruce and Ayala, and Nozawa et al. being 0.83, 0.41, and 0.71, respectively (Nozawa et al. 1982). This indicates that c is not the same for all laboratories. In a study of ground squirrels, Smith and Coss (1984) also speculated that the c value for their data was probably smaller than 0.25, and obtained $k = 6.7 \times 10^6$.

Sampling Errors

As mentioned earlier, a large number of loci should be used to obtain a reliable estimate of genetic distance, particularly when genetic distance is small. In practice, however, many authors use a relatively small number of loci for various reasons. Consider an extreme case where two populations are virtually monomorphic and fixed for different alleles at 10 percent of the loci. The expected genetic identity for this case is 0.9 ($D = 0.11$), but the observed identity can be substantially larger or smaller than 0.9. For example, if r loci are examined, the observed value of I will be 1 (or D will be 0) with the probability of $(0.9)^r$. In the case of $r = 20$, this probability is 0.12, which is not very small. To make this probability smaller than 0.05, r must be equal to or greater than 29, whereas the minimum value of r that makes the probability less than 0.01 is 44. This is a substantial number of loci.

In a study of genetic distances between different species of sea urchins from the Pacific and Atlantic coasts of Panama, Lessios (1979) observed an unusually low distance value for one of the four species pairs examined, though the Panamanian Isthmus is known to have been above sea level for the last 2 to 5 million years. From this observation, he concluded that electrophoretic data cannot be used for dating evolutionary time. However, he studied only 18 protein loci, so that the chance that this unusual result is due to sampling error is quite high (Vawter et al. 1980). Actually, if we consider the average genetic distance (0.39) for all the four species pairs, the divergence time (2.4 MY) estimated is consistent with the one from geological data (table 9.4).

Nonlinear Relationship of D with Time

When the time since divergence between two species is long, the relationship between D and t is no longer linear, as mentioned earlier. This factor has not been properly taken into account in most of the empirical studies cited in table 9.4. Correction for this nonlinearity seems to improve the agreement between the estimates of divergence times from genetic distance and other sources in some cases. For example, King and Wilson (1975) obtained $D = 0.62$ between man and chimpanzee using equation (9.24). If we use equation (9.55) with $\alpha = 10^{-7}$, this gives $t = 3.1 \times 10^6$ years, which seems to be too low compared with the estimate (5×10^6) obtained from Sarich and Wilson's (1967) immunological distance. However, if we estimate D_v by (9.60) and put it into (9.55), we have $t = (4 \pm 1) \times 10^6$ years, which is no longer incompatible with the estimate from other sources. If we use equation (9.62), t becomes even larger. Similarly, Kim et al.'s (1976) estimate of t can be improved by using (9.60), as shown in table 9.4. The same property was noted by Smith and Coss (1984) in their study of ground squirrels, though they used a proportionality constant of $k = 6.7 \times 10^6$.

The largest proportionality constant k in table 9.4 is that of Gorman and Kim (1977), Adest (1977), and Vawter et al. (1980). This value was obtained by comparing Sarich and Wilson's (1967) immunological distance (d_I) with D (Maxson and Wilson 1974). In this comparison, many D values larger than 1 were used. Therefore, the k value obtained is expected to be an overestimate. However, Gorman and Kim's (1977) $k = (18 \times 10^6)$ seems to be better than $k = 5 \times 10^6$ in explaining Vawter et al.'s (1980) and Adest's (1977) electrophoretic data. One can use equation (9.60) to compute the expected evolutionary time under the assumption of $k = 5 \times 10^6$, but the values obtained are much smaller than the estimates obtained from other sources. The only data that can be accommodated with the value of $k = 5 \times 10^6$ are those of Gorman and Kim (1977) (table 9.4). Since Vawter et al.'s data are based on 10 pairs of species, their results cannot be dismissed as a special case. This suggests that the fishes and reptiles studied by Vawter et al. and Adest do not show the same evolutionary rate as that for many other organisms or that the electrophoretic technique used for these organisms did not detect protein differences as efficiently as in other organisms.

Bottleneck Effects

One of the troublesome problems in dating evolutionary time from genetic distance data is the bottleneck effect. As we have seen earlier, the bottleneck effect accelerates the increase of D temporarily. In the presence of bottleneck effects, therefore, equation (9.55) is expected to give an overestimate of evolutionary time. On the other hand, if one tries to calibrate the evolutionary clock by using D values which have been affected by bottlenecks, a relatively small value of k would be obtained.

It is generally difficult to know whether or not a particular population or a particular pair of populations has gone through bottlenecks. In some cases, however, there is clear-cut evidence for the bottleneck effect, and we can make a correction for it. For example, the cave populations of the fish *Astyanax mexicanus* in Mexico are very small, and they are apparently derived from the nearby surface populations (Avise and Selander 1972). At the present time, the average homozygosities in the cave populations are virtually 1, and if we use equation (9.64), the bottleneck effect is eliminated. The genetic distance for this species in table 9.4 has been estimated in this way.

In some other cases, the bottleneck effect (or the effect of population size reduction) can be inferred from the level of heterozygosity even if we do not know the history of the populations. For example, the pocket gopher *Thomomys talpoides* consists of a number of subspecies which have a low average heterozygosity for enzyme loci and are fixed for different chromosome numbers (Nevo et al. 1974). It is, therefore, possible that the genetic distances among them are affected by bottleneck effects. This could be another reason why the k value estimated for these subspecies is lower than 5×10^6.

Some Remarks

The above considerations suggest that if we use equation (9.55), taking into account various factors that affect the relationship between D and t, electrophoretic data are useful for obtaining a rough idea about evolutionary time. It should be noted that D is intended to measure the number of amino acid substitutions that are detectable by electrophoresis. Therefore, as long as amino acid substitution occurs at a constant rate, D should increase as evolutionary time increases.

It should also be noted that the electrophoretic clock is used mainly for closely related species. If D is too large (say, $D \geq 1$), its variance becomes very large even if a substantial number of loci are studied, so that the reliability of dating declines. The high frequency of backward and parallel mutations at the electrophoretic level in the case of $D \geq 1$ also makes the clock unreliable. According to Peetz et al. (1986), the frequency of backward mutations at the electrophoretic level is higher than that expected under random substitution, apparently because of purifying selection. Nozawa et al. (1982) noted that the estimate of the time since divergence between humans and macaques was much smaller than that indicated by the fossil records (about 30 million years). I believe that humans and macaques are too distantly related for the electrophoretic clock to be applicable.

A number of authors (Aquadro and Avise 1981; Ohnishi et al. 1983; Goldman et al. 1986) have used two-dimensional electrophoresis to study the evolutionary relationship of related species. Although this method does not detect as much variation as ordinary electrophoresis does, it permits an examination of a large number of polypeptides on one gel. Because of this advantage, it may become a useful tool for the study of evolution in the future. Recently, Goldman et al. (1986) improved this method by introducing ^{35}S methionine labeling and surveyed 383 fibroblast polypeptides in humans and apes. Using these data, they constructed a phylogenetic tree of humans and apes, which has the same topology as that of figure 2.4.

In chapters 4 and 5, we discussed various methods of measuring protein or DNA divergence between different species. As expected, there are high correlations among different measures of genetic divergence. The electrophoretic distance (D) is also highly correlated with other measures of genetic divergence (Sarich 1977; Highton and Larson 1979). Recently, O'Brien et al. (1985a) examined the genetic relationship of the giant panda with various species of bears and raccoons by using albumin distance, DNA hybridization data, electrophoretic distance, and chromosomal banding pattern. All these data have indicated that the panda is a bear rather than a raccoon, closing a long-standing controversy among zoologists (see also Sarich 1973).

In recent years, many authors have used mitochondrial DNA to study evolutionary relationships of organisms. It should be noted, however, that the resolving power of mitochondrial DNA is not necessarily higher than that of protein electrophoresis. This is particularly so when the restriction enzyme technique is used. To see the difference in resolving power between different methods, let us consider how many nucleotides are assayed by electrophoresis. As discussed in chapter 3, electrophoresis detects only those nucleotide substitutions that change the net charge of the protein encoded. Under the assumption of random nucleotide substitution, 71 percent of nucleotide changes result in amino acid substitutions. In practice, however, the number of silent nucleotide substitutions per silent site is five times more frequent than that of amino acid altering substitutions per nonsilent site (table 5.4). Therefore, about 33 [$= 71/(71 + 29 \times 5)$] percent of nucleotide substitutions seem to result in amino acid changes. Since about 25 percent of amino acid changes are detectable by electrophoresis, about 8.3 percent of nucleotide substitutions are assayed by electrophoresis. On the other hand, the number of amino acids in an average protein is about 400, so that the number of nucleotides in the coding regions of an average gene is 1,200. Therefore, electrophoresis is expected to survey about $1,200 \times 0.083 \simeq 100$ nucleotides per locus. If we examine 60 loci by electrophoresis, it is equivalent to studying 6,000 nucleotides. This is much larger than the number of nucleotides (895) sequenced by Brown et al. (1982) for human and ape mitochondrial DNAs (see chapter 11).

When the restriction enzyme technique is used, the number of nucleotides surveyed is given by $\Sigma_i m_i r_i$, where m_i and r_i are the number of restriction sites per sequence and number of nucleotides in the recognition sequence of the ith enzyme used, respectively. In their study of the evolution of human and ape mitochondrial DNAs, Ferris et al. (1981) used eighteen 6-base enzymes and one 4-base enzyme. The average number of restriction sites per sequence for all 6-base enzyme was 42, whereas the number for the 4-base enzyme was 7. Therefore, the total number of nucleotides assayed is $42 \times 6 + 7 \times 4 = 280$. This number is even smaller than the number of nucleotides sequenced by Brown et al.

Of course, the detectability of protein differences by electrophoresis seems to be affected by various electrophoretic conditions, as discussed earlier. Therefore, the above comparison of resolving power should be regarded only as a general guideline.

CHAPTER TEN

DNA POLYMORPHISM WITHIN AND BETWEEN POPULATIONS

The study of protein polymorphism has indicated that the extent of genetic variation in natural populations is enormous. However, the total amount of genetic variation cannot be known unless it is studied at the DNA level. The study of DNA polymorphism is still in its infancy, but the results so far obtained indicate that the extent of DNA polymorphism is far greater than that of protein polymorphism. In this chapter, I would like to discuss statistical methods for studying DNA polymorphism and some general features of the results so far obtained.

DNA Sequence Polymorphism Due to Nucleotide Substitution

Number of Polymorphic Sequences

When nucleotide sequences are known for a number of genes (alleles) sampled from a population, there are several ways of measuring the extent of DNA polymorphism. A simple measure is the number of different (polymorphic) sequences (k) in the sample. This measure has the same statistical property as that of the number of alleles in the sample discussed in chapter 8 and is highly dependent on sample size. Therefore, to compare the extent of polymorphism between two populations or loci by using this measure, one has to have the same or nearly the same sample size. In the case of DNA sequences, however, there is another problem. That is, when a long DNA sequence is studied, all sequences examined may be different from each other. In this case, the number of polymorphic sequences is no longer an informative measure of polymorphism.

Number of Polymorphic Nucleotide Sites

The second measure of DNA polymorphism is the number of polymorphic (segregating) nucleotide sites per nucleotide site. It is given by $p_n = s_n/m_T$, where s_n and m_T are the number of polymorphic sites per sequence and the total number of nucleotides examined, respectively. This measure is superior to the number of different sequences because it is applicable even when all sequences examined are different. However, this is also dependent on sample size.

If we use the infinite-site model of neutral mutations (see chapter 13), the expected value of p_n over independent populations (over the stochastic process) is

$$E(p_n) = L[1 + 1/2 + 1/3 + \cdots + (n-1)^{-1}], \tag{10.1}$$

where n is the number of sequences examined and $L = 4N\mu$ (Watterson 1975). Here, N and μ are the effective population size and mutation rate per nucleotide site per generation, respectively. When n is large, say $n > 20$, equation (10.1) is approximately given by

$$E(p_n) = L[0.577 + \log_e(n-1)]. \tag{10.1a}$$

When there is no recombination, the variance of p_n over the stochastic process is given by

$$V(p_n) = E(p_n)/m_T + L^2 \sum_{i=1}^{n-1} 1/i^2. \tag{10.2}$$

Therefore, it is possible to estimate $L = 4N\mu$ by

$$\hat{L} = p_n/A, \tag{10.3}$$

where $A = 1 + 1/2 + \cdots + (n-1)^{-1}$. This $\hat{L}$ can be used as a measure of polymorphism. However, note that equation (10.3) is valid only when neutral mutations are considered and the population is in equilibrium with respect to the effects of mutation and genetic drift.

Nucleotide Diversity

A more appropriate measure of DNA polymorphism than the above two measures is the average number of nucleotide differences per site between two sequences or *nucleotide diversity*. This is defined by

$$\pi = \sum_{ij} x_i x_j \pi_{ij},$$ (10.4)

where x_i is the population frequency of the ith type of DNA sequence, and π_{ij} is the proportion of different nucleotides between the ith and jth types of DNA sequences. In a randomly mating population, π is simply heterozygosity at the nucleotide level. It can be estimated either by

$$\hat{\pi} = \frac{n}{n-1} \sum_{ij} \hat{x}_i \hat{x}_j \pi_{ij}$$ (10.5)

or by

$$\hat{\pi} = \sum_{i<j} \pi_{ij}/n_c,$$ (10.6)

where n, $\hat{x}_i$, and n_c are the number of DNA sequences examined, the frequency of the ith type of DNA sequence in the sample, and the total number of sequence comparisons $[n(n-1)/2]$, respectively. In equation (10.6), i and j refer to the ith and jth sequences rather than to the ith and jth types of sequences. Assuming that π_{ij}'s are constant, the variance of $\hat{\pi}$ obtained by (10.5) is given by

$$V(\hat{\pi}) = \frac{4}{n(n-1)} \left[(6-4n) \left(\sum_{i<j} x_i x_j \pi_{ij} \right)^2 \right.$$

$$\left. + (n-2)\Sigma x_i x_j x_k \pi_{ij}\pi_{ik} + \sum_{i<j} x_i x_j \pi_{ij}^2 \right].$$ (10.7)

(Nei and Tajima 1981). This is the variance generated at the time of allele frequency survey and does not include the variance due to stochastic errors.

When all mutations are neutral, the expectation of $\hat{\pi}$ over the stochastic process is given by

$$E(\hat{\pi}) = L/[1 + (4/3)L]$$

$$\approx L \qquad \text{for } L \ll 1 \qquad\qquad (10.8)$$

(Kimura 1968b). On the other hand, the variance of $\hat{\pi}$ over the sto-chastic process is

$$V(\hat{\pi}) = \frac{n+1}{3(n-1)m_T}L + \frac{2(n^2+n+3)}{9n(n-1)}L^2 \qquad (10.9)$$

[modified from Tajima (1983a)]. As n increases, this approaches

$$V(\hat{\pi}) = \frac{1}{3m_T}L + \frac{2}{9}L^2. \qquad (10.10)$$

EXAMPLE

Kreitman (1983) sequenced the alcohol dehydrogenase (Adh) gene re-gion for eleven alleles of Drosophila melanogaster. The gene region studied contained 2,379 nucleotides (m_T), excluding deletions and insertions. Nine of the 11 sequences were different from each other, indicating a high degree of sequence polymorphism. There were 43 polymorphic nu-cleotide sites, so that p_n was $43/2379 = 0.0181$. Under the assumption of neutrality, therefore, we have $\hat{L} = 0.0062$ from equation (10.3), since A is 2.929 for $n = 11$. From Kreitman's data, it is possible to compute the proportion of different nucleotides for all pairs of alleles. The results obtained are presented in table 10.1. The nucleotide diversity for this gene region is therefore estimated to be 0.0065 ± 0.0017 from either equation (10.5) or (10.6). Note that $\hat{x}_i = 1/11$ for all but one haplotype (sequence type). The one haplotype has a sample frequency of 3/11 (see table 10.1). The standard error of the estimate was obtained by using equation (10.7). It is interesting to see that the estimate of nucleotide diversity is close to the one estimated from the number of polymorphic sites under the assumption of neutrality.

The variance of $\hat{\pi}$ due to stochastic factors (including sampling vari-ance) under the assumption of neutrality can be computed by using equation (10.9) and becomes 12.61×10^{-6}. The total standard error is therefore 0.0036. This is much larger than the standard error due to sampling error alone (0.0017). Previously, we obtained an estimate of $\hat{L}$ from $\hat{p}_n$. The standard error of this $\hat{L}$ under the assumption of neutrality

Table 10.1   Percent nucleotide differences between 11 alleles of
the *Adh* locus in *Drosophila melanogaster*. The total number of
nucleotide sites compared is 2,379. Data from Kreitman (1983).

Allele	(1)	(2)	(3)	(4)	(5)	(6)	(7)	(8)	(9)	(10)
(1) Wa-S										
(2) Fl-1S	0.13									
(3) Af-S	0.59	0.55								
(4) Fr-S	0.67	0.63	0.25							
(5) Fl-2S	0.80	0.84	0.55	0.46						
(6) Ja-S	0.80	0.67	0.38	0.46	0.59					
(7) Fl-F	0.84	0.71	0.50	0.59	0.63	0.21				
(8) Fr-F	1.13	1.10	0.88	0.97	0.59	0.59	0.38			
(9) Wa-F	1.13	1.10	0.88	0.97	0.59	0.59	0.38	0.00		
(10 Af-F	1.13	1.10	0.88	0.97	0.59	0.59	0.38	0.00	0.00	
(11) Ja-F	1.22	1.18	0.97	1.05	0.84	0.67	0.46	0.42	0.42	0.42

becomes 0.0028 from equations (10.2) and (10.3). This is slightly smaller
than that of $\hat{\pi}$.

At the *Adh* locus of *D. melanogaster*, there are two electromorphs (alleles) that are detectable by electrophoresis, i.e., *S* and *F*. The alleles
studied by Kreitman consist of six *S*'s and five *F*'s. Therefore, we can
estimate nucleotide diversity for these two electromorphs separately. It
becomes 0.0056 for *S* and 0.0029 for *F*. Thus, the nucleotide diversity
for *S* is twice as high as that for *F*. Table 10.1 shows that the nucleotide
differences between alleles of electromorphs *S* and *F* are considerably larger
than the intra-electromorph nucleotide diversity. This suggests that the
S and *F* alleles diverged a long time ago.

DNA Polymorphism Estimated from Restriction Site Data

Although DNA sequences give a complete picture of genetic variation
in populations, it is time-consuming to sequence DNA for a large number of individuals. The extent of DNA polymorphism is therefore usually studied by using the restriction enzyme technique.

DNA Regions to Be Studied

Unlike the case of protein polymorphism, the unit of study of DNA
polymorphism is not always well defined. Except in small genomic DNAs

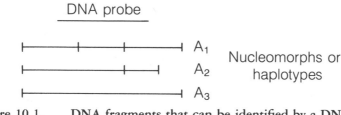

Figure 10.1.   DNA fragments that can be identified by a DNA
probe.

such as mitochondrial and chloroplast DNA, DNA polymorphism is
usually studied for a given fragment of DNA sequence. This fragment
may include any region of DNA such as amino acid coding regions,
introns, flanking regions or intergenic regions. It could be a short piece
of DNA including only one type of DNA region or a long piece of DNA
encompassing many structural genes. When polymorphism is studied by
the restriction enzyme technique, the distinction between different gene
regions is not always possible, and polymorphism is usually studied by
using the entire DNA fragment as a unit. Furthermore, in the case of
nuclear DNA a particular DNA probe is used, and the restriction fragments that hybridize with the probe are identified (figure 10.1). In this
case, polymorphism for the DNA probe region as well as the outside
regions may be detected (e.g., sequences A_1 and A_2 in figure 10.1). Nei
and Tajima (1981) called the entire DNA segment under investigation
a *nucleon* and a particular DNA sequence (or restriction site sequence) for
the segment studied a *nucleomorph*. Nucleon and nucleomorph correspond
to gene (locus) and allele in classical genetics, respectively. In recent
literature, however, a nucleomorph is often called a *haplotype*. In the
following, we shall use both terminologies, depending on the situation.
    In this connection, it should be mentioned that human geneticists
often use the phrase *restriction fragment length polymorphism* (RFLP) (Botstein et al. 1980). In practice, this cumbersome phrase is unnecessary;
it can be replaced either by "DNA polymorphism" or "restriction site
polymorphism."

Number of Haplotypes and Nucleon (Haplotypic) Diversity

Suppose that one is interested in the extent of DNA polymorphism
for a particular region of nuclear DNA or the entire mitochondrial (or
chloroplast) DNA in a population. In this case, a certain number of

individuals are sampled from this population, and their DNAs are examined by using restriction enzymes. These DNAs are classified into different nucleomorphs or haplotypes according to their restriction site patterns. For example, in the case of figure 10.1 there are three different haplotypes. Once different haplotypes are identified, their relative frequencies in the sample are computed just like the allele frequencies at an electrophoretic locus. We designate the population frequency of the ith haplotype by x_i and the sample frequency by $\hat{x}_i$.

DNA polymorphism can be studied at two different levels, i.e., at the nucleon and nucleotide levels. At the nucleon level, DNA polymorphism due to both nucleotide substitution and deletion/insertion may be studied without distinction, whereas the study of polymorphism at the nucleotide level is usually done by considering only nucleotide substitutions. The simplest measure of DNA polymorphism at the nucleon level is the number of haplotypes (k) observed in the sample. As in the case of the number of different DNA sequences, this number depends on sample size so that the same sample size must be used to compare the extents of polymorphism of different populations. A more appropriate measure of nucleon polymorphism is *nucleon* or *haplotypic diversity* (h) (Nei and Tajima 1981). This is equivalent to heterozygosity or gene diversity used in the study of protein polymorphisms. It is defined by the same equation as (8.1) and can be estimated by (8.4). The sampling property and the expectation of h under the infinite-allele model are also the same as those of gene diversity discussed in chapter 8.

Mean Number of Restriction-Site Differences

The extent of DNA polymorphism can also be measured by the mean number of restriction-site differences between two randomly chosen haplotypes (Nei and Tajima 1981). The number is defined as

$$\nu = \Sigma_{ij} x_i x_j \nu_{ij}, \qquad (10.11)$$

where ν_{ij} is the number of restriction-site differences between the ith and jth haplotypes and summation is taken over all combinations of haplotypes. The estimate of ν and its sampling variance are given by equations (10.5) and (10.7), respectively, if we replace π_{ij} by ν_{ij}.

For neutral haplotypes due to nucleotide substitution, the expectation of ν under the infinite-site model of neutral mutations is

$$E(\nu) = M, \qquad\qquad (10.12)$$

where $M = 4N\nu$ and ν is the mutation rate per nucleon at the restriction site level (see chapter 13). On the other hand, the stochastic variance of $\hat{\nu}$ is approximately given by (10.9) if we replace L by M and eliminate m_T. Elimination of m_T is necessary because we are now studying variation per nucleon rather than variation per nucleotide site [see Tajima (1983a)].

Number of Polymorphic Restriction Sites

A quantity closely related to ν is the number of polymorphic restriction sites (s_r), i.e., the number of restriction sites that are polymorphic in a sample of n nucleons. Like the number of haplotypes in a sample, this quantity is also highly dependent on sample size. If we assume the neutrality of haplotypes without deletion and insertion, the expectation of s_r is given by (10.1) if we replace L by M. Similarly, the variance of s_r is given by (10.2) if we eliminate m_T and replace L and p_n with M and s_r, respectively.

Nucleotide Diversity

One problem with the above measures of DNA polymorphism is that they all depend on the size of the nucleon studied and generally increase as the size increases. In practice, nucleon size varies greatly with the gene or gene region studied, so that they cannot be used as a general measure. This problem can be avoided if we measure variation at the nucleotide level. If most restriction-site polymorphisms are due to nucleotide substitution, this can be done by using nucleotide diversity defined by (10.4). When the restriction enzyme technique is used, π_{ij} in (10.4) may be estimated by (5.41) or (5.50) in chapter 5. Therefore, if sample haplotype frequencies are available, π can be estimated by (10.5) or (10.6).

As mentioned earlier, the expected value of π under the infinite-site model of neutral mutations is $L = 4N\mu$, whereas the expected value of ν is $M = 4N\nu$. Therefore, π can be estimated from ν if we know the relationship between μ and ν. Nei and Tajima (1981) have shown that

$$\nu = 2 \sum_{i=1}^{s} m_i r_i \mu, \qquad\qquad (10.13)$$

where m_i and r_i are the number of restriction sites and the number of nucleotides in the recognition sequence of the ith enzyme, respectively, and s is the number of enzymes used. Therefore, π can be estimated by

$$\hat{\pi} = \hat{v}/R, \tag{10.14}$$

where $R = 2\Sigma_{i=1}^{s} m_i r_i$. If we use (10.3), π can also be estimated from s_r. That is,

$$\hat{\pi} = s_r/(AR) \tag{10.15}$$

where $A = 1 + 1/2 + \cdots + (n-1)^{-1}$.

Applying Wright's (1931) theory for a pair of two neutral alleles, Ewens et al. (1981) showed that $\pi \equiv L$ can be estimated by

$$\hat{\pi} = (s_{r4} + s_{r6})/[(8m_4 + 12m_6)\log_e n], \tag{10.16}$$

where s_{r4} and s_{r6} are the numbers of polymorphic restriction sites for 4-base and 6-base enzymes, respectively, and m_4 and m_6 are the total numbers of restriction sites for 4-base and 6-base enzymes, respectively. This equation has been derived under the condition that n is large. Actually, (10.16) is regarded as an approximation to (10.15), since A is approximately $\log_e n$ when n is large, say $n > 100$.

Under the assumption of random arrangement of nucleotides in nucleons but without assuming neutrality, Engels (1981a) and Hudson (1982) derived an approximate formula for the proportion of polymorphic sites ($p_r = s_r/m$, where m is the total number of restriction sites) for *a given sample*. To estimate L from p_r, however, it is necessary to assume neutrality of haplotypes. Under this assumption, they derived essentially the same formula as (10.16). Engels (1981a) also derived a formula for estimating π. It is given by

$$\hat{\pi} = \frac{nc - \Sigma c_i^2}{rc(n-i)}, \tag{10.17}$$

where c_i is the number of members of the sample that have a cleavage at site $i (i = 1, \cdots, s_r)$ and $c = \Sigma c_i$. This formula does not require the assumption of neutrality as emphasized by Engels, but depends on the assumption of linkage equilibrium among restriction sites. By contrast,

$\hat{\pi}$ given by (10.5) or (10.14) requires the assumption of neither neutrality nor linkage equilibrium. However, when the number of DNA sequences (nucleons) examined is large (say, $n > 100$), (10.17) seems to be easier to compute than (10.5) or (10.14). Engels also presented a formula for computing the variance of $\hat{\pi}$ obtained by (10.17), but since it depends on the assumption of linkage equilibrium, it is expected to give an underestimate.

EXAMPLE

Shah and Langley (1979) studied the genetic variability of mtDNA in three species of *Drosophila (melanogaster, simulans, and virilis)* by using restriction endonucleases *Hae*III (GGCC), *Hpa*II (CCGG), *Eco*RI (GAATTC), and *Hind*III (AAGCTT). They identified seven haplotypes, and the frequencies of these haplotypes are given for each species in table 10.2. In *D. simulans,* only five nucleons (DNA sequences) were sampled, and no polymorphism was found. In *D. melanogaster,* there are four haplotypes, and the estimate of nucleon diversity ($\hat{h}$) is 0.71 ± 0.04 (table 10.3). From this value, we can estimate $M \equiv 4Nv$ by $\hat{M} = \hat{h}/(1 - \hat{h})$ under the assumption of no selection. It becomes 2.46. On the other hand, the estimate of M obtained from the number of haplotypes (k) by using (8.16) is 1.95, which is considerably smaller than the estimate obtained from nucleon diversity. This difference apparently occurred because nucleon diversity (heterozygosity) tends to give an overestimate of M when it is based on a single locus (Zouros 1979).

To compute the mean number of restriction site differences, it is convenient to make a table of the numbers of restriction-site differences (v_{ij}) for all pairs of haplotypes (table 10.4). These numbers are obtainable from Shah and Langley's (1979) figure 1. From the values of v_{ij} for *D. melanogaster* in table 10.4, we obtain $\hat{v} = 1.22$ by using (10.5) with π_{ij} replaced by v_{ij}. Under the assumption of neutral haplotypes, this is an

Table 10.2   Haplotype frequencies in samples of mtDNAs from three species of *Drosophila*. $n =$ number of mtDNAs sampled. Data from Shah and Langley (1979).

Haplotype	m	m_a	m_b	m_c	s	v	v_d	(n)
D. melanogaster	0.1	0.3	0.5	0.1				10
D. simulans					1.0			5
D. virilis						0.6	0.4	10

Table 10.3 Haplotypic diversity *(ĥ)*, number of haplotypes *(k)*, average number of restriction-site differences *(v̂)*, number of segregating sites *(sᵣ)*, and nucleotide diversity *(π̂)* in *Drosophila melanogaster* and *D. virilis*.

	$\hat{h}$	k	$\hat{v}$	s_r	$\hat{\pi}$
D. melanogaster					
Estimate	0.71 ± 0.12	4	1.22 ± 0.27	3	0.008 ± 0.002
$M\equiv4Nv$	2.46	1.95	1.22	1.06	
$L\equiv4N\mu$	0.017	0.013	0.008	0.007	0.008
D. virilis					
Estimate	0.53 ± 0.09	2	0.53 ± 0.09	1	0.004 ± 0.001
$M\equiv4Nv$	1.144	0.43	0.53	0.35	
$L\equiv4N\mu$	0.0088	0.003	0.004	0.002	0.004

Table 10.4 Restriction-site differences *(vᵢⱼ)* and *Ŝᵢⱼ* values for pairs of haplotypes. The figures above the diagonal are *vᵢⱼ*'s and those below the diagonal are *Ŝᵢⱼ*'s. The upper *Ŝᵢⱼ* value for each pair of haplotypes is for *Hae*III and *Hpa*II *(r=4)*, whereas the lower *Ŝᵢⱼ* value is for *Eco*RI and *Hin*dIII *(r=6)*.

Haplotype	m	m_a	m_b	m_c	s	v	v_d
m		1	1	1	11	13	14
m_a	0.93		2	2	10	12	13
	1.00						
m_b	1.00	0.93		2	12	14	15
	0.94	0.94					
m_c	1.00	0.93	1.00		12	14	15
	0.94	0.94	0.89				
s	0.43	0.46	0.43	0.43		14	15
	0.84	0.84	0.80	0.80			
v	0.18	0.20	0.18	0.18	0.22		1
	0.71	0.71	0.67	0.67	0.59		
v_d	0.18	0.20	0.18	0.18	0.22	1.00	
	0.67	0.67	0.63	0.63	0.56	0.92	

estimate of $4Nv$. It is still smaller than the estimate obtained from the number of haplotypes. The mtDNA polymorphism in *D. melanogaster* is caused by the polymorphism at three restriction sites. That is, $s_r=3$ in this species. If we equate this number to the expectation in (10.1) with

L replaced by M, we have $3 = 2.83M$ since $n = 10$. Therefo,
another estimate of $\hat{M} = 1.06$, which is close to the estimates

Let us now relate $4Nv$ to $4N\mu$ under the assumption that
tionary changes in these species occurred by nucleotide substit\
that they are neutral. The average number of restriction sites fⱴ. *Hae*II,
*Hpa*II, *Hind*III, and *Eco*RI for the four haplotypes of *D. melanogaster* are
3.7, 4, 4.6, and 4, respectively. Therefore, we have the relationship
$v = 2\Sigma m_i r_i \mu = 2 \times [(3.7 + 4) \times 4 + (4.6 + 4) \times 6]\mu = 164.8\mu$. Similarly,
we obtain $v = 180\mu$ for *D. simulans* and $v = 100.8\mu$ for *D. virilis*. The
average of these estimates is 149μ. In this connection, it should be
noted that the $G + C$ content of mtDNA in *Drosophila* is about 0.22
even if the A-T rich region is excluded (Kaplan and Langley 1979), and
thus *Hae*III (GGCC) and *Hpa*II (CCGG) do not produce many restriction
sites. At any rate, if we use the relationship $v = 149\mu$, an estimate of
$4N\mu$ can be obtained from the $4Nv$ value. It ranges from 0.007 to
0.017, but the latter value which was obtained from $\hat{h}$ is probably an
overestimate for the reason mentioned earlier.

For estimating nucleotide diversity, π, we must first compute the
proportion of shared restriction sites for each pair of haplotypes by using
the formula $\hat{S}_{ij} = 2m_{ij}/(m_i + m_j)$, where m_i and m_j are the numbers of restriction sites for the ith and jth haplotypes, respectively, and m_{ij} is the
number of shared restriction sites (chapter 5). The estimate of π_{ij} is then
given by $\hat{\pi}_{ij} = (-\log_e \hat{S}_{ij})/r$. When two or more enzymes with the same
r value are used, $\hat{S}_{ij}$ should be computed by pooling m_i, m_j, and m_{ij} over
all enzymes. If r is not the same, they should be computed separately.
(The maximum likelihood method in chapter 5 can also be used.) In
table 10.4, $\hat{S}_{ij}$'s are given separately for *Hae*III and *Hpa*II ($r = 4$) and for
*Eco*RI and *Hind*III ($r = 6$). From these values, $\hat{\pi}$ can be computed in
the same way as $\hat{v}$ is computed. In *D. melanogaster*, it becomes 0.0080
for the enzymes with $r = 4$ and 0.0076 for the enzymes with $r = 6$, the
average being 0.008. This value is another estimate of $4N\mu$ and is close
to the estimates obtained from $\hat{v}$ and s_r.

In *D. virilis*, the same computations were done, and the results obtained are presented in table 10.3. All estimates of genetic variability in
this species are smaller than those in *D. melanogaster*, and the estimates
of $4N\mu$ obtained by different methods are again more or less the same,
except the one from nucleon diversity. This result suggests that mtDNA
is less variable in *D. virilis* than in *D. melanogaster*. However, since the
number of mtDNAs sampled and the number of restriction enzymes

used are both small, a more extensive study should be done before any definite conclusion is drawn. In this connection, it should be noted that the standard error generated by stochastic errors is much larger than the sampling error. For example, the $\hat{v}$ for *D. melanogaster* is 1.22. Therefore, the stochastic standard error under the assumption of neutral mutations is 0.96 from (10.9) with the modification mentioned earlier. This is considerably larger than the sampling standard error (0.27). To reduce the standard error relative to the estimate, it is necessary to use a large number of restriction enzymes and many independent genes.

DNA Length Polymorphism

Recent data indicate that a substantial portion of DNA polymorphism is caused by deletion and insertion (e.g., Langley et al. 1982; Bell et al. 1982; Chapman et al. 1986). DNA length polymorphism is also generated by unequal crossover (Coen et al. 1982). This polymorphism is usually found by the presence of gaps between two DNA or restriction-site sequences compared. A simple method of measuring this type of polymorphism is to compute nucleon diversity, treating each polymorphic DNA type as a haplotype. However, this quantity usually depends on the length of DNA studied, as in the case of polymorphism due to nucleotide substitution. Another measure is the average number of gap nucleotides (nucleotides in the gaps) per nucleotide site between two randomly chosen haplotypes. Let m_{gij} be the number of gap nucleotides between the ith and jth haplotypes, and m_{Tij} be the total number of nucleotides compared, including gap nucleotides. The number of gap nucleotides per nucleotide site (g_{ij}) between the two haplotypes is then given by $g_{ij} = m_{gij}/m_{Tij}$, whereas the average of g_{ij} weighted with m_{Tij} for all combinations of haplotypes is

$$g = \bar{m}_g / \bar{m}_T, \tag{10.18}$$

where $\bar{m}_g$ and $\bar{m}_T$ are the means of m_{gij} and m_{Tij}, respectively.

Note that $\bar{m}_g$ in (10.18) is equivalent to (10.5), so that the variance of $\bar{m}_g$ can be obtained by replacing π_{ij} by m_{gij} in (10.7). The approximate variance of g may be obtained by assuming that $\bar{m}_T$ is constant, since $\bar{m}_T$ is usually much larger than $\bar{m}_g$.

EXAMPLE

Langley et al. (1982) examined the restriction site polymorphism of the alcohol dehydrogenase gene region of *Drosophila melanogaster* and identified six DNA-length haplotypes (table 10.5). These haplotypes are apparently caused by deletions from the most common haplotype (N1), which is about 12 kb long. All deletions were observed in the noncoding regions of DNA. The nucleon diversity for this polymorphism becomes

Table 10.5 Deletion/insertion polymorphisms in the alcohol dehydrogenase gene region of *Drosophila melanogaster*. Data from Langley et al. (1982).

		Deletion/insertion				
Haplotype	Observed number	Δa 20 bp	Δb 550 bp	Δc 900 bp	Δd 180 bp	Δf 30 bp
N1 (Standard)	10	−	−	−	−	−
N2	2	+	−	−	−	−
N3	1	−	+	−	−	−
N4	1	−	−	+	−	−
N5	2	−	−	−	+	−
N6	2	−	−	−	−	+

Table 10.6 Estimates of nucleotide diversity (π) or the proportion of nucleotide differences between a selected pair of DNA sequences (π_{ij}).

DNA or gene region	Organism	Method	n	bp	π or π_{ij}
		Nucleotide diversity (π)			
mtDNA	Man	R	100	16,500	0.004
mtDNA	Chimpanzee	R	10	16,500	0.013
mtDNA	Gorilla	R	4	16,500	0.006
mtDNA	Peromyscus	R	19	16,500	0.004
mtDNA	Fruitfly	R	10	11,000	0.008
β-globin	Man	R	50	35,000	0.002
Growth hormone	Man[a]	R	52	50,000	0.002
Adh gene region	Fruitfly	R	18	12,000	0.006
Adh coding region	Fruitfly[b]	S	11	765	0.006
H4 gene region	Sea urchin[c]	S	5	ca 1,300	0.019
Hemagglutinin	Influ. virus	S	12	320	0.510
		Selected pair of DNA sequences (π_{ij})			
Insulin	Man	S	2	1,431	0.003
Immuno. C_κ	Rat	S	2	1,172	0.018
Immuno. IgG2a	Mouse	S	2	1,114	0.100

n: sample size.   *bp:* base pairs.   *R:* restriction enzyme technique.   *S:* DNA sequencing.
Source: [a]Chakravarti et al. (1984); [b]Kreitman (1983); [c]Yager et al. (1984); see Nei (1983) for others.

0.682 ± 0.111, which is a little lower than the nucleon diversity (0.853) obtained by Langley et al. (1982) for non-deletion restriction site polymorphism for the same gene region. To estimate g, we must first compute g_{ij}'s for all pairs of haplotypes. For example, g_{12} is given by $m_{g12}/m_{T12} = 20/12000 = 0.002$. In the present case, all deletions/insertions are nonoverlapping, so that $m_T = 12,000$ for all g_{ij}'s. Once g_{ij}'s are computed, g can be obtained by (10.18). It becomes $g = 0.0175 \pm 0.009$. This is about three times the nucleotide diversity (0.006) for the same DNA region (see table 10.6).

Some Observations on DNA Polymorphism

Nucleotide Diversity

In the preceding sections, we considered several ways of measuring DNA polymorphism. If we disregard the effect of deletions and insertions, the most fundamental measure is nucleotide diversity. This measure does not depend on the length of DNA and sample size so that it can be used for comparing the extents of polymorphism of different genes or populations.

In the last few years, nucleotide diversity has been studied for a substantial number of genes. Some examples are presented in table 10.6. Most of these estimates were obtained by an indirect method, i.e., the restriction enzyme technique. This table also includes examples of the proportion of nucleotide differences (π_{ij}) between a selected pair of alleles (DNA sequences). Nucleotide diversity varies from 0.002 to 0.019 in eukaryotic organisms and is nearly the same for both mitochondrial DNA and nuclear genes. If we note that a gene usually consists of several thousand nucleotide pairs, this result suggests that the nucleotide sequences of two genes randomly chosen from a population are rarely identical. Mitochondrial DNAs (mtDNAs) are maternally inherited and exist in the haploid form. Therefore, the effective population size for mtDNAs is expected to be about one-fourth that of nuclear genes. However, the mutation rate for mtDNA is apparently considerably higher than that for nuclear genes (chapter 3). These two compensating factors probably make π for mtDNA nearly equal to that of nuclear genes. In some cases, however, nucleotide diversity seems to be very low. For example, mtDNA in American Indians from Venezuela is virtually monomorphic (Johnson et al. 1983). Apparently, this population went through

a bottleneck relatively recently. Unlike eukaryotic genes, the hemagglutinin gene of the influenza A virus shows an extremely high nucleotide
diversity. As will be discussed later, this high nucleotide diversity is
apparently due to an unusually high mutation rate in this organism.

Although π_{ij} for a selected pair of alleles would not be a good estimate of nucleotide diversity, data in table 10.6 indicate that the nucleotide diversity estimated by the restriction enzyme technique is in rough
agreement with that obtained by the sequencing method. There is, however, one exception. Namely, the π_{ij} for the immunoglobulin $\gamma 2a$ heavy-
chain constant region gene in the mouse (see chapter 6) is one order of
magnitude higher than the usual π value for eukaryotic genes. This high
value of π_{ij} has been suspected to be due to a gene conversion that might
have occurred between this gene and its neighboring gene (Schreier et
al. 1981).

If the population is in equilibrium, the expected value of π is a function of $N\mu$. Therefore, it is possible to study the relationship between
the observed value of π and $N\mu$ as in the case of average heterozygosity.
However, there are not enough data to conduct such a statistical study
at the present time.

Silent Polymorphism

From the standpoint of the neutral theory, it is interesting to examine
the extent of DNA polymorphism that is not expressed at the amino
acid level. Under the neutral theory, this silent polymorphism is expected to be high compared with the polymorphism expressed at the
amino acid level, because silent mutations are subject to purifying selection less often than nonsilent mutations (chapter 5). On the other hand,
if polymorphism is actively maintained by natural selection and the effect of genetic drift is unimportant, one would expect that silent polymorphism is lower than nonsilent polymorphism. One way of testing
this hypothesis is to examine the polymorphism at the first, second, and
third positions of codons in the coding regions. All nucleotide changes
at the second position of nuclear codons lead to amino acid replacement,
whereas at the third position only about 28 percent of changes are expected to affect amino acids because of degeneracy of the genetic code.
At the first position, about 95 percent of nucleotide substitutions result
in amino acid changes. Therefore, if the neutral theory is valid, the
extent of DNA polymorphism is expected to be highest at the third

position and lowest at the second position. Available data indicate that this is indeed the case except for some immunoglobulin genes. For example, the F and S alleles at the alcohol dehydrogenase locus in *Drosophila melanogaster* are electrophoretically distinguishable because of the amino acid difference (threonine vs. lysine) at the 192nd residue (codon change from ACG to AAG). The nucleotide sequences of about a dozen alleles at this locus indicate that there is no other amino acid substitution in the enzymes but that there are many third position substitutions that are silent (Benyajati et al. 1981; Kreitman 1983). Similar examples of a high proportion of silent substitutions are found in the tryptophan operon genes in *Escherichia coli* (Milkman and Crawford 1983), the β globin gene complex in man (Poncz et al. 1983), and the histone 4 gene in the sea urchin *Strongylocentrotus purpuratus* (Yager et al. 1984).

However, this is not necessarily true in immunoglobulin genes. Sheppard and Gutman (1981) showed that two alleles (*LEW* and *DA*) at the rat κ light-chain constant region gene show twelve nucleotide differences in the coding region. Eleven of these twelve nucleotide differences (92%) have resulted in amino acid differences. This is considerably higher than the expected value of 74 percent under random nucleotide substitution, though the difference is not statistically significant. In the case of the constant region of the mouse immunoglobulin γ heavy-chain gene (γ2a), two alleles, *IgG2a^a* and *IgG2a^b*, show a total of 111 nucleotide differences plus 15 additional differences due to insertion and deletion when 1,114 nucleotides are examined (Schreier et al. 1981). Of the 111 nucleotide differences, 18 (16%) were silent and the rest were amino acid altering substitutions. These two observations suggest that in immunoglobulin genes amino acid altering nucleotide substitutions might be favored by selection because variability is needed in immunoglobulins. Similar selection seems to be involved in the mouse MHC loci, where extremely high variability is observed (Klein and Figueroa 1986).

Nucleotide Substitutions with Adaptive Significance

While most silent substitutions seem to be neutral or nearly neutral, amino acid altering substitutions may have drastic phenotypic effects. A well known example is the substitution of codon GTG for GAG at the 6th codon of the β globin gene in man. This substitution results in the replacement of glutamic acid by valine, which in turn causes sickle cell anemia in homozygous condition. However, the heterozygotes for this mutant gene are resistant to malaria, so that the frequency of the mutant gene has increased substantially in malarial areas.

Another example of adaptive change by a single nucleotide substitution is that of a chloroplast gene in plants. Many commercially important herbicides kill plants by inhibiting photosynthesis. This inhibition of photosynthesis occurs because herbicides bind to a thylakoid-membrane protein encoded by a chloroplast gene, *psbA*. Several plant species have mutant strains that are resistant to herbicides. Hirschberg and McIntosh (1983) sequenced the *psbA* genes from a normal and a mutant strain of the pigweed *Amaranthus hybridus* and found that there is only one codon difference between the two genes; the normal gene has a serine codon (AGT) at the 264th codon, whereas the mutant gene has a glycine codon (GGT). Clearly, the A → G mutation has occurred. Interestingly, the serine codon AGT is evolutionarily conserved and is shared by at least three quite different species, the pigweed, spinach, and tobacco. Furthermore, Erickson et al. (1984) and Golden and Haselkorn (1985) showed that the serine residue at the same position is shared even by chlamydomonas and cyanobacteria. They also showed that the herbicide-resistant genes in these organisms are caused by a single amino acid substitution [TCT (Ser) → GCT (Ala) in chlamydomonas and TCG (Ser) → GCG (Ala) in cyanobacteria]. These findings suggest that the serine at the 264th residue is responsible for herbicide binding and that the replacement of this serine by some other amino acid prevents herbicides from binding to the chloroplast thylakoid membranes.

The above examples indicate that a single nucleotide (or amino acid) substitution may have a drastic effect on gene function and consequently on the fitness of an individual. It is expected that the number of such examples will increase as more polymorphic alleles are studied at the DNA level. However, it should be noted that there are many examples of nucleotide or amino acid substitutions that are of little consequence to the survival of an individual (chapters 8 and 14).

Influenza Viruses and Other RNA Viruses

Influenza remains an uncontrolled disease in man, mainly because of antigenic variation of the two surface proteins, hemagglutinin and neuraminidase. From year to year, the influenza virus becomes progressively more resistant to antibodies made against older viruses. This resistance is caused by the accumulation of amino acid changes in the surface proteins (antigenic drift). Every so often, however, a new type of virus with a major change of antigenicity appears and causes a pandemic (antigenic shift). There are 13 subtypes of hemagglutinin identified for the influenza A virus. This virus is an RNA virus, but cDNA can be produced in the laboratory. Air (1981) sequenced the cDNA from a portion of the hemagglutinin gene (a total of about 320 nucleotides for the signal peptide and the HA1 polypeptide) for 12 subtypes. Some parts of her results are presented in figure 10.2. It is clear from this figure that the subtypes are quite different from each other at both nucleotide and amino acid levels. Particularly in the signal peptide region, sequence variation is extremely high. However, there are a number of amino acid sites at which the same amino acid is present for all subtypes (two of them are shown in figure 10.2). This is apparently due to the functional constraint of the protein. We note that although amino acids remain the same at these sites, silent substitutions occur with a high frequency.

Table 10.7 gives estimates of nucleotide diversities among the 12 subtypes for the three nucleotide positions of codons in the hemagglutinin gene. The nucleotide diversity (π) in the signal peptide region is extremely high. It is close to the maximum expected value of 0.75 at the third position. The π value is considerably lower at the second position than at the third position, apparently reflecting the functional constraint of the protein. The π value for the HA1 polypeptide region of the hemagglutinin gene is smaller than that for the signal peptide at

```
H1   ATG AAG GCA AAC CTA CTG GTC CTG TTA TGT GCA CTT GCA GCT GCA GAT GCA GAC ACA ATA|TGT|ATA|GGC|TAC CAT GCG AAC
     Met Lys Ala Asn Leu Leu Val Leu Leu Cys Ala Leu Ala Ala Ala Asp Ala Asp Thr Ile|Cys|Ile|Gly|Tyr His Ala Asn
                                                                        ↑
H2           ATG GCC ATC ATT TAT CTC ATT CTC CTG TTC ACA GCA GTG AGA GGG GAC CAG ATA|TGC|ATT|GGA|TAC CAT GCC AAT
             Met Ala Ile Ile Tyr Leu Ile Leu Leu Phe Thr Ala Val Arg Gly Asp Gln Ile|Cys|Ile|Gly|Tyr His Ala Asn
                                                                        ↑
H5       ATG GAG AGA GTA GTG CTT CTT CTT GCA ATG ATC AGT CTT GTC AAA AGT GAC CAG ATT|TGC|ATT|GGT|TAC CAT GCA AAC
         Met Glu Arg Val Val Leu Leu Leu Ala Met Ile Ser Leu Val Lys Ser Asp Gln Ile|Cys|Ile|Gly|Tyr His Ala Asn
                                                                        ↑
H11      ATG AAG AAA GTA CTG CTT TTT GCA GCA ATC ATC ATC TGT ATT CGA GCA GAC GAA ATC|TGC|ATT|GGA|TAC CTG AGC AAC
         Met Lys Lys Val Leu Leu Phe Ala Ala Ile Ile Ile Cys Ile Arg Ala Asp Glu Ile|Cys|Ile|Gly|Tyr Leu Ser Asn
                                                                        ↑
H6       ATG ATT GCA ATC ATT GTA GTA GCG ATA CTG GCA ACA GCC GGA AGG TCT GAC AAG ATC|TGC|ATT|GGA|TAT CAT GCC AAC
         Met Ile Ala Ile Ile Val Val Ala Ile Leu Ala Thr Ala Gly Arg Ser Asp Lys Ile|Cys|Ile|Gly|Tyr His Ala Asn
                                                                        ↑
H8   ATG GAG AAA TTC ATC GCA ATA GCA ATG CTC TTG GCG AGC ACA AAT GCA TAC GAT AGG ATA|TGC|ATT|GGG|TAC CAA TCA AAC
     Met Glu Lys Phe Ile Ala Ile Ala Met Leu Leu Ala Ser Thr Asn Ala Tyr Asp Arg Ile|Cys|Ile|Gly|Tyr Gln Ser Asn
                                                                        ↑
```

Figure 10.2. Partial nucleotide and amino acid sequences for the hemagglutinin gene in six subtypes (H1, H2, H5, H11, H6, H8) of the influenza A virus. The hemagglutinin gene codes for the signal peptide and the HA1 and HA2 polypeptides. This figure shows the sequences for the signal peptide and the first ten amino acids of HA1. The arrow sign indicates the end of the signal peptide. The amino acids in boxes are identical for all 12 subtypes examined. After Air (1981).

Table 10.7 Estimates of nucleotide diversities (π) for the three nucleotide positions of codons for the signal peptide and the HA1 polypeptide of hemagglutinin for the influenza A virus. These results were obtained from Air's (1981) sequence data (12 subtypes). Only the codons shared by all subtypes were used.

Position in codon	Signal peptide (14 codons)	HA1 (81 codons)	Total (95 codons)
First	0.705	0.440	0.480
Second	0.588	0.351	0.386
Third	0.732	0.651	0.663
Total	0.675	0.481	0.510

all three nucleotide positions, but it is still much higher than the value for most eukaryotic genes.

The high values of π are due to an extremely high rate of mutation in RNA viruses (Holland et al. 1982). In the influenza A virus, the mutation rate can be estimated by using virus strains that have been kept in refrigerators. For example, the Asian flu (subtype H2) was first isolated in 1957, and its later variant strains were isolated in 1967, 1968, 1972, 1977, etc. Therefore, comparing the nucleotide sequences of these strains, one can determine the mutation rate. Using this method, Nei (1983), Hayashida et al. (1985), and Saitou and Nei (1986a) have estimated that the mutation rate for the hemagglutinin gene is about 0.01 per nucleotide site per year (see figure 5.2). However, the rate of nucleotide substitution is one order of magnitude lower in many influenza genes apparently because of purifying selection.

The extraordinarily high degree of genetic polymorphism is not confined to the influenza A virus but also occurs in all kinds of RNA viruses, including retroviruses discussed in chapter 6 (Holland et al. 1982; Rabson and Martin 1985).

Linkage Disequilibria Among Restriction Sites

In the study of DNA polymorphism, the restriction enzyme method is used more frequently than DNA sequencing because of its simplicity. This technique has already uncovered many restriction-site polymorphisms for various genes. One prominent feature of restriction-site polymorphism is that there is strong nonrandom association of polymorphic restriction sites among themselves or with neighboring genetic loci (Kan and Dozy 1978; Kazazian et al. 1983; Aquadro et al. 1986). This nonrandom association or linkage disequilibrium occurs mainly because restriction sites and their neighboring genetic loci are tightly linked (see chapter 12).

Although restriction sites are generally in linkage disequilibrium, there are a few exceptions in which two closely linked restriction sites show little nonrandom association. At the top of figure 10.3, a linkage map of polymorphic restriction sites is shown for the β globin gene complex in man. In the region of the ϵ to the $\psi\beta$ genes there are five polymorphic restriction sites. Therefore, $2^5 = 32$ different haplotypes may be generated if all sites are randomly combined. In the Mediterranean population, however, only three common types (frameworks 1, 2, 3) have been observed. The 3' region starting from the β gene also includes five polymorphic restriction sites, but there are only three haplotypes (frameworks A, B, C) in this region. This is a clear case of strong nonrandom association of restriction sites. However, frameworks 1, 2, 3 and frameworks A, B, and C are more or less randomly combined. Thus, all nine possible types for the two groups of frameworks are observed with appreciable frequencies (figure 10.3). This result suggests that there is a "hot spot" of recombination between the two groups of frameworks. Kazazian et al. (1983) speculate that the hot spot exists near the restriction site ($Hinf$I) between the δ and β genes. Interestingly, the 5' side of the β globin gene is known to have a χ sequence (5'GCTGGTGG3'), which is presumably a promoter of recombination (Chakravarti et al. 1984).

It should be noted, however, that although there seem to be recombination hot spots in eukaryotic genomes, the recombination frequency at a particular hot spot is apparently quite low. Using information on the linkage disequilibria within and between the two linkage blocks in the β globin gene complex, Chakravarti et al. (1984) have estimated that the recombination value between the two blocks is $0.0003 \sim 0.003$ (however, see Weir and Hill 1986). It should be noted that the recombination value (r) does not have to be very high to generate linkage equilibrium. Even if r is as low as 0.001, the linkage disequilibrium would become very close to 0 in several thousand generations if the effective population size (N) is large and random mating occurs. However, note that if $4Nr$ is smaller than 1, the story is different. In this

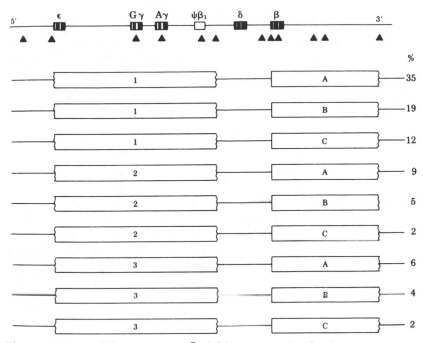

Figure 10.3.   Nine common β globin gene region haplotypes
observed in Mediterranean populations of man. Two nonrandomly
associated sequence clusters are shown; the 5' cluster includes the ϵ,
$^G\gamma$, $^A\gamma$, and $\psi\beta1$ genes, and the 3' cluster includes the β gene and at
least 18 kb downstream to it. The three common sequence types in the
5' cluster are designated 1, 2, and 3, and the three common sequence
types in the 3' cluster are called A, B, and C. The percentage of each
haplotype among normal Mediterranean chromosomes is shown on the
right. A recombination "hot spot" is tentatively located around the
$Hinf$I site between the δ and β genes. The triangle signs indicate the
polymorphic restriction sites. From Kazazian et al. (1983).

case, linkage disequilibrium may be generated by genetic drift even if it
is initially 0 (Ohta and Kimura 1971b). Kreitman's (1983) DNA sequence data for the *Adh* locus of *D. melanogaster* suggest that intragenic
recombination or gene conversion occurs even outside the recombination
hot spots. However, these events occur with an extremely low frequency
so that one can identify the events by comparing extant DNA sequences.
Hudson and Kaplan (1985) and Stephens (1985) have devised statistical
methods to detect these events.

DNA Divergence Between Populations

In chapter 5, we discussed methods of estimating the number of nucleotide substitutions between two DNA sequences obtained from different species, ignoring the effect of polymorphism. To estimate the extent of DNA divergence between two closely related species, however, we must consider the effect of polymorphism. This can be done in a way similar to the estimation of genetic distance.

Nucleotide Differences Between Populations

Suppose that there are m different haplotypes for a particular DNA region (nucleon) in populations X and Y, and let $\hat{x}_i$ and $\hat{y}_i$ be the sample frequencies of the ith haplotype for populations X and Y, respectively. The average number of nucleotide substitutions for a randomly chosen pair of haplotypes (d_X) in population X can be estimated by

$$\hat{d}_X = \frac{n_X}{n_X - 1} \Sigma_{ij} \hat{x}_i \hat{x}_j d_{ij}, \tag{10.19}$$

where n_X is the number of sequences sampled and d_{ij} is the number of nucleotide substitutions per site between the ith and jth haplotypes. d_{ij} may be estimated by one of the methods discussed in chapter 5 [e.g., equation (5.3) or (5.42)]. When all DNA sequences are different, $\hat{x}_i = 1/n_X$. The average number of nucleotide substitutions (d_Y) for Y can be estimated in the same way.

On the other hand, the average number (d_{XY}) of nucleotide substitutions between DNA haplotypes (nucleomorphs) from X and Y can be estimated by

$$\hat{d}_{XY} = \Sigma_{ij} \hat{x}_i \hat{y}_j d_{ij}, \tag{10.20}$$

where d_{ij} is the nucleotide substitutions between the ith haplotype from X and the jth haplotype from Y. The number of net nucleotide substitutions between the two populations (d_A) is then estimated by

$$\hat{d}_A = \hat{d}_{XY} - (\hat{d}_X + \hat{d}_Y)/2. \tag{10.21}$$

If the rate of nucleotide substitution is constant and is λ per site per year and the time since divergence between the two populations is T, then the expected value (d_A) of $\hat{d}_A$ is

$$d_A = 2\lambda T. \tag{10.22}$$

The rationale of equations (10.21) and (10.22) is that in the presence of polymorphic sequences the average time since divergence between the alleles (nucleomorphs) from X and the alleles from Y is longer than the time of population splitting, as seen from figure 10.4. Therefore, to estimate the population splitting time (T), we must subtract the average nucleotide difference for polymorphic alleles at the time of population splitting from d_{XY}. This average nucleotide difference is estimated by $(\hat{d}_X + \hat{d}_Y)/2$ under the assumption that the expected value of $\hat{d}_X$ or $\hat{d}_Y$ is the same for the entire evolutionary process (chapter 13). The sampling variance of $\hat{d}_A$ generated at the time of nucleon sampling is given by

$$V(\hat{d}_A) = V(\hat{d}_{XY}) + \frac{1}{4}\,[V(\hat{d}_X) + V(\hat{d}_Y)]$$

$$-\frac{1}{2}\,[Cov(\hat{d}_{XY},\ \hat{d}_X) + Cov(\hat{d}_{XY},\ \hat{d}_Y)], \tag{10.23}$$

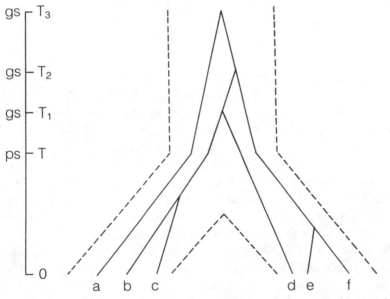

Figure 10.4.   Diagram showing that the time of gene splitting (gs) is usually earlier than the time of population splitting (ps) when polymorphism exists.

where

$$V(\hat{d}_{XY}) = \frac{1}{n_X n_Y}[(1 - n_X - n_Y)(\Sigma\ x_i y_i d_{ij})^2$$

$$+ (n_X - 1)(\Sigma\ x_i^2 y_j d_{ij}^2 + \Sigma\ y_i x_j x_k d_{ij} d_{ik})$$

$$+ (n_Y - 1)(\Sigma\ x_i y_j^2 d_{ij}^2 + \Sigma\ x_i y_j y_k d_{ij} d_{ik})$$

$$+ \Sigma\ x_i y_j d_{ij}^2], \tag{10.24}$$

$$Cov(\hat{d}_{XY}, \hat{d}_X) = 2\left[\sum_{i \neq j} \sum_k x_i x_j y_k d_{ij} d_{ik} \right.$$

$$\left. - (\Sigma\ x_i x_j d_{ij})(\Sigma\ x_i y_j d_{ij}) \right]/n_X, \tag{10.25}$$

$$Cov(\hat{d}_{XY}, \hat{d}_Y) = 2\left[\sum_{i \neq j} \sum_k y_i y_j x_k d_{ij} d_{ik} \right.$$

$$\left. - (\Sigma\ x_i y_j d_{ij})(\Sigma\ y_i y_j d_{ij}) \right]/n_Y, \tag{10.26}$$

and $V(\hat{d}_X)$ and $V(\hat{d}_Y)$ are given by (10.7) with π_{ij} replaced by d_{ij}. Here, x_i and y_i are population frequencies, but they are replaced by sample frequencies when we estimate $V(\hat{d}_A)$. The variance in (10.23) does not include stochastic variance, and if we include this component, the total variance is quite large (chapter 13).

EXAMPLE

Bodmer and Ashburner (1984) and Cohn et al. (1984) sequenced the alcohol dehydrogenase gene for two sibling species of *D. melanogaster*, i.e., *D. simulans* and *D. mauritiana*. The sequences published by the two groups are slightly different from each other, and this difference is apparently due to polymorphism within species. Stephens and Nei (1985) studied the interspecific and intraspecific differences of these sequences as well as those for *D. melanogaster* (Kreitman 1983), considering the shared region of 822 nucleotides. The results obtained are presented in table 10.8. The d_A values suggest that *simulans* and *mauritiana* are more closely related to each other than to *melanogaster* and that speciation occurred first between *melanogaster* and the *simulans-mauritiana* group and then between *simulans* and *mauritiana*.

If we assume a constant rate of nucleotide substitution, we can estimate the time of divergence between these species. The average rate of nucleotide substitution for coding regions of genes seems to be about

Table 10.8 Estimates of interpopulational (d_{XY}),
intrapopulational (d_X or d_Y), and net (d_A) nucleotide differences
among three species of *Drosophila* for the alcohol dehydrogenase
gene (822 nucleotides examined). The figures on the diagonal
refer to d_X (or d_Y), and those below the diagonal d_{XY}. The figures
above the diagonal represent the values of $d_A = d_{XY} - (d_X + d_Y)/2$.
All values should be divided by 100. From Stephens and Nei
(1985).

	D. melanogaster	D. simulans	D. mauritiana
D. melanogaster	0.70 ± 0.41	1.73 ± 0.28	2.36 ± 0.24
D. simulans	2.45 ± 0.49	0.73 ± 0.30	0.86 ± 0.22
D. mauritiana	2.96 ± 0.54	1.47 ± 0.38	0.49 ± 0.24

2×10^{-9} per site per year $(0.71 \times 0.88 \times 10^{-9} + 0.29 \times 4.65 \times 10^{-9} \approx 2 \times 10^{-9})$ (see chapter 3 and table 3.6). If we use this value, the estimate of the time of divergence between *simulans* and *mauritiana* becomes $T = 0.0086/(2 \times 2 \times 10^{-9}) = 2.1 \times 10^6$ years from equation (10.22). This divergence time is only slightly larger than the divergence time (1.8 MY) of two polymorphic alleles ($d_X = 0.0073$) in *simulans*. On the other hand, the divergence times between *melanogaster* and *simulans* and between *melanogaster* and *mauritiana* are estimated to be 4.4×10^6 and 5.8×10^6 years, respectively, the average being 5.1 MY. Therefore, the divergence between *melanogaster* and the *simulans-mauritiana* group seems to have occurred about three times earlier than the divergence between *simulans* and *mauritiana*.

Restriction Site Differences

Equation (10.21) was presented in the context of DNA sequence data. However, it can also be used for estimating the number of net restriction-site differences per nucleon (ν_A) between two populations. The only thing necessary for this case is to replace d_{ij} by ν_{ij} in equations (10.19) and (10.20). If the two populations studied are in mutation-drift equilibrium, the expectations of ν_X and ν_Y (corresponding to d_X and d_Y, respectively) are $M = 4N\nu$, as mentioned earlier. On the other hand, the expectation of ν_{XY} is

$$E(\nu_{XY}) = M + (2m_T a - M)(1 - e^{-2r\mu t}),$$ (10.27)

which becomes

$$E(\nu_{XY}) = M + 2\nu t,$$ (10.31)

approximately if $2r\mu t \ll 1$, since $\nu = 2m_T a r\mu$ (Nei and Tajima 1981). Here, m_T is the total number of nucleotides in the nucleon and a is given by (5.31). Therefore, $\nu_A = \nu_{XY} - (\nu_X + \nu_Y)/2$ is expected to increase linearly with time as long as $2r\mu t \ll 1$.

Genealogical Relationships of Genes Within and Between Populations

One advantage of DNA sequencing or restriction-site mapping over electrophoretic study of protein polymorphism is that the phylogenetic relationship of different genes (nucleomorphs) can be studied. Phylogenetic analysis of genes often gives detailed information of the evolutionary history of the genes studied.

Early in this chapter we estimated the number of nucleotide substitutions for all pairs of 11 alleles at the *Adh* locus in *D. melanogaster*. Using this set of data and additional information, Stephens and Nei (1985) conducted a phylogenetic analysis of these alleles (see also Kreitman 1983). Their result indicates that the *S* alleles are on the average much older than the *F* alleles and that the latter were derived from the former about one million years ago. This indicates that the *S–F* polymorphism is very old. To explain this ancient polymorphism, however, balancing selection is not required because if the effective size of this species is of the order of two million, it can be accommodated with the neutral theory (see chapter 13).

When a phylogenetic tree is constructed for genes sampled from several closely related populations, the genes from different populations are genealogically mixed. Figure 10.5 shows one such example in which the phylogenetic tree of 10 mtDNAs from each of three major races of man, Caucasoid, Negroid, and Mongoloid, is given. MtDNAs from the three major races do not cluster well. A similar result was obtained by Cann et al. (1982) and Cann (1982). Cann (1982) interpreted this observation as being an indication of migration that occurred among the major races.

However, genealogical mixing of mtDNAs from different populations

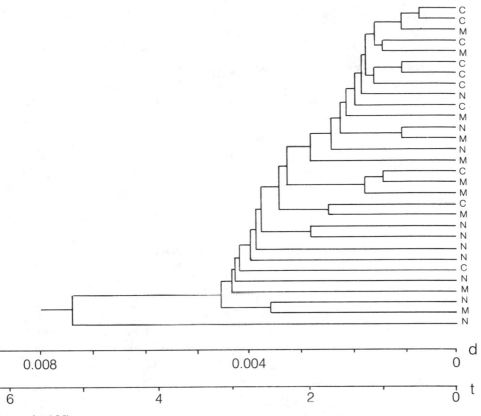

Figure 10.5.   Phylogenetic tree of mtDNAs from 10 individuals from each of Caucasoid (C), Negroid (N), and Mongoloid (M). From Nei (1985).

is expected to occur even without migration if the ancestral population was polymorphic and the time since divergence between the populations is relatively short. This is because many of the polymorphic alleles in the current populations are expected to have diverged before population splitting (see figure 10.4). It should also be noted that the time of gene splitting is usually much earlier than the time of population splitting and thus the former cannot be used for estimating the latter, as discussed earlier. Figure 10.5 shows that the oldest Mongoloid mtDNA diverged from an old Negroid mtDNA about 300,000 years ago, but this time of DNA divergence is almost certainly earlier than the time of racial splitting, since many other Caucasoid and Mongoloid genes diverged from Negroid genes at later times.

Theoretically, one can study the evolutionary relationship of the three major races by using the d values given by equation (10.21). However, when the populations studied are closely related, the variance of d is so large that the relationship obtained is not very reliable (chapter 13). Furthermore, in the presence of bottleneck effects, the d value in equation (10.21) would not be proportional to evolutionary time, since $(\hat{d}_X + \hat{d}_Y)/2$ will change with time. In these cases, $\hat{d}_{XY}$ seems to be more useful than $\hat{d}$ for finding the order of population splitting, since $\hat{d}_{XY}$ has a smaller variance than $\hat{d}$ and is not influenced by bottleneck effects (Nei 1985).

Differentiation of Haplotype Frequencies

When the populations studied are closely related, the extent of genetic differentiation can also be studied by examining the frequencies of nucleomorphs or haplotypes. Wainscoat et al. (1986) studied the frequencies of 14 polymorphic haplotypes for the ϵ-γ-$\psi\beta$ region of the β-globin gene complex in eight different human populations (see figure 10.3). The results obtained are presented in table 10.9. In European and Oriental populations, haplotypes $+ - - - -$ and $- + - + +$ are predominant, whereas in African populations haplotypes $- + - - +$ and $- - - - +$ show high frequencies. Furthermore, Asians and Pacific Islanders have several haplotypes which are not possessed by either Europeans or Africans. From these results, Wainscoat et al. concluded that Europeans and Orientals are genetically closer to each other than to Africans. This conclusion is in agreement with that obtained from gene frequency data for protein loci (Nei and Roychoudhury 1974b, 1982). However, since the genetic relationships derived from gene or haplotype frequencies are subject to a large sampling error, the above conclusion must be regarded as tentative.

Table 10.9 shows that the Melanesian population has two chromosomes of the African haplotype $- - - - +$. Wainscoat et al. speculate that these chromosomes were generated by a recombination event in the past because they are different from the African haplotype at another polymorphic restriction site near the $^G\gamma$ gene. Note also that none of the three most common haplotypes in Europeans and Asians can be derived from any of the other two by a single mutation. It seems

Table 10.9   Frequencies of the 14 restriction-site haplotypes for the ϵ-γ-$\psi\beta$ gene region in 8 different human populations. Wainscoat et al. (1986).

Haplotypes[a]	British	Cypriot	Italian	Indian	Thai	Mela-nesian	Poly-nesian	African
+ – – – –	16	59	32	58	29	117	43	3
– + – + +	15	14	15	28	0	29	6	6
– + + – +	5	8	2	15	2	3	0	1
+ – – + +	0	0	0	0	0	11	0	0
– + + – –	1	1	1	3	0	0	0	0
– + + + +	0	0	0	0	0	5	1	0
+ + – + +	0	0	0	3	0	0	0	0
– – – – –	0	0	0	1	0	?	1	0
+ + + – +	0	0	0	1	0	0	0	0
– – – + +	0	0	0	1	1	0	0	0
+ + – – –	0	0	0	1	0	0	0	0
– + – – –	0	0	0	0	0	0	0	2
– + – – +	0	0	0	0	0	4	4	12
– – – – +	0	0	0	0	0	2	0	37
Total	37	82	50	111	32	173	55	61

[a] + and – represent the presence and absence of a restriction site, respectively.

that several recombination events or mutations occurred to generate these haplotypes.

Avise et al. (1979) studied the differentiation of mtDNA among geographic populations of the pocket gopher, *Geomys pinetis*, in the southeastern United States. They found that this species can be divided into two distinct groups of individuals in terms of mtDNA haplotypes. One group inhabits Florida and southeastern Georgia and the other the region west of the Apalachicola River, including Alabama and northwestern Georgia. The average number of nucleotide differences per site between the mtDNAs from the two regions was estimated to be 0.034. Interestingly, similar geographic subdivision exists in several freshwater fish species, such as *Amia calva*, *Lepomis punctatus*, *L. gulosus*, and *L. microlophus* (Bermingham and Avise 1986). It seems that this congruent pattern of geographic differentiation of mtDNA in several different vertebrate species was produced by repeated episodes of sea level changes that isolated Florida from the continent periodically during the last 20 MY (Riggs 1984; Bermingham and Avise 1986).

Another interesting genetic "break" that distinguishes between two geographical populations is observed in the horseshoe crab *Limulus poly-phemus*. This species inhabits the Atlantic coast of North America and the Gulf of Mexico coast. Figure 10.6 shows the distributions of poly-morphic haplotypes for four restriction enzymes in this region. All four restriction enzymes show essentially the same pattern of haplotype dis-tribution, though *Bcl*I and *Hinc*II reveal three haplotypes rather than

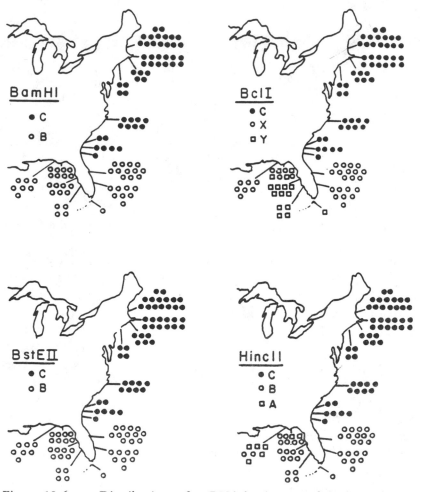

Figure 10.6.   Distributions of mtDNA haplotypes of the horseshoe crab in the coast of the eastern United States. From Saunders et al. (1986).

two; the haplotype existing in the Atlantic coast north of Florida is different from that existing in the coast of Florida and the Gulf of Mexico. Essentially the same pattern of haplotype distribution was observed for three other restriction enzymes (Saunders et al. 1986). This result is consistent with that obtained from an electrophoretic study of protein polymorphism. Selander et al. (1970) examined the gene frequencies for 8 polymorphic protein loci in two populations of the Atlantic coast north of Florida and two populations of the Gulf of Mexico coast. Their data indicate that the Atlantic and Gulf of Mexico coast populations are genetically differentiated, though the extent of differentiation is much less than that for mtDNA. This area of genetic divergence corresponds to a long-recognized zoogeographic boundary between warm-temperate and tropical marine faunas. It is, therefore, possible that the genetic differentiation between the two geographical populations is caused by gene flow barriers or selection associated with water mass differences (Saunders et al. 1986).

Introgression of Mitochondrial DNA

Ordinarily different species are reproductively isolated, and there is no gene flow between them. Recently, however, a number of authors have reported that mtDNA can cross species boundaries.

In southern Denmark, there is a hybrid zone between two species of mice, *Mus musculus* and *M. domesticus* (formerly two subspecies), northern Denmark being inhabited by *M. musculus*. Examining the mtDNAs of the two species, Ferris et al. (1983) showed that, unlike *musculus* mice from eastern Europe, *musculus* mice from north of the hybrid zone have the *domesticus* mtDNA, which shows a 5 percent sequence difference from the *musculus* mtDNA. This *domesticus* mtDNA is confined to northern Denmark and some parts of Sweden. Therefore, the introgression of mtDNA from *domesticus* to *musculus* seems to have occurred relatively recently. Interestingly, nuclear genes show no evidence of introgression. Possibly, introgression of nuclear genes is prevented by hybrid sterility or inviability which is caused by nuclear genes themselves, whereas mtDNA can cross species boundaries because it is inherited independently of nuclear genes. [See Takahata (1985) for a mathematical model for this problem.]

Spolsky and Uzzell (1984) reported a similar introgression of mtDNA from *Rana lessonae* to *R. ridibunda* in frogs. *R. ridibunda* has two types of mtDNA, and one of them is very close to that of the related species *R. lessonae,* suggesting introgression of the *R. lessonae* mtDNA into *R. ridibunda.* In the present case, introgression of mtDNA seems to have been facilitated by a special reproductive system of hybrids (hybridogenesis) between the two species. Nuclear genes again show no tendency of introgression. A case of introgression of mtDNA was also reported between *Drosophila pseudoobscura* and *D. persimilis* (Powell 1983).

At the present time, it is not clear how often introgression of mtDNA occurs. In general, it seems to be a very rare event. However, once it occurs, the phylogenetic relationship of mtDNAs will be quite different from the phylogenetic tree of the species involved. Therefore, caution should be exercised when one wants to infer the evolutionary relationship of different species from mtDNA data.

☐ CHAPTER ELEVEN ☐
PHYLOGENETIC TREES

One of the most important achievements in the study of molecular evolution is the discovery of the approximate constancy of the rate of amino
acid or nucleotide substitution. This discovery has provided evolutionists
with a new tool of constructing phylogenetic trees. As discussed in chapter 4, the constancy of the rate of amino acid or nucleotide substitution
holds only approximately, but compared with morphological or physiological characters, molecular data show a much more regular pattern of
evolutionary change. It is therefore expected that they give a clearer
picture of evolutionary relationship of organisms than morphological
characters. While it is very difficult to give an evolutionary time scale
for a morphological tree, it is routinely done for a molecular tree.

In the past two or three decades, many statistical methods have been
developed for constructing a quantitative tree. A comprehensive treatment of this subject is given in Sneath and Sokal's (1973) and Clifford
and Stephenson's (1975) books. The tree-making methods discussed in
these books are primarily concerned with morphological characters, but
many of them are applicable to molecular data as well. However, molecular data often pose new problems of their own, so that some special
consideration is necessary. For example, as we have already seen, stochastic factors are very important in molecular evolution whether or not
there is selection, and the effect of these factors must be taken into
account in the reconstruction of phylogenetic trees. In classical numerical taxonomy, little consideration has been given to this effect.

At the present time, a great deal of controversy is going on among
the workers of numerical taxonomy, particularly between pheneticists
and cladists (see Sneath and Sokal 1973; Eldredge and Cracraft 1980).
A large part of this controversy is concerned with the philosophy and
methodology of classification of organisms. Since our primary interest is
the construction of phylogenetic trees rather than taxonomy, this controversy is beyond the scope of this book. In the following, we shall discuss
various tree-making methods, some of which are favored by one school
of numerical taxonomy and the others by another school. The sole cri-

terion of inclusion of a method is its utility for molecular data. The discussion presented in this chapter is primarily for molecular data and does not necessarily apply to morphological characters. Before discussing tree-making methods, I shall first consider the types of phylogenetic trees molecular evolutionists are interested in.

Types of Phylogenetic Trees

Gene Trees vs. Species Trees

Evolutionists are usually interested in a phylogenetic tree which represents the evolutionary pathways of a group of species or populations. This type of tree is called a *species* or *population* tree. In a species tree, the time of divergence between two species refers to the time when the two species were reproductively isolated from each other (see figure 11.1). When a phylogenetic tree is constructed from one gene from each species, however, the tree estimated does not necessarily agree with the species tree. Indeed, in the presence of polymorphism the times of divergence of genes sampled from different species are expected to be longer than those of species divergence (figures 10.4 and 11.1). To distinguish this tree from the species tree, we call it a *gene tree* (Tateno et al. 1982). Futhermore, even the branching pattern (topology) of a tree constructed from genes may be different from that of the species tree. Figure 11.1 shows three different possible relationships between the species and gene trees for the case of three species. In relationships (a) and (b), the topologies of the species and gene trees are the same, but in relationship (c) they are different. As will be discussed in chapter 13, the probability of occurrence of relationship (c) is quite high when the time interval between the first and second species splitting ($t_1 - t_0$) is short.

Even if the pattern of gene splitting agrees with that of species splitting, the topology of a reconstructed gene tree still may not agree with that of the species tree if the number of nucleotides or amino acids examined is small. This is because nucleotide or amino acid substitution occurs stochastically, so that the number of substitutions in lineage Z in figure 11.1(a) or (b) may become smaller than that in lineage X or Y. To avoid this type of error, we must examine a large number of nucleotides or amino acids.

When the gene studied belongs to a multigene family, another problem occurs; it is not easy to identify the homologous genes for all species

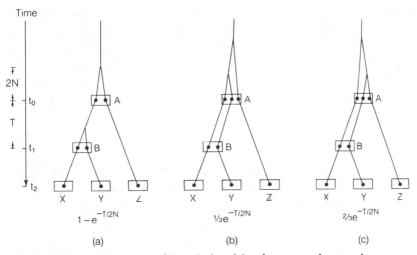

Figure 11.1. Three possible relationships between the species and gene trees for the case of three species in the presence of polymorphism. t_0 and t_1 are the times of the first and second species splitting, respectively. The probability of occurrence of each tree is given underneath the tree (see chapter 13). $T = t_1 - t_0$, and N is the effective population size.

studied. When the number of duplicate genes involved is small, the identification is relatively simple, but when the number is large, as in the case of immunoglobulin V_H genes (see figure 6.11), it is virtually impossible. We should, therefore, exercise great caution in the estimation of species trees from gene trees.

Of course, gene trees are not always produced just to estimate species trees. As in the case of the study of concerted evolution, it is often important to know the evolutionary change of genes themselves rather than species. In this case, we must study a gene tree.

Expected vs. Realized Distance Tree

In the estimation of species trees, one is generally interested in the time since divergence between each pair of species. In this type of tree, the lengths of the two branches leading to two extant species from the common ancestral species must be the same. Even in a gene tree, the time since divergence between a pair of genes must be the same. If the rate of gene substitution is constant, the expected evolutionary distance should also be the same. Thus, one can construct a tree each branch of which is proportional to evolutionary time or expected evolutionary distance. This type of tree will be called an *expected distance tree*. On the other hand, the number of mutational changes or gene substitutions that actually occur in each of the two evolutionary lineages may not necessarily be the same because of stochastic error or nonconstancy of the rate of gene substitution. In some cases, it is important to estimate the actual number of mutational changes that occurred in each branch of a tree. To distinguish this tree from the previous expected distance tree, we call it a *realized distance tree*. Note that a species tree is always an expected distance tree, whereas a gene tree may be either an expected distance tree or a realized distance tree.

Rooted vs. Unrooted Trees

Phylogenetic relationships of genes or organisms are usually presented in a treelike form with a root, as those in figure 11.2(a). This type of tree is called a *rooted tree*. On the other hand, it is possible to draw a tree without a root, as those in figure 11.2(b). This type of tree is called an *unrooted tree* or *network*, though the latter word has a different meaning in graph theory. The branching pattern of a tree is called a *topology*. There are many possible rooted and unrooted trees for a given number of species (n). In the case of $n = 4$, there are 15 possible rooted trees and 3 possible unrooted trees, as shown in figure 11.2. This number of possible trees rapidly increases with increasing n. In general, the number of bifurcating rooted trees for n species is given by

$$1 \cdot 3 \cdot 5 \cdots (2n-3) = \frac{(2n-3)!}{2^{n-2}(n-2)!} \qquad (11.1)$$

for $n \geq 2$ (Cavalli-Sforza and Edwards 1967). This indicates that when $n = 10$, the number is 34,459,425. Only one of these trees is the true tree. The number of bifurcating unrooted trees for n species is given by replacing n by $n-1$ in equation (11.1). This becomes more than two million for $n = 10$ [See Felsenstein (1978a) for the number of multifurcating trees.] In many cases, of course, a majority of these possible trees can be excluded because of obviously unlikely genetic relationships or of other biological information. Nevertheless, it is a very difficult task to find the true phylogenetic tree from observed data on extant species when n is large.

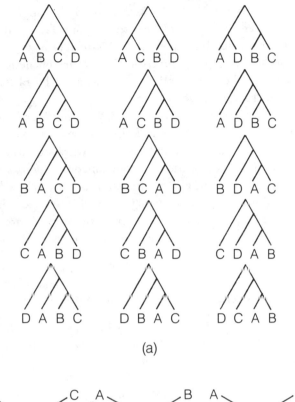

(a)

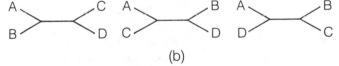

(b)

Figure 11.2.   Fifteen possible rooted trees (a) and three possible unrooted trees (b) for four species.

Phenetic vs. Cladistic Trees

In numerical taxonomy, there are two different approaches to the construction of phylogenetic trees. One is the *phenetic* approach, in which a tree is constructed by considering the phenotypic similarities of the species without trying to understand the evolutionary pathways of the species. When the rules of evolutionary change of the characters used are not well understood, this approach is reasonable because it is virtually impossible to know the real evolutionary pathways. This approach is also justified when one is interested in the taxonomy based on phenotypic resemblance only. Since the tree constructed by this approach does not necessarily represent the evolutionary relationship, it is often called a *phenogram*. In this case, tree construction is merely a way of grouping organisms [see Farris (1977) for criticisms of the phenetic approach]. However, some of the statistical methods invented by pheneticists can be used for constructing phylogenetic trees.

In the second approach, which is called the *cladistic* approach, a phylogenetic tree is constructed by considering various possible ways of evolution, following certain rules and choosing the best possible tree. A tree constructed by this approach is often called a *cladogram* in contrast with a phenogram. One might think that this approach is better than the phenetic approach in constructing phylogenetic trees. In practice, however, the assumptions required for cladistic methods are not always satisfied with molecular data, so that the superiority of this approach is not guaranteed.

Reconstruction of Phylogenetic Trees

There are many different methods that are used for reconstructing phylogenetic trees from molecular data. Two most popular types are the *distance matrix methods* and the *maximum parsimony methods.* In distance matrix methods, evolutionary or genetic distance is computed for all pairs of species (or populations), and a phylogenetic tree is constructed by considering the relationships among these distance values. Once distance values are obtained, there are several different ways of constructing a tree. In maximum parsimony methods, the nucleotide or amino acid sequences of ancestral species are inferred from those of extant species, and a tree is produced by minimizing the number of evolutionary changes for the entire tree. There are several different algorithms for constructing a maximum parsimony tree. In addition to the above two types, the so-called compatibility method and the maximum likelihood method are also used. Furthermore, in the case of gene frequency data, several other methods such as the minimum evolution method, cluster analysis, and Cavalli-Sforza and Piazza's (1975) multivariate analysis method have been proposed. In the next section, I shall discuss some of these methods in detail with emphasis on long-term evolution.

In general, it is an extremely difficult task to reconstruct the true evolutionary tree through which the extant species or populations evolved. There are two types of errors in the reconstruction of trees. They are *topological errors* and *branch length errors*. The former errors refer to the errors in the branching pattern of a tree or topology and the latter to the errors in branch lengths. In the case of species trees, a branch length means the time interval between two branching points or between one branching point and one extant species. In the case of gene trees, it means either the expected or realized number of gene substitutions for a tree branch. Clearly, there is only one true topology for a group of organisms, and evolutionists want to find this topology from existing data. One might think that if the topology of a tree is wrong, it is useless to estimate branch lengths. If a large part of the topology is wrong, this is certainly true. However, if only a minor part is wrong, estimation of branch lengths is still meaningful. In practice, there is usually no way to know whether or not a tree obtained is the correct one. The only thing we can do is to use a reliable tree-making method, but the reliability of a method depends on the type of data used and the purpose of the investigator, as will be discussed below.

Distance Matrix Methods

Average Distance Method (UPGMA)

The simplest method in this category is the average distance method or unweighted pair-group method with arithmetic mean (UPGMA). This method was originally developed for constructing a phenogram (Sokal and Michener 1958), but it can be used for constructing a phylogenetic tree as well, particularly when a distance measure of which the *expected value* is approximately proportional to evolutionary time is used. In their computer simulation studies, Tateno et al. (1982), Nei et al. (1983a), and Sourdis and Krimbas (1987) have shown that when distance estimates are subject to large stochastic errors, UPGMA is often superior to other distance matrix methods in recovering the true tree. This method is intended to estimate a species tree or expected gene tree. For this purpose, it is recommended that linear distance measures such as the number of amino acid substitutions or standard genetic distance be used. Note that these distance measures do not necessarily follow the principle of triangle inequality, but this principle is not required for this method.

Algorithm.   In UPGMA, a certain measure of evolutionary distance is computed for all pairs of *operational taxonomic units* (OTUs), i.e., species or populations, and the distance values obtained are presented in a matrix similar to the following one.

OTU	1	2	3
2	d_{12}		
3	d_{13}	d_{23}	
4	d_{14}	d_{24}	d_{34}

Here, d_{ij} stands for the distance between the ith and jth OTUs. Clustering of OTUs starts from the two OTUs with the smallest distance, and more distantly related OTUs are gradually added to the cluster. Suppose that the distance between OTUs 3 and 4 (d_{34}) is smallest among all distance values in the above matrix. These two OTUs are then clustered with a branching point located at distance $d_{34}/2$. Here, we have assumed that the lengths of the branches leading from the branching point to OTUs 3 and 4 are the same. OTUs 3 and 4 are then combined into a single OTU, (34). New distances between this combined OTU and the other OTUs are calculated:

OTU	1	2
2	d_{12}	
(34)	$d_{1(34)}$	$d_{2(34)}$

Here, $d_{1(34)}$ and $d_{2(34)}$ are given by $(d_{13}+d_{14})/2$ and $(d_{23}+d_{24})/2$, respectively. We again search for the smallest value in the new distance matrix. Suppose that $d_{2(34)}$ is smallest. OTU 2 then joins the (34) cluster with a branching point located at distance $d_{2(34)}/2$. In this case, OTU 1 is the last to be clustered. The branching point at which this last OTU

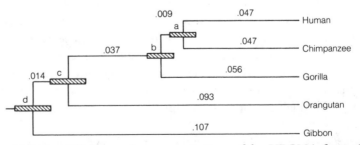

Figure 11.3.   Phylogenetic tree reconstructed by UPGMA from the distance matrix in table 11.1. The hatched bar represents one standard error on each side of the branching point.

joins the others is $d_{1(234)}/2 = (d_{12} + d_{13} + d_{14})/(3 \times 2)$. If $d_{1(34)}$ is smallest among the three distance values above, OTU 1 joins the (34) cluster first, and then OTU 2. On the other hand, if d_{12} is smaller than any of $d_{1(34)}$ and $d_{2(34)}$, OTUs 1 and 2 are clustered, and then the two clusters (12) and (34) are joined into the final single family. When more than four OTUs are involved, the above procedure is continued until all OTUs are clustered into a single family (e.g., figure 11.3).

Statistical Properties.   The underlying assumption of UPGMA when applied to molecular data is that the *expected* rate of gene substitution is constant. If the distance measure used is exactly linear with evolutionary time without error, UPGMA gives the correct topology and correct branch lengths. When the distance measure is subject to stochastic errors, however, both the topology and branch lengths may be incorrect even if the expected rate of gene substitution is constant. In actual data, we usually do not know the correct topology, so that it is difficult to evaluate the extents of topological errors and branch length errors. Computer simulations, however, have shown that UPGMA is often better than other distance matrix methods in recovering the true species tree unless the expected rate of gene substitution varies substantially with evolutionary lineage, as will be discussed later. It is also known that once the correct topology is obtained, UPGMA gives least-squares estimates of branch lengths (Chakraborty 1977). Namely, UPGMA minimizes the quantity

$$S = \Sigma_{ij}(d_{ij} - 2\lambda t_{ij})^2, \qquad (11.2)$$

where λ is the rate of gene substitution, and t_{ij} is the time since divergence between OTUs i and j.

One way to examine the accuracy of the topology of a phylogenetic tree from existing data is to compute the standard error of each branching point of the tree. In the case of a UPGMA tree, this standard error can be evaluated easily if we assume a constant rate of gene substitution. Obviously, the standard error of a branching point leading to two extant OTUs (e.g., a in figure 11.3) is equal to half the standard error of the distance between the two OTUs. In general, however, a branching point connects two clusters consisting of various numbers of OTUs. Suppose that the two clusters (A and B) have r and s OTUs, respectively. The mean distance of these two clusters in a UPGMA tree is given by

$$d_{AB} = \Sigma_{ij} d_{ij}/(rs), \qquad (11.3)$$

where d_{ij} is the intercluster distance between the ith OTU in cluster (A) and the jth OTU in cluster (B). Therefore, the variance of d_{AB} is

$$V(d_{AB}) = [\Sigma V(d_{jk}) + \Sigma \, Cov(d_{jk}, \, d_{lm})]/(rs)^2. \qquad (11.4)$$

There are rs variances and $rs(rs-1)$ covariances. [$Cov(d_{ij}, d_{kl})$ is equal to $Cov(d_{kl}, d_{ij})$.] The variances involved are directly obtainable from equations (4.6), (5.4), (5.51), (9.39), etc., depending on the data used. The covariance $Cov(d_{ij}, d_{kl})$ also depends on the data. When amino acid sequence data are used, it is given by

$$Cov(d_{ij}, d_{kl}) = (1-q)/(qn), \qquad (11.5)$$

where n is the number of amino acids examined, and

$$q = e^{-d_{AB} + [d_{(ij)} + d_{(kl)}]/2}. \qquad (11.6)$$

Here $d_{(ij)}$ stands for the intracluster distance between i and j, and $d_{(ij)} = 0$ if $i = j$. The quantity $d_{AB} - [d_{(ij)} + d_{(kl)}]/2$ is equal to the expected distance between the branching point of OTUs i and j and the branching point of OTUs k and l. In the case of the number of nucleotide substitutions estimated by Jukes and Cantor's formula, the covariance is given by (5.4) with

$$q = \frac{1}{4} + \frac{3}{4} \, e^{-\frac{4}{3}\{d_{AB} - [d_{(ij)} + d_{(kl)}]/2\}} \qquad (11.7)$$

[For the derivation of (11.6) and (11.7), see Nei et al. (1985).] Therefore, all variances and covariances on the right-hand side of (11.4) are computable, and thus $V(d_{AB})$ is obtainable. Once $V(d_{AB})$ is obtained, the variance of the branching point b_{AB} is given by

$$V(b_{AB}) = V(d_{AB})/4. \qquad (11.8)$$

The above method is easily adapted to distance values estimated from restriction-site data or electrophoretic data. The details of the procedures are given by Nei et al. (1985).

EXAMPLE

Brown et al. (1982) sequenced a segment (895 nucleotides) of mitochondrial DNA (mtDNA) for the human, chimpanzee, gorilla, organutan, and gibbon. From these data, we can estimate the number of nu-cleotide substitutions and their standard errors by using equations (5.3) and (5.4), respectively. The results obtained are presented in table 11.1. It is seen that the value between the human and chimpanzee is smallest, so that the human and chimpanzee are the first to be clustered with a branching point (a) at $b_{HC} = 0.094/2 = 0.047$ (figure 11.3). The human and chimpanzee are now combined into a single OTU, (HC). The d values between this OTU vs. the gorilla, orangutan, and gibbon become (0.111 + 0.115)/2 = 0.113, (0.180 + 0.194)/2 = 0.187, and (0.207 + 0.218)/2 = 0.212, respectively. The other distance values remain un-changed. The smallest d value in the new d matrix is that (0.113) be-tween (HC) and the gorilla. Thus, the gorilla joins (HC) with a branch-ing point of $b_{G(HC)} = 0.113/2 = 0.056$. If this type of computation is repeated, we finally obtain the phylogenetic tree given in figure 11.3.

The variances or standard errors of branching points in this phyloge-netic tree can be obtained by using the method described above. In the case of branching point b, $r = 1$ and $s = 2$ in equation (11.3), and we have $d_{AB} = (0.111 + 0.115)/2 = 0.113$, $V(d_{HG}) = (0.012)^2$, and $V(d_{CG}) = (0.012)^2$ from table 11.1, where the subscripts of d refer to the organ-isms concerned. On the other hand, the expected distance between the gorilla and the branching point (a) between the human and chimpanzee is $d = d_{AB} - b_{HC} = 0.066$, so that the corresponding q value is $1/4 + (3/4)\exp[-(4/3) \times 0.066] = 0.937$ from equation (11.7). Thus $Cov(d_{HG}, d_{CG}) = V(d)$ can be computed from this q value by using (5.4). It becomes 7.833×10^{-5}. Therefore, the standard error of branching point b is $[V(d_{(HC)G})/4]^{1/2} = 0.005$. Similarly, the standard errors of

Table 11.1   Proportion of different nucleotides (p) (above the diagonal) and estimates of the number of nucleotide substitutions (d) per site and their standard errors (below the diagonal) obtained from nucleotide sequence data for five primate species. Data from Brown et al. (1982).

	Human	Chimpanzee	Gorilla	Orangutan	Gibbon
Human		.088	.103	.160	.181
Chimpanzee	.094 ± .011		.106	.170	.189
Gorilla	.111 ± .012	.115 ± .012		.166	.189
Orangutan	.180 ± .016	.194 ± .016	.188 ± .016		.188
Gibbon	.207 ± .017	.218 ± .017	.218 ± .017	.216 ± .017	

branching points c and d in figure 11.3 become 0.0071 and 0.0074, respectively. The difference between branching points a and b and that between branching points c and d are not statistically significant. However, the difference between branching points b and c is significant. For the details of the significance test, see Nei et al. (1985).

Fitch and Margoliash's Method

As mentioned earlier, the purpose of UPGMA is to construct a species tree or gene tree with expected branch lengths. However, it is often important to know the realized number of gene substitutions for each branch of a tree. When the rate of gene substitution varies extensively from evolutionary lineage to lineage, UPGMA is likely to give an incorrect topology, and the branch lengths estimated would not be very meaningful. In this case, if we use a method which allows different rates of gene substitution for each branch, we would obtain a more reliable topology than that of the UPGMA tree. There are several methods that can be used for this purpose, i.e., for obtaining a realized distance tree. One of the methods is that of Fitch and Margoliash (1967a). The principle of this method is the same as that of Cavalli-Sforza and Edwards' (1967) *additive method,* but the actual procedure is simpler.

In this method, three OTUs are considered at a time. Let A, B, and C be the three OTUs as considered, and assume that they are related by the unrooted tree given in figure 4.4. Here, x, y, and z denote the numbers of nucleotide (or amino acid) substitutions that occurred in the branches concerned. In chapter 4, we have seen that x, y, and z can be estimated by equations (4.12a)–(4.12c) if we know the distances between the three OTUs. Thus, it is possible to know the realized numbers of nucleotide substitutions for each branch of the tree. In practice, of course, the d values are usually estimates of the numbers of substitutions rather than the sum of the exact numbers of substitutions. Therefore, x, y, and z are also estimates of the realized numbers of substitutions.

When there are more than three OTUs, the basic procedure is the same except that C now represents a composite OTU consisting of all OTUs other than A and B. We first choose two OTUs which show the smallest distance and designate them by A and B. The distance between A and C is the average of the distances between A and all OTUs in C. The distance between B and C is computed in the same way. For example, in the data given in table 11.1, the distance between the human and chimpanzee is smallest, so we denote the human and chimpanzee by A and B, respectively. We then have $d_{AB} = 0.094$, $d_{AC} = (0.111 + 0.180 + 0.207)/3 = 0.166$, and $d_{BC} = (0.115 + 0.194 + 0.218)/3 = 0.176$. The values of x, y, and z are therefore given by 0.042, 0.052, and 0.124, respectively, from (4.12). Here, x and y represent the numbers of substitutions (a and b) for the human and chimpanzee lines, whereas z is the distance between the composite OTU C and the branching point between the human and chimpanzee (see figure 11.4).

We now combine A and B and regard them as a single OTU (AB). We then recompute the distances between this combined OTU (AB) and all other OTUs and choose two OTUs which show the smallest value among all distances, including those which do not involve OTU (AB). These two OTUs are again designated by A and B, whereas C represents the composite OTU consisting of all the remaining OTUs. The new x, y, and z values are computed by the same procedure. In the construction of the UPGMA tree, the distances between OTU (AB) and the other OTUs (gorilla, orangutan, and gibbon) have already been computed (0.113, 0.187, and 0.212, respectively), and the smallest distance in the new distance matrix is that between (AB) and the gorilla. Therefore, (AB) and the gorilla are now designated as the new A and B, respectively, whereas C represents the orangutan and gibbon. We now have $d_{AB} = 0.113$, $d_{AC} = (0.180 + 0.194 + 0.207 + 0.218)/4 = 0.200$, and $d_{BC} = (0.188 + 0.218)/2 = 0.203$. Therefore, $x = 0.055$, $y = 0.058$, and $z = 0.145$. Branch lengths c and d of the evolutionary tree in figure 11.4 are then estimated from

$$d_{AB} = (a+b)/2 + c + d,$$

$$d_{AC} = (a+b)/2 + c + z,$$

$$d_{BC} = d + z.$$

We know that $(a+b)/2 = 0.047$ and $z = 0.145$. Therefore, $c = 0.008$ and $d = 0.058$.

The above procedure is repeated until all OTUs are clustered into a single family. In the present example, this method gives the same unrooted tree as that of figure 11.3. Unlike UPGMA, however, the present method does not give the root for the tree. To determine the root, we need some additional information or assumption. When there is a clear-cut outgroup organism, the root can be determined by using sequence data for this organism. In the present case, the baboon or macaque would serve as an outgroup organism. Unfortunately, however, sequence data are not available for these organisms. Another method is to assume the constant rate of evolutionary change and to put the root at the point of $(x + y + 2z)/4$ in the final triplet in tree-making. The topology of a tree obtained by this method is always the same as that of a UPGMA tree, but the estimates of branch lengths are obviously different. In the present example, the difference in branch lengths between the two trees is rather small (figure 11.4).

The above procedure is, however, only one part of Fitch and Margoliash's method. They consider the possibility that the topology of the tree obtained above is wrong, and suggest that many other topologies should be examined before a tree is accepted as the final one. Since the number of possible trees is enormously large for a moderately large number of OTUs, they suggest that only topologies which are close to the first one should be examined. For example, in the tree of figure 11.4 one may interchange the chimpanzee and gorilla branches and recompute all branch lengths under this condition. This tree is then compared with the first tree in terms of the so-called percent "standard deviation." This percent standard deviation is defined as follows:

$$s_{FM} = \left[\frac{2\Sigma_{ij}\{(d_{ij} - e_{ij})/d_{ij}\}^2}{n(n-1)} \right]^{1/2} \times 100 \qquad (11.9)$$

for $i < j$, where n is the number of OTUs used, and d_{ij} and e_{ij} are the observed distance and patristic distances between OTUs i and j, respectively. The *patristic distance* between OTUs i and j is the sum of the lengths of all branches connecting the two OTUs. For example, the patristic distance between the human and gorilla in the tree of figure 11.4 is $a + c + d = 0.108$. If the s_{FM} value is smaller for the second tree than for the first, then the second tree is judged to be better.

In the tree given in figure 11.4, s_{FM} becomes 1.30. In this tree, the chimpanzee and gorilla are relatively close to each other. We can, therefore, change this part of the topology so as to make the chimpanzee and gorilla cluster first and then the human join this cluster. If we estimate all branch lengths with this topology and compute s_{FM}, it becomes 3.11. We can also try another topology in which the human and gorilla are

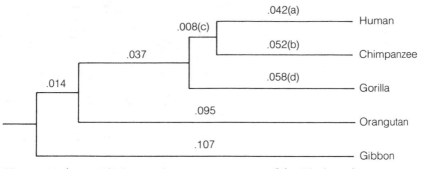

Figure 11.4. **Phylogenetic tree reconstructed by Fitch and Margoliash's method from the distance matrix in table 11.1.**

clustered first and then the chimpanzee is joined. In this case, we have $s_{FM} = 1.88$. Therefore, the first topology is judged to be the best among the three topologies considered. In the present case, no other topologies seem to be better than the first one.

When the number of OTUs used is small, say, six or less, it is possible to examine all possible topologies. However, if this number is large, we must choose a limited number of alternative topologies for testing. Fitch and Margoliash (1967a) suggest that alternative topologies should be produced automatically by choosing a different pair of OTUs for A and B whenever the distance between them is larger than that between the initial pair of OTUs by a given amount or less. This scheme is incorporated in the computer program developed by W. M. Fitch (personal communication).

It should be noted that Fitch and Margoliash's algorithm does not necessarily give the minimum value of s_{FM} for a given topology. If necessary, one can adjust the branch lengths of a given topology so that s_{FM} is minimized. In practice, however, the amount of reduction in s_{FM} obtained by this procedure is usually very small.

Note also that equation (11.9) is only one of several measures for evaluating the difference between d_{ij} and e_{ij}. Numerical taxonomists have used the correlation coefficient between d_{ij} and e_{ij}, which is often called the *cophenetic correlation*. However, this correlation is very close to 1 in most trees obtained from molecular data and does not have much discriminatory power. The other measures that have been used are:

$$s_F = \Sigma_{ij} |d_{ij} - e_{ij}|$$
(11.10)

(Farris 1972), and

$$s_O = \left[\frac{2\Sigma_{ij}(d_{ij} - e_{ij})^2}{n(n-1)} \right]^{1/2} \tag{11.11}$$

(Tateno et al. 1982). These quantities give nearly the same result as that of equation (11.9) in discriminating among different trees.

One of the problems in Fitch and Margoliash's method is the low power of s_{FM}, s_F, and s_O in identifying the correct topology. According to Tateno et al.'s (1982) computer simulation, these quantities do not have a high correlation either with the extent of topological error or with that of branch length error. Nevertheless, they are often used because there is no other observable quantity by which one can measure the closeness of an estimated tree to the true tree. In my opinion, the differences in these quantities among alternative trees are meaningful only when they are quite large. In general, much confidence cannot be given to them.

In Fitch and Margoliash's method, estimates of branch lengths can be negative. Theoretically, this is acceptable because when the true branch lengths are close to 0, their estimates may become negative because of sampling error. However, it is possible to impose the condition that all branch lengths be non-negative. That is, when x in $d_{AB} = x + y$ becomes negative, we can set $x = 0$ and define $y = d_{AB}$. If this rule is adopted, there will be no negative estimates of branch lengths.

Transformed Distance Method

Farris (1977) noted that when the rate of gene substitution varies from evolutionary lineage to evolutionary lineage, the following distance transformation may give an improved topology for the average distance method.

$$d'_{ij} = (d_{ij} - d_{ir} - d_{jr})/2, \tag{11.12}$$

where r refers to one of the OTUs used (reference OTU). This property was also discussed by Farris et al. (1970) in relation to discrete characters, and by Klotz et al. (1979), Klotz and Blanken (1981), and Li (1981b) in relation to molecular data. Consider an artificial evolutionary tree given in figure 11.5(a). If we assume that all nucleotide substitutions are detectable, this tree gives the following distance matrix.

	A	B	C	D
B	18			
C	24	18		
D	20	14	6	
E	21	15	9	5

If we use UPGMA, this matrix produces the tree given in figure 11.5(b). This is clearly an incorrect phylogenetic tree. However, suppose that the ancestral DNA sequence (X) is known so that we can count the total number of nucleotide substitutions for each lineage from X. This number is 16, 10, 12, 8, and 9 for OTUs A, B, C, D, and E, respectively. We now regard this ancestral sequence as a reference OTU and compute the following transformed distance.

$$d'_{ij} = (d_{ij} - d_{ir} - d_{jr})/2 + \bar{d}_r, \qquad (11\ 13)$$

where $\bar{d}_r$ is the average of d_{ir} over all i's. Here, $\bar{d}_r$ is introduced to make d'_{ij} positive. (Actually, any other positive value can be used.) In the present case, $\bar{d}_r = 11$. Therefore $d'_{AB} = (18 - 16 - 10)/2 + 11 = 7$, $d'_{AC} = (24 - 16 - 12)/2 + 11 = 9$, etc. Thus, the transformed distance matrix becomes

	A	B	C	D
B	7			
C	9	9		
D	9	9	4	
E	9	9	5	5

It is now obvious that if we apply UPGMA to this new distance matrix, the correct topology [figure 11.5(a)] is obtained.

The rationale for the above method can be seen from figure 11.5(a). It is obvious that $d'_{AB} = -a + \bar{d}_r - 2$, $d'_{AC} = d'_{AD} = d'_{AE} = d'_{BC} = d'_{BD} = d'_{BE} = \bar{d}_r - 2$, $d'_{CD} = -(b + c) + \bar{d}_r - 2$, and $d'_{CE} = d'_{DE} = -b + \bar{d}_r - 2$, where $a = 2$, $b = 4$, and $c = 1$. That is, transformed distances are composed of ($\bar{d}_r - 2$) minus shared branch lengths, where 2 is the length of the stem in figure 11.5(a). Therefore, it always leads to a correct topology. This property holds for any number of OTUs (Klotz et al. 1979).

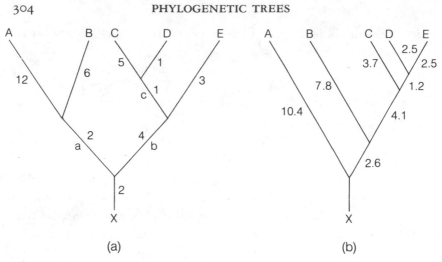

Figure 11.5.   An artificial model tree (a) and the tree (b)
reconstructed by UPGMA.

In the above method, we assumed that the ancestral sequence X is
known. In practice, this is usually not the case. However, if there is a
clearcut outgroup organism in the OTUs examined, it can be used as a
reference organism. As long as all nucleotide substitutions are detectable, the outgroup organism gives the same result as that obtained by
using the ancestral organism. The problem is that there are usually a
substantial number of undetectable substitutions when an outgroup organism is used.

When one is interested in unrooted trees, neither ancestral nor outgroup organism is necessary. In this case, we can regard one of the
OTUs as the reference organism and determine the topology and the
numbers of nucleotide substitutions for all branches. As long as the
pairwise distances are the sums of the nucleotide substitutions for all
branches concerned, this method gives the correct topology and correct
branch lengths. This is obvious from figure 11.5(a), since when the
ancestral OTU X is removed, any of A, B, C, D, and E can serve as the
reference OTU to obtain the correct unrooted tree. If one is willing to
accept the constant rate of evolutionary change, a rooted tree can also be
produced by the method discussed in the previous section.

As mentioned above, most molecular data suffer from backward and
parallel mutations, so that a distance estimate between two OTUs is not
really the sum of all nucleotide substitutions that have occurred after the

two OTUs diverged. This introduces both topological errors and branch length errors. Furthermore, the unrooted tree obtained by using one OTU as the reference OTU will not necessarily be the same as that obtained by using another OTU as the reference OTU. For this reason, it is recommended that in the application of the transformed distance method an unrooted tree be produced for each reference OTU, and the most consistent tree or a consensus tree be chosen.

It should be noted that the transformed distance method is for constructing a topology and does not provide estimates of branch lengths. Once a topology is obtained, however, the branch lengths can be estimated by using Fitch and Margoliash's, distance-averaging, or some other method. In the case of tree (a) in figure 11.5, the correct branch lengths can be obtained by using Fitch and Margoliash's branch-length estimation method after the topology is obtained.

Farris' Distance-Wagner Method

This method is also designed to estimate a realized gene tree. In this method, however, a distance measure satisfying triangle inequality (a metric) is supposed to be used (Farris 1972). As in the case of Fitch and Margoliash's method, many possible trees are examined, but this process is built into the method so that a unique final tree is produced.

Consider a distance matrix for five OTUs, similar to that in table 11.1, and let d_{ij} be the distance between the ith and jth OTUs. Suppose that distance d_{12} is smallest in the matrix. OTUs 1 and 2 are then combined first. The distance between this combined OTU (1,2) and each of the three remaining OTUs is computed by taking the average of the distance between OTU 1 and a third OTU and that between OTU 2 and the third OTU. Suppose that the distance between OTU (1,2) and OTU 3 is smallest among the distances thus obtained. Then, OTU 3 is combined with OTU (1,2) as shown in figure 11.6(a). This figure represents a network (unrooted tree) rather than a rooted tree, because this method cannot determine the root. In this figure, X is the branching point. Each branch length of the network is computed by the following formulas.

$$L(3,X) = (d_{13} + d_{23} - d_{12})/2, \qquad (11.14a)$$

$$L(1,X) = d_{13} - L(3,X), \qquad (11.14b)$$

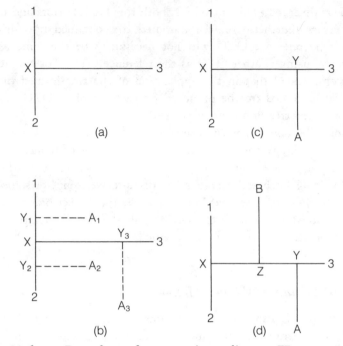

Figure 11.6. Procedure of constructing a distance-Wagner tree. See text.

$$L(2,X) = d_{23} - L(3,X), \qquad (11.14c)$$

where $L(a,b)$ represents the length between points a and b. In practice, these values are computed for every pair of OTU (1,2) and the remaining OTUs, and the OTU which shows the smallest length to X is chosen.

We now proceed to the next step, where one more OTU is added to the network. There are three possibilities for one (say, A) of the remaining OTUs (4 and 5) to be connected to the network; OTU A may be connected at point Y_1, Y_2, or Y_3 in figure 11.6(b). The subscripts of A in the figure correspond to the three possibilities. The branch lengths $L(A_1, Y_1)$, $L(A_2, Y_2)$, and $L(A_3, Y_3)$ are then computed. This computation is done for all remaining OTUs (4 and 5), and the OTU which gives the smallest branch length is chosen to be connected to the network. In practice, $L(A_i, Y_i)$'s are computed by the following formulas.

$$L(A_1, Y_1) = [L(A_1, 1) + L(A_1, X) - L(1, X)]/2, \qquad (11.15a)$$

$$L(A_2, Y_2) = [L(A_2, 2) + L(A_2, X) - L(2, X)]/2, \qquad (11.15b)$$

$$L(A_3, Y_3) = [L(A_3, 3) + L(A_3, X) - L(3, X)]/2. \qquad (11.15c)$$

In these formulas, $L(A_1, 1)$, $L(A_2, 2)$, and $L(A_3, 3)$ are directly obtained from the distance matrix, whereas $L(1, X)$, $L(2, X)$, and $L(3, X)$ have already been computed by using equation (11.14). On the other hand, $L(A_1, X)$, $L(A_2, X)$, and $L(A_3, X)$ are computed by the following formulas.

$$L(A_1, X) = L(A_1, 2) - L(2, X) = L_1, \text{ or}$$
$$= L(A_1, 3) - L(3, X) = L_2, \qquad (11.16a)$$

$$L(A_2, X) = L(A_2, 1) - L(1, X) = L_3, \text{ or}$$
$$= L_2, \qquad (11.16b)$$

$$L(A_3, X) = L_3 \text{ or } L_1. \qquad (11.16c)$$

Among L_1, L_2, and L_3, Farris chooses the largest value and uses it for all of the $L(A_i, X)$'s in equations (11.15a)–(11.15c). Suppose that $L(A_3, Y_3)$ is smallest among $L(A_1, Y_2)$, $L(A_2, Y_2)$, and $L(A_3, Y_3)$. Then OTU A is connected to the branch 3–X, as shown in figure 11.6(c). The last OTU (B, which is either OTU 4 or 5) is then added to the network as in figure 11.6(d). The connecting procedure is the same as the above, except that there are five possible ways of connection in this case. This procedure is continued until all OTUs are clustered.

The present method does not give the root. To place the root for the tree obtained, one can either use an outgroup organism or assume a constant rate of evolutionary change. In the latter case, the root is given at the middle point of the longest branch that connects two OTUs.

Figure 11.7 shows the distance Wagner tree obtained from data in table 11.1. In this case, the proportions of different nucleotides (p) rather than the d values are used because the latter are not metrics. The topology of the tree is the same as that of the tree obtained by UPGMA or Fitch and Margoliash's method, but the branch lengths near the root are considerably shortened compared with those of the trees in figure 11.3 and 11.4. This difference occurred mainly because the distance-Wagner tree is based on the proportions of different nucleotides (p).

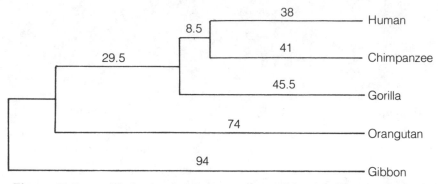

Figure 11.7.   Phylogenetic tree reconstructed by the distance-Wagner
method from the distance matrix in table 11.1. Branch lengths are
measured in $1,000 \times p$.

Although Farris (1972) recommended that a metric be used for his
distance-Wagner method, it is not essential. As will be discussed later,
the use of a metric does not necessarily improve the accuracy of the
method compared with the case where a nonmetric is used. Actually,
the linear distance d or D in equation (9.24) seems to yield more accurate branch lengths than π or I in (9.25). The only problem is that
when a nonmetric is used, some estimates of branch lengths may become
negative. Of course, if necessary, these negative estimates may be avoided
by the method mentioned earlier.

Modifications of Farris' Method

When Farris' method is applied to molecular data, there arises one
problem. It concerns the estimation of branch lengths $L(A_i,X)$'s by the
largest value of L_1, L_2, and L_3 in equations (11.16a)–(11.16c). Since
estimates of genetic distances are generally subject to large sampling
errors, this procedure is expected to lead to an overestimate of $L(A_i,X)$
and, subsequently, an overestimate of $L(A_i,Y_i)$. In a computer simulation, Tateno et al. (1982) have shown that Farris' method often gives
overestimates of branch lengths because of this property. They have then
suggested that the average of L_1, L_2, and L_3 be used as $L(A,X)$. This
modification of Farris' method gives substantially better estimates of branch
lengths when a linear distance measure is used.

Faith (1985) modified Farris' method in another direction, regarding the method as an approximation to the maximum parsimony method for discrete characters. His interest is in reducing topological errors rather than branch length errors. The reader who is interested in this problem may refer to his original paper.

Relative Merits of Different Distance Matrix Methods

It is now clear that different tree-making methods depend on different assumptions, so that their performance in recovering the correct tree would not be the same. UPGMA is based on the assumption of a constant rate of evolution, whereas the others do not require this assumption. Therefore, one might expect that UPGMA is inferior to the others, since the other methods are supposed to work for the case of varying rate as well as for the case of constant rate. The actual situation is not so simple, because molecular data are usually subject to large stochastic error and this introduces another source of error in tree making. Because of this error, the phylogenetic tree reconstructed from sequence or genetic distance data is often incorrect in any tree-making method.

It is generally very difficult to study the relative efficiencies of different tree-making methods by using actual data, because we almost never know the true tree for the group of species under investigation. For this reason, a number of authors have used computer simulations to study this problem. In computer simulations, we can set up a model tree and compare a reconstructed tree obtained from simulated molecular data with it. We can thus determine how often the true (model) tree is recovered by using each tree-making method.

Tateno et al. (1982) compared the efficiencies of the UPGMA, Fitch and Margoliash, Farris, and their modified Farris methods in recovering the correct topology of the model tree as well as in estimating the branch lengths. They considered the cases of both constant rates and varying rates of nucleotide substitution, using a sequence of 300 nucleotides for eight OTUs. Their results indicate that when substitution rate is constant and the number of nucleotide substitutions per site is relatively small (d for the most distantly related pair of species $= 0.09$) the Farris and the modified Farris methods tend to be better than the other two methods in recovering the correct topology (rooted tree). For estimating the number of substitutions for each branch of the tree, however, the modified Farris method shows a better performance than the Farris method. When the number of substitutions is relatively large (d for the most distantly related pair of species $= 0.37$), UPGMA shows the best performance among the four methods examined in both recovering the correct rooted tree and estimating branch lengths. When substitution rate varies with nucleotide position or with evolutionary lineage to the extent observed for the hemoglobin or cytochrome c gene, the Farris and modified Farris methods tend to perform slightly better than the other two methods in recovering the correct tree, even if the number of nucleotide substitutions is quite large. (The Farris method is usually slightly better than the modified Farris method in obtaining an unrooted tree.) For estimating expected branch lengths, however, UPGMA is almost always best. Similar conclusions were obtained by Sourdis and Krimbas (1987) in a more extensive computer simulation.

The effect of a high degree of rate heterogeneity among different branches on the recovery rate of the correct topology was examined by Blanken et al. (1982). Their results indicate that in the presence of extensive rate heterogeneity the transformed distance method is much better than the Fitch and Margoliash method, which is in turn better than UPGMA. On the other hand, if the rate of substitution is constant, UPGMA is better than the transformed distance method or any other method examined.

Nei et al. (1983a) studied the accuracies of three distance matrix methods (UPGMA, Farris, and modified Farris) in the reconstruction of phylogenetic trees from simulated gene frequency data, using four measures of genetic distance (D, D_R, f_θ, D_A; see chapter 9). Their method of computer simulation was briefly as follows. (1) An ancestral population, which was in equilibrium with respect to the effects of mutation and genetic drift, split into two populations. At a later time, one of the two descendant populations again split into two populations. This process was continued until the model tree given by figure 11.8(a) was completed. (2) The expected number of gene substitutions per locus was proportional to evolutionary time. In figure 11.8(a), the expected number of gene substitutions for the shortest branch ($b \equiv D/2$) is 0.1. (3) Starting from the ancestral population, the gene frequency change in each population was followed by using the infinite-allele model of mutations (chapter 13), and at the end of the evolutionary change the gene frequencies for all populations were recorded. The number of loci examined for each replication was 100. (4) Ten sets of distance matrices were computed for each distance measure by using the first 10 loci, first 20 loci, . . . , and all 100 loci. (5) For each of these distance matrices, phylogenetic trees were reconstructed by using the UPGMA, Farris, and modified Farris methods, and each reconstructed tree was compared with the model (true) tree. (6) This simulation was repeated ten times for each of $b = 0.1$ and $b = 0.004$.

Figure 11.8 shows one example of the comparison of the trees reconstructed by the three tree-making methods. Here, the results from only one replication for the case of 50 loci used are presented. In this particular example, the tree obtained from UPGMA is quite close to the true tree when Nei's distance D is used. However, the trees produced by the other two methods are not as good as the UPGMA tree. When a tree is constructed by using Rogers' distance D_R, the branches of the tree near the root are condensed. This is because D_R is not proportional to the number of gene substitutions and there is an upper limit to it. The trees produced by using f_θ and D_A also have a similar property.

Nei et al. (1983a) constructed trees for all ten sets of different numbers of loci, four distance measures used, and ten replications. Their general conclusions are as follows. In all tree making methods, the accuracies of both the topology and branch lengths of a reconstructed tree (rooted tree) are very low when the number of loci used is less than 20 but gradually increases with increasing number of loci. When the expected number of gene substitutions per locus for the shortest branch (b) is 0.1 or more per locus and 30 or more loci are used, the topological error as measured by Robinson and Foulds' (1981) distortion index (d_T) is not great, but the probability of obtaining the correct topology (P) is less than 0.5 even with 60 loci. (d_T is roughly twice the number of branch interchanges that are required for the topology of a reconstructed tree to be converted to that of the model tree.) When b is as small as 0.004, P is substantially lower. In obtaining a good topology (small d_T and high P) UPGMA and the modified Farris method generally show a better performance than the Farris method. The poor performance of the Farris method is observed even when Rogers' distance, which obeys the triangle inequality principle, is used. For estimating the expected branch lengths of the true tree, UPGMA generally shows a better performance than the other methods, particularly when a distance measure proportional to the number of gene substitutions is used.

The unexpectedly good performance of UPGMA seems to be due to the fact that genetic distance estimates are subject to large stochastic error and the procedure of distance averaging in UPGMA reduces the effect of this error on the estimates of branch lengths. This suggests that

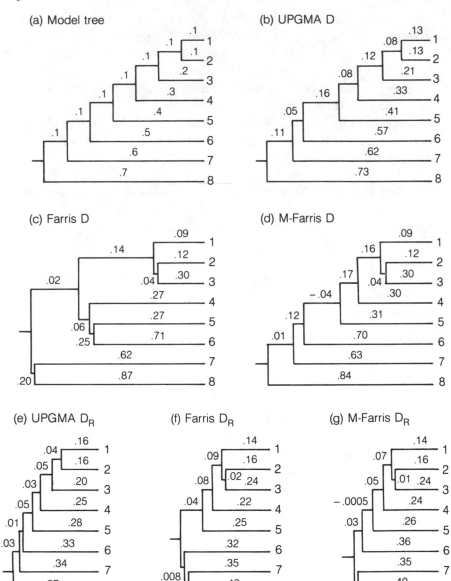

Figure 11.8.   Model tree (a) and reconstructed trees (b, c, d) obtained by using Nei's distance, D, and those (e, f, g) obtained by using Rogers' distance, D_R. These are based on the same set of gene frequency data from one replication of computer simulation with 50 loci. From Nei et al. (1983a).

as long as genetic distance is subject to large stochastic error but has an expectation that is approximately linear with time UPGMA is a good tree-making method. Nei et al.'s computer simulation was conducted by considering neutral mutations, but the above conclusion is expected to hold even in the presence of selection, provided that stochastic error is very large and the rate of gene substitution remains more or less constant. Of course, if the rate varies greatly among evolutionary lineages, then the transformed distance or the Farris method should produce better trees.

Before leaving this subject, I should emphasize that UPGMA is useful only for constructing a species tree or an expected gene tree. To estimate the number of realized gene substitutions for each branch, other methods such as the transformed distance and the Farris methods are better. Note also that the relative merits of different tree-making methods depend on many factors, such as the shape of the model tree, the extent of rate heterogeneity, the extent of genetic divergence, the kind of data used, etc (e g , Fiala and Sokal 1985). The details of the effects of these factors are still under investigation.

Parsimony Methods

There are several versions of parsimony methods (Felsenstein 1982), but here we consider only the one that is routinely used for molecular data. This method was first used by Eck and Dayhoff (1966) for amino acid sequence data, but it can be used for nucleotide sequence data as well (Fitch 1971a, 1977). The principle of this method is to infer the amino acid or nucleotide sequences of the ancestral species and choose a tree that requires the minimum number of mutational changes. A tree obtained by this method is called a maximum parsimony tree. This method is primarily for constructing a topology, and branch lengths are not obtainable except under certain assumptions. It is intended to construct a realized tree rather than an expected tree.

Algorithm

We first consider a particular topology for a group of OTUs and infer the ancestral sequences for this topology. We then count the minimum number of substitutions that are required for explaining the evolutionary changes of sequences. Once this number is obtained, another topology is tried, and the minimum number of substitutions for this topology is determined. This process is continued for all reasonable topologies, and the topology that requires the smallest number of substitutions is chosen as the final tree.

The minimum number of substitutions required for a given unrooted tree is computed for each nucleotide (or amino acid) site, and the numbers of substitutions for all sites are added. I suggest that the first topology to be examined be obtained by the Farris, the modified Farris, or the transformed distance method. Consider six OTUs which are assumed to be related by the rooted tree given in figure 11.9(a), and suppose that at a given nucleotide position the OTUs have the nucleotides listed at the tips of the tree. From these nucleotides, we can infer the nucleotides for the five ancestral sequences (nodes) 1, 2, 3, 4, and 5. The nucleotide at node 1 must be either C or T if we consider the minimum number of substitutions. Similarly, the nucleotide at node 2 is inferred to be either G or T, whereas the nucleotide at node 3 must be A, G or T. On the other hand, node 4 is expected to have T because its immediate descendant nodes (1 and 3) both have T. Finally, we infer the nucleotide at node 5 to be either A or T. It is now clear that the minimum number of nucleotide substitutions for this set of OTUs is obtained by assuming that all the ancestral nodes had nucleotide T. The number becomes four. However, this is not the only possible way of explaining the evolutionary change of the OTUs with four substitutions.

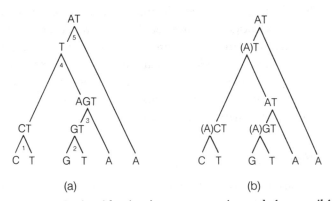

(a)   (b)

Figure 11.9.   Nucleotides in six extant species and the possible nucleotides in five ancestral species. From Fitch (1971a).

If we assume that all intermediate nodes have A, the number of substitutions required is again 4 [see figure 11.9(b)]. Actually, there are nine more possibilities with the same minimum number of substitutions. Four of them are as follows: (5–A, 4–T, 3–T, 2–T, 1–T), (5–A, 4–A, 3–A, 2–A, 1–T), (5–A, 4–A, 3–A, 2–A, 1–C), and (5–A, 4–A, 3–A, 2–G, 1–A). It is clear from these that the nucleotides at the ancestral node cannot always be determined uniquely, and all the nucleotides listed in figure 11.9(b) are parsimonious ones. However, it is possible to count the minimum number of substitutions required.

In the above example, we considered a rooted tree. However, the tree can be transformed into an unrooted tree if we eliminate the apex (node 5) of the tree. Elimination of this node does not change the minimum number of substitutions, but the number of possible ways of substitution is reduced. For example, the two possibilities (5–T, 4–T, 3–T, 2–T, 1–T) and (5–A, 4–T, 3–T, 2–T, 1–T) are no longer distinguishable. Except under certain conditions, the root of a tree cannot be placed unconditionally, so most workers produce unrooted trees.

Alternative topologies for a given set of data can be produced by various methods. One is to use Fitch and Margoliash's (1967a) method mentioned earlier. Another one is Dayhoff and Park's (1969) method, in which a cut is made for each of the branches of the first tree, and each resultant part is grafted onto all branches of the other part. The minimum number of substitutions is examined for all topologies produced in this way, and the best one is chosen. This process is continued several times, and the best of all topologies examined is chosen as the maximum parsimony tree.

Informative Sites for Determining Topology

One might assume that any polytypic (polymorphic) site in which two or more different types of nucleotides (or amino acids) exist is useful for determining the topology. In the construction of maximum parsimony trees, this is not the case, and any nucleotide that exists uniquely in an OTU (*singular nucleotide;* Fitch 1977) is not informative. This is because such a nucleotide can always be assumed to have arisen by a single mutation in the immediate branch leading to the OTU in which the nucleotide exists, and the mutational change is compatible with any topology. A nucleotide site is *informative* only when there are at least two different kinds of nucleotides, each represented at least two times. For example, the first nucleotide site in figure 11.10 is noninformative, whereas the fourth nucleotide site is informative.

EXAMPLE

Figure 11.10 shows Brown et al.'s (1982) nucleotide sequences of mitochondrial DNAs from the human (H), chimpanzee (C), gorilla (G), orangutan (O), and gibbon (B). In this figure, only sites in which at least two different kinds of nucleotides exist are given. There are 282 such polytypic sites, but only 90 sites are informative for determining topology. They are marked with dots in figure 11.10. Brown et al. considered six different topologies which had previously been proposed by various authors for the evolution of this group of organisms. The four best are presented in figure 11.11.

In the construction of a maximum parsimony tree, it is convenient to classify the informative sites into different polytypic or polymorphic patterns. In the present case, there are 15 different polytypic patterns as given in table 11.2. Polytypic pattern (HC)–(GOB) indicates that OTU H and C share the same nucleotide, whereas G, O, and B share a different nucleotide. Symbol H–(CG)–(OB) indicates that OTU H has a unique

```
           1                                    50                                              100
Human      CCCGTCATTCACCGTACATACAACCCTTCCTAATATCAATTTTCTAGTCTCCTTCAGAACTGCAAACGCATATTCTCCCTTAACTCAGCAGACCCATCCT
Chimpanzee     A T C    A     T TT C T C       C  C     C TC TCAG  C     T A      C       C T  A   G  G
Gorilla        TG T T A  A   T    ACCT   C C   CC    G C   C   A G CA     A    CC C TT     TCT AT  G TT
Orangutan      AC  CC GTTAC  CC  G   ACC    C     C CCC    A TCA CC TA   CAA G TAT   TC  CA   A      A A T T C  C
Gibbon     TATAC G C    AACTCCG   T   A C A GGCGCT GC CCGC       C      CAT TCAAGGT C GG CCT ATACCCG    GA G   T    TTC
             ....  .       ..   .     ....      ...          .  .         ..  ..      .. ...  .  .      .  ..  .

           101                                  150                                             200
Human      GCTTCCTACACATCACCCCCGTCAAACATCCTCCGACCTACGTTCTCTTCCGCCAGTTACGGCATTCACACATGCCATCATACCTACCACATTACACCCAA
Chimpanzee A    TG     T T A TT    CT  TT  T C TACA    C         C      T T T AT C  CG     T T  C     T
Gorilla    A C  ACC     T T  A  G   C      TTC   CA   C    T    C A     GTG  A GCT  A   TT  G C      T G
Orangutan  ATAC AC   TT  CT     C     C T   AG  CGT  CG  AC  TATTGCCC  A       TCA T  GG CATCG TTT CC TG    C
Gibbon     A A TA  A    C    TATAACT GCTCC  A TA     TC ATC CCTTA T   CGAATGCC      CTAT  A   A  GAT   CCG TT
             .  ..       .    ...  .     .. .. .     .    ...         .  .

           201                                  250
Human      CCGTGCTCCCCTTACCCGTCTCAATCCCTTATCCATTGCACATTCTCTCCCCAGCTACGTTTACTGAACAGCAAAAACCCGC
Chimpanzee A CT T   C    A       CT  T   G       A G  CTCT       A    C     A G  A
Gorilla    A  T T      T ACT   G    C  T C       C C    TCA      C     AG   A    TT A
Orangutan  TA    C  TACCGT  A  C   C  C A GGCCA   GC  CTA A T C  GAACCCG AA  TGAA CG CTA AA
Gibbon     G AC      TACAG  TA  CTG  T GCC ATG CCATCT  A CA T T   AC  ACC  T A      ATG  GCTAGA
             .  ..      .....       .   .   .          .  .         ..   .
```

Figure 11.10.   Nucleotide sequences of mitochondrial DNAs from humans and apes. Here, only polytypic sites are presented. In chimpanzees, gorillas, orangutans, and gibbons, the nucleotides that are identical with those of humans are not shown. Nucleotide sites with dots are informative sites. Data from Brown et al. (1982).

nucleotide, whereas C and G share a nucleotide and O and B share another nucleotide. We note that some polytypic patterns are compatible with topology (A) in figure 11.11 under the assumption of a single nucleotide substitution, whereas others are not. For example, polytypic pattern a is compatible with topology (A) under the assumption of a single nucleotide substitution because the single substitution involved can be assumed to have occurred between nodes α and β in figure 11.11(a). However, polytypic pattern b requires at least two substitutions to be accommodated with topology (A), whereas it requires only one to be accommodated with topology (C). Therefore, pattern b is compatible with topology (C) but not with (A). Table 11.3 lists compatible polytypic patterns for each of the four topologies. To compute the minimum number of substitutions required for a topology, we must consider the minimum numbers for all compatible and incompatible polytypic patterns and sum them up. The results obtained are given in table 11.3. This computation indicates that topology (B) requires 145 substitutions and that this number is smallest among the four topologies examined. In the present case, there are 11 more unrooted trees, but none of them has as few substitutions as 145. Therefore, topology (B) is chosen as the most parsimonious tree.

Incidentally, it is interesting to note that the number of compatible sites in this example is negatively correlated with the minimum number of substitutions, the correlation being perfect. Therefore, the maximum parsimony tree is identical with the tree having the largest number of compatible sites. In general, however, the correlation between the two quantities is not necessarily perfect, so that the two criteria may lead to different trees. Le Quesne (1969) and Estabrook et al. (1975) proposed that a phylogenetic tree be constructed by considering the number of

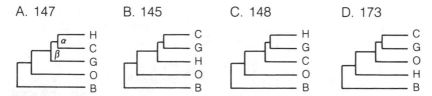

A. 147   B. 145   C. 148   D. 173

Figure 11.11.   Four possible phylogenetic trees (topologies) for the human (H), chimpanzee (C), gorilla (G), orangutan (O), and gibbon (B). The number given to each topology is the minimum number of nucleotide substitutions required for explaining the sequence differences.

Table 11.2   Polytypic patterns and their frequencies in informative sites for determining the parsimonious tree for the human (H), chimpanzee (C), gorilla (G), orangutan (O), and gibbon (B). Singular mutations are listed separately.

Polytypic pattern	Frequency	Polytypic pattern	Frequency
I. Informative sites			
a. (HC)–(GOB)	10	i. (GB)–(HCO)	8
b. (HG)–(COB)	5	j. (OB)–(HCG)	29
c. (HO)–(CGB)	2	k. G–(HC)–(OB)	3
d. (HB)–(CGO)	4	l. O–(HG)–(CB)	2
e. (CG)–(HOB)	10	m. O–(HB)–(CG)	1
f. (CO)–(HGB)	2	n. B–(HG)–(CO)	2
g. (CB)–(HGO)	4	o. B–(HO)–(CG)	1
h. (GO)–(HCB)	7	Total	90
II. Singular mutations			
p. Human	20	s. Orangutan	59
q. Chimpanzee	26	t. Gibbon	79
r. Gorilla	26	Total	210

compatible characters (sites). This method is called the *compatibility method*. In the case of molecular data, the parsimony method and the compatibility method often give the same topology. Particularly when the number of OTUs used is 5 or less, they give the same result.

Note that the maximum parsimony tree obtained above is different from that [topology (A)] obtained by various distance matrix methods discussed earlier. However, the difference in the number of substitutions between topologies (A) and (B) is only two. It is, therefore, difficult to decide which topology is likely to be correct. It should be noted that

Table 11.3   Compatible polytypic patterns with the four topologies given in figure 11.11 and the minimum number of nucleotide substitutions required for each topology.

Topology	Compatible polytypic patterns	Number of compatible sites	Minimum number of substitutions
(A) HCGOB	a, j, k	42	147
(B) CGHOB	e, j, k, m, o	44	145
(C) HGCOB	b, j, k, l, n	41	148
(D) CGOHB	d, e, m, o	16	173

Sibley and Ahlquist's (1984) tree estimated from DNA hybridization data (see chapter 2) is in agreement with topology (A) rather than with topology (B). Sibley and Ahlquist's tree seems to be quite reliable because their data are apparently subject to low sampling error. Topology (A) is also in agreement with the topology of the trees reconstructed from protein sequences (Goodman et al. 1982), chromosome banding patterns (Yunis and Prakash 1982), and two-dimensional electrophoresis (Goldman et al. 1986).

Some Remarks

Some numerical taxonomists are of the opinion that parsimony methods do not require the assumption of approximate constancy of nucleotide or amino acid substitution. This is true if the number of substitutions per site is very small. However, if this number is so large that parallel and backward mutations occur with an appreciable frequency, parsimony methods tend to make serious errors, unless the substitution rate is constant or nearly constant. This is true even if the number of nucleotides or amino acids examined is very large (Felsenstein 1978b). Furthermore, if the number of nucleotides or amino acids examined is small and there are many backward and parallel mutations, parsimony methods have a high probability of producing an erroneous tree even when substitution rate is constant (Peacock and Boulter 1975).

Since parsimonious trees are based on *minimum* estimates of realized numbers of substitutions, it is not easy to develop a statistical method for testing the difference in the parsimonious number of substitutions between two different topologies. The reliability of the difference depends on the tree structure, total number of substitutions, heterogeneity of substitution rate among different evolutionary lineages, etc. To circumvent this problem, Templeton (1983) proposed a nonparametric test for the case of restriction site data. As shown by Nei and Tajima (1985) and Li (1986), however, parsimonious estimates of the number of restriction site changes for a polytypic site are subject to systematic error depending on the tree structure and the number of changes. Furthermore, when four or more OTUs are involved, it is not clear what kind of null hypothesis is being tested (Cavender 1981; Felsenstein 1985). A more rigorous theoretical study is necessary on this important subject.

As mentioned earlier, parsimony methods usually do not provide estimates of branch lengths. However, it is possible to infer all ancestral nucleotides (or amino acids) at each informative site and compute the probabilities of having each nucleotide by using Dayhoff and Park's (1969), Fitch's (1971a), or Goodman et al.'s (1974) method. Therefore, the average number of substitutions for each branch can be computed for all informative sites. We can then add singular substitutions to obtain the total number of substitutions for each branch. Another method for estimating branch lengths is to use Fitch and Margoliash's branch-length estimation method after the parsimony tree is determined (see previous section).

When the number of substitutions per site is small and the number of OTUs used is small, a large proportion of the mutational changes are singular and uninformative for constructing a parsimonious tree. This is one disadvantage of parsimony methods compared with distance matrix methods. In the latter, all sites are used for computing distances. Saitou and Nei (1986b) have studied the probabilities of obtaining the correct topology from nucleotide sequence data for the parsimony method described above and several distance matrix methods under the assumption of constant rate of substitution. Although the study was done for special model trees with consideration of the evolution of humans and apes, the results obtained indicate that the probability (P) of obtaining the correct topology is generally higher for the transformed distance and Farris methods than for the parsimony method, which in turn shows a higher P value than UPGMA. However, to make a P value higher than 0.95, a large number of nucleotides must be examined. For example, when mitochondrial DNA is used and information on outgroup species (orangutans and gibbons) is available, at least about 2,600 nucleotides seem to be necessary to determine the branching order of the human-chimpanzee-gorilla divergence. When nuclear genes are used, a much larger number of nucleotides is required because the rate of nucleotide substitution is lower.

Maximum Likelihood Method

The maximum likelihood method of tree-making was first studied by Cavalli-Sforza and Edwards (1967). These authors attempted to construct a tree from gene frequency data using the model of Brownian motion (Gaussian process). Later, Felsenstein (1973a) and Thompson (1975) developed algorithms for constructing a maximum likelihood tree, extending Cavalli-Sforza and Edwards' approach. However, the evolutionary change of gene frequency is typically non-Gaussian, so this method has not been used extensively. In the case of amino acid or nucleotide sequence data, the evolutionary change can be followed relatively easily by using the Poisson process or some other probability model. Felsenstein (1973b, 1981) took advantage of this and developed maximum likelihood algorithms for constructing a tree. Langley and Fitch (1974) also used the maximum likelihood method for estimating expected branch lengths (or evolutionary times associated) for a given topology when the observed number of substitutions for each branch is known. The latter problem is obviously much simpler than the problem of finding both topology and branch lengths. Let us first discuss Langley and Fitch's method and then consider Felsenstein's approach.

Langley and Fitch's Method

Consider the phylogenetic tree given in figure 11.12, where x_i stands for the observed number of substitutions for the ith branch and t_i refers to the time interval specified in the figure. We are interested in esti-

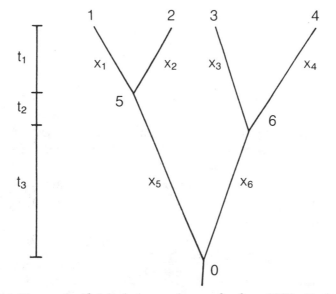

Figure 11.12.   Artificial phylogenetic tree for four OTUs (1, 2, 3, and 4). t_i represents the evolutionary time, whereas x_i stands for the observed number of substitutions.

mating t_1, t_2, and t_3 by using the maximum likelihood method. To apply this method, we must have a specific mathematical model for the change of genes or proteins. We assume that the rate of substitution (λ) is constant per year or per generation throughout the evolutionary process and that the number of substitutions for the gene or protein considered follows the Poisson distribution for any time interval. That is, the probability that x substitutions occur during evolutionary time t in one lineage is $\exp(-\lambda t)(\lambda t)^x/x!$. Therefore, the likelihood of observing x_1, x_2, $\cdots$, x_6 for the tree in figure 11.12 is

$$L = e^{-v_1}\frac{v_1{}^{x_1}}{x_1!} \cdot e^{-v_1}\frac{v_1{}^{x_2}}{x_2!} \cdot e^{-(v_1+v_2)}\frac{(v_1+v_2)^{x_3}}{x_3!}$$

$$\cdot e^{-(v_1+v_2)}\frac{(v_1+v_2)^{x_4}}{x_4!} \cdots, \tag{11.19}$$

where $v_i = \lambda t_i$. Since λ is usually unknown, we seek the maximum likelihood estimates of v_i's rather than t_i's. They are given by the following simultaneous equations.

$$\frac{\partial \log_e L}{\partial v_1} = -4 + \frac{x_1+x_2}{v_1} + \frac{x_3+x_4}{v_1+v_2} = 0, \tag{11.20a}$$

$$\frac{\partial \log_e L}{\partial v_2} = -3 + \frac{x_3+x_4}{v_1+v_2} + \frac{x_5}{v_2+v_3} = 0, \tag{11.20b}$$

$$\frac{\partial \log_e L}{\partial v_3} = -2 + \frac{x_5}{v_2+v_3} + \frac{x_6}{v_3} = 0. \tag{11.20c}$$

Solutions to these equations can be obtained numerically. The same method can be used for any number of OTUs, since the number of parameters to be estimated is always smaller than the number of observed values.

In the estimation of v_i, we assumed that λ is constant in all evolutionary lineages. Once v_i's are estimated, however, the assumption of constant λ can be tested by

$$\chi^2 = \sum_{i=1}^{m} \frac{(x_i - \hat{v}_i)^2}{\hat{v}_i}, \tag{11.21}$$

with $m-p$ degrees of freedom, where $\hat{v}_i$ is the estimate of v_i, and m and p are the number of observed quantities (x_i's) and the number of parameters estimated (v_i's), respectively. In the above example, $m = 6$ and $p = 3$. In practice, x_i's are usually unobservable quantities. However, they can be estimated by one of the parsimony or distance matrix methods discussed earlier. In most studies (e.g., Langley and Fitch 1974), these estimates are directly used in place of x_i's, though they are often underestimates of real values. Therefore, some caution should be exercised in the test of constancy of λ by the above χ^2.

The present method can easily be extended to the case where data for several genes or proteins are available. The rate of substitution may vary from gene to gene, but the evolutionary time for a pair of OTUs is the same for all genes. The reader may refer to Langley and Fitch (1974) for details of the procedure.

Using amino acid substitution data for hemoglobins α and β, cytochrome c, and fibrinopeptide A, Langley and Fitch constructed a composite evolutionary tree for 18 vertebrate species. This tree is presumably more reliable than the trees based on individual genes. Furthermore, their χ^2 equivalent to (11.21) was highly significant. This suggests that the rate of amino acid substitution is not strictly constant. Nevertheless, the number of amino acid substitutions appeared to increase approximately linearly with evolutionary time when long-term evolution was considered.

Felsenstein's Method

Unlike Langley and Fitch's method, Felsenstein's (1981) method is intended to obtain both topology and branch lengths. In this method, the maximum likelihood of obtaining observed nucleotide sequences for a group of OTUs is computed for many different topologies, and the topology which shows the highest likelihood value is chosen as the final tree.

Let us again consider the tree given in figure 11.12 as an example. In Felsenstein's method, information on x_i is not required. Instead, we consider the nucleotides of all extant OTUs (1, 2, 3, 4) at each polytypic site and evaluate the likelihood of having these observed nucleotides under a particular probability model. Let s_1, s_2, s_3, and s_4 be the nucleotides of OTUs 1, 2, 3, and 4 at a given nucleotide site, respectively, and assume that the nucleotides at nodes 0, 5, and 6 are s_0, s_5, and s_6, respectively. We denote nucleotides A, T, C, and G by 1, 2, 3, and 4, respectively. Therefore, s_i takes the value of 1, 2, 3, or 4. Suppose we know the probability, $P_{ij}(v)$, that nucleotide i at time 0 becomes nucleotide j at time v. As before, time is measured in terms of $v = \lambda t$, where λ and t are the rate of substitution and the absolute time, respectively. However, Felsenstein considers the possibility that λ varies with branch, so that he designates by v_i the expected number of substitutions for the ith branch. Then the likelihood of having the observed nucleotides for OTUs 1, 2, 3, and 4 and given nucleotides for s_0, s_5, and s_6 is

$$L = g_{s_0} P_{s_0 s_5}(v_5) \; P_{s_5 s_1}(v_1) \; P_{s_5 s_2}(v_2) \; P_{s_0 s_6}(v_6)$$

$$\cdot P_{s_6 s_3}(v_3) \; P_{s_6 s_4}(v_4),$$

where g_{s_0} is the probability that node 0 has nucleotide s_0. In practice, we do not know s_0, s_5, and s_6, so the likelihood will be the sum of the above quantity over all possible nucleotides for these nodes. That is,

$$L = \sum_{s_0} \sum_{s_5} \sum_{s_6} g_{s_0} P_{s_0 s_5}(v_5) P_{s_5 s_1}(v_1) P_{s_5 s_2}(v_2) \; P_{s_0 s_6}(v_6) \; P_{s_6 s_3}(v_3) P_{s_6 s_4}(v_4)$$

$$= \sum_{s_0} g_{s_0} \left[\sum_{s_5} P_{s_0 s_5}(v_5) P_{s_5 s_1}(v_1) P_{s_5 s_2}(v_2) \right] \left[\sum_{s_6} P_{s_0 s_6}(v_6) P_{s_6 s_3}(v_3) P_{s_6 s_4}(v_4) \right]. \qquad (11.22)$$

To compute L, we must specify $P_{ij}(v)$. Felsenstein suggests the following random substitution model.

$$P_{ij}(v) = e^{-v} \delta_{ij} + (1 - e^{-v}) g_j, \qquad (11.23)$$

where δ_{ij} is 1 for $i = j$ and 0 for $i \neq j$. We are now in a position to search for the maximum value of L by varying all v_i's. This usually requires a large amount of computer time even for a moderate number of OTUs. Felsenstein has introduced an algorithm (pulley principle) to reduce this computer time. This procedure is repeated for many topologies, and then the topology which shows the highest L value is chosen as the maximum likelihood tree.

In practice, however, this method seems to have some problems. First, the equation for $P_{ij}(v)$ is unrealistic, since the actual transition probabilities are quite asymmetric and vary from site to site (see chapters 3 and 5). The actual value of v for a given branch also varies with nucleotide position, because there are usually conserved and nonconserved regions in a DNA sequence. One can, of course, modify $P_{ij}(v)$ taking into account these factors, but the computation will be very complicated. Second, the likelihood computed in this method is conditional for each topology, so that it is not clear whether or not the topology showing the highest likelihood has the highest probability of being the true topology when a relatively small number of nucleotides are examined. Felsenstein justifies his method by considering the limiting case of infinitely many nucleotides, but the number of nucleotides actually used is usually quite small. Nevertheless, it is possible that this method is insensitive to these complicating factors, as is usually the case with many statistical methods. It is worth conducting some computer simulation to clarify this point and investigate the merits and demerits of the method relative to others.

Role of Population Genetics Theory in Tree Construction

Molecular phylogeny is based on the number of nucleotide substitutions in DNA or the number of gene substitutions in the genome. Nucleotide or gene substitution does not occur instantly but is a product of a complicated process of mutation, selection, and genetic drift, as will be discussed in the next two chapters. It is therefore important to take into account this process in the construction of phylogenetic trees.

At the present time, most molecular trees are constructed by using a single gene from each species, and they are often regarded as species trees. As mentioned earlier, however, a gene tree is sometimes quite different from a species tree, and to evaluate the magnitude of difference between them it is necessary to understand the pattern of gene substitution in populations (chapter 13).

Some authors have attempted to construct trees from electrophoretic data, treating each allele as a frequency-less cladistic character. They have paid no attention to the process of allele frequency change or gene substitution in populations. To construct a phylogenetic tree of populations or species, however, we must take into account this process. Nei et al.'s (1983a) recommendation of the use of UPGMA together with Nei's standard genetic distance for electrophoretic data is based on this consideration. Of course, if there are bottleneck effects or inbreeding effects on genetic distances, as in the case of human populations (Nei and Roychoudhury 1982), UPGMA is expected to produce an erroneous tree. In this case, the transformed distance or the distance-Wagner method is preferable to infer the topology of the tree (Li 1981b).

It is often stated that the parsimony method is better than other methods because no assumption is necessary about the process of evolutionary change (e.g., Sober 1985). This statement may be correct when we do not know the mechanism of evolutionary change of the characters used. When this mechanism is known, however, the knowledge should be used effectively for tree reconstruction. When long-term evolution is considered, backward and parallel mutations occur with a high frequency. In this case, there is no reason that we should not use corrected distances for tree-making. Indeed, as mentioned earlier, Saitou and Nei's computer simulation indicates that at least under certain conditions the transformed distance method with corrected distances recovers the true tree more often than the parsimony method.

At the present time, however, the theoretical foundations of many tree-making methods are not well established from the point of view of population genetics. It is hoped that more study will be made on this problem in the future.

CHAPTER TWELVE

POPULATION GENETICS THEORY: DETERMINISTIC MODELS

The basic process of evolution is gene substitution or gene frequency change in a population. Gene substitution or gene frequency change is caused by mutation, selection, and random genetic drift, and the actual process is quite complicated. It is, therefore, necessary to know some mathematical theory to understand the process of gene substitution. In this and the following chapters, I present a brief discussion of mathematical theory that is important for the study of molecular evolution. Although some current mathematical theory is quite sophisticated, I shall avoid higher mathematics as far as possible and present only the essential aspects. The reader who is interested in more mathematical detail may refer to Wright (1969), Crow and Kimura (1970), Kimura and Ohta (1971a), Nei (1975), and Ewens (1979). A more elementary exposition of the mathematical theory of population genetics is found in the books by Hartl (1980) and Hedrick (1983).

The mathematical theory of population genetics can be divided into two categories, i.e., the deterministic theory and the stochastic theory. In the former, the gene frequency change from one generation to the next is assumed to occur uniquely from one value to another, whereas in the latter it is assumed to occur probabilistically. In this chapter, we consider mainly the deterministic theory.

Mutation

Single Locus

When mutation occurs repeatedly from one allele to another, it causes gene frequency change. Consider a pair of alleles, A_1 and A_2, at a locus, and let x and $1-x$ be the frequencies of A_1 and A_2, respectively. We denote the mutation rate from A_1 to A_2 by u and that from A_2 to A_1 by v. The gene frequency of A_1 in the next generation (x') is then given by

$$x' = (1-u)x + v(1-x).$$ (12.1)

Therefore, the amount of change per generation ($\Delta x = x' - x$) is

$$\Delta x = v - (u+v)x$$
$$= (u+v)(\hat{x} - x),$$ (12.2)

where $\hat{x} = v/(u+v)$.

If there are no other evolutionary forces operating, an equilibrium is reached when $\Delta x = 0$. The equilibrium gene frequency is given by

$$\hat{x} = v/(u+v).$$ (12.3)

The gene frequency at the tth generation can be obtained from (12.2). Since the mutation rate is usually much smaller than 1, Δx may be written as dx/dt, where t is the time measured in generations. Therefore,

$$\frac{dx}{dt} = (u+v)(\hat{x} - x).$$ (12.4)

Solution of (12.4) gives

$$x = \hat{x} - (\hat{x} - x_0)e^{-(u+v)t},$$ (12.5)

where x_0 is the initial gene frequency at $t = 0$.

If we note that u and v are of the order of 10^{-5} or less per gene per generation, it is clear that the gene frequency change due to mutation is a very slow process. Because of this finding, early population geneticists such as Wright (1931) and Haldane (1932) concluded that mutation is an unimportant factor in determining the speed of evolution, though it is certainly the ultimate source of variation. This finding had a strong influence in the formation of neo-Darwinism, as will be discussed in chapter 14.

Although equation (12.5) is given in most population genetics books, its utility is quite limited. There are two reasons for this. First, at the molecular level mutation usually creates a new allele (nucleotide sequence), so that the assumption of forward and backward mutation between a pair of alleles is not satisfied. Second, even in the absence of selection the gene frequency is affected by random genetic drift, and this effect is often far greater than the effect of mutation. We shall discuss these problems in detail in the next chapter.

Two Loci

When two or more loci are considered simultaneously, the genotype frequencies cannot be expressed by gene frequencies alone except when all chromosome frequencies are in linkage equilibrium (chapter 7). In general, we need linkage disequilibrium parameters to determine genotype frequencies for multiple loci.

Let us now examine the effect of mutation on linkage disequilibrium, considering two loci each with two alleles A_1, A_2, and B_1, B_2. As in chapter 7, we denote the chromosome frequencies of A_1B_1, A_1B_2, A_2B_1, and A_2B_2 by X_1, X_2, X_3, and X_4, respectively. The linkage disequilibrium is then given by $D = X_1X_4 - X_2X_3$ in equation (7.45). We also denote the mutation rate per generation from A_1 to A_2 and that from A_2 to A_1 by u_A and v_A, respectively, and the corresponding mutation rates at the B locus by u_B and v_B. Therefore, the chromosome frequency of A_1B_1 after mutation is given by $X_1' = (1 - u_A)(1 - u_B)X_1 + (1 - u_A)v_BX_2 + v_A(1 - u_B)X_3 + v_Av_BX_4$, which becomes $(1 - u_A - u_B) X_1 + v_BX_2 + v_AX_3$ approximately, if we ignore quadratic terms of mutation rates. Similarly, the chromosome frequencies of A_1B_2, A_2B_1, and A_2B_2 after mutation are approximately $X_2' = u_BX_1 + (1 - u_A - v_B)X_2 + v_AX_4$, $X_3' = u_AX_1 + (1 - v_A - u_B)X_3 + v_BX_4$, and $X_4' = u_AX_2 + u_BX_3 + (1 - v_A - v_B)X_4$. Thus, the linkage disequilibrium after mutation is given by

$$X_1'X_4' - X_2'X_3' = (1 - u_A - v_A - u_B - v_B)(X_1X_4 - X_2X_3), \qquad (12.6)$$

where the terms involving products of mutation rates are neglected. Therefore, mutation reduces the amount of linkage disequilibrium in the same way as does recombination (see chapter 7).

If we combine the effects of recombination and mutation, the linkage disequilibrium in the next generation is given by

$$D' = (1 - r - u)D, \qquad (12.7)$$

approximately, where r is the recombination value and $u = u_A + v_A + u_B + v_B$.

In the above formulation, we assumed that mutation occurs recurrently between the two pairs of alleles at each locus. At the molecular level, however, mutations are often unique and nonrecurrent as mentioned earlier, and a new mutant allele at a locus is initially associated with a particular allele at a second locus that happened to exist on the same chromosome when the mutation occurred. Therefore, if we consider a particular set of alleles at the molecular level, equation (12.7) does not apply. Rather, mutation is expected to generate linkage disequilibrium.

A good example of this type of linkage disequilibrium is the association of electromorphs with inversion chromosomes mentioned in chapter 7. A new mutant inversion chromosome carries a particular allele, and this allele may increase in frequency with the inversion chromosome because the recombination value between an inversion and a noninversion chromosome is extremely low. Therefore, inversion chromosomes and electromorphs are expected to be generally in linkage disequilibrium. This is indeed the case as discussed earlier.

Another example of this type of linkage disequilibrium is the association of the sickle-cell anemia gene with a variant *Hpa*I restriction site in the β globin gene complex in man. In most individuals, the β globin gene is contained in a 7.6 kb *Hpa*I restriction fragment (see figure 12.1). However, there are two different variant types of restriction fragments with lengths of 7.0 kb and 13.0 kb in the American black population. (The frequency of the 7.0 kb fragment is very low.) As is well known, the American black population has a high frequency of the sickle-cell anemia gene (*S*) at the β globin locus. Kan and Dozy (1978, 1980) have

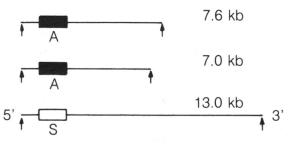

Figure 12.1.   Diagram of three different *Hpa*I restriction fragments at the β globin locus region in man. Arrows, *Hpa*I sites; A and S, normal and sickle β globin genes, respectively. From Kan and Dozy (1978).

shown that the S genes are associated with the 13 kb fragment in 67.9 percent of American black chromosomes, whereas the association of the normal β globin gene (A) with this fragment is only 3.4 percent. They proposed that this nonrandom association was generated by an $A \to S$ mutation that occurred in the 13 kb fragment and that the subsequent increase in frequency of the S gene in the malarial environment in West Africa helped to increase the frequency of the 13 kb fragment by "hitch-hiking." In this hypothesis, the A gene with the 13 kb fragment or the S gene with the 7.6 kb fragment is considered to have been produced by occasional recombination.

In a large population, this type of linkage disequilibrium will eventually disappear if there is any recombination. But it would take a long time for the disequilibrium to disappear when the recombination value is low. In the case of inversion chromosomes, the recombination value (double crossing-over frequency) between the inversion and noninversion chromosomes seems to be about 10^{-4} to 10^{-5} (Ishii and Charlesworth 1977). Therefore, linkage disequilibrium is expected to last for a long time. This problem can be studied by using equation (7.57).

Let us consider the association of the β globin and restriction fragment polymorphism in the American black population. As mentioned earlier, the frequency of the 13 kb fragment (A_1) among the sickle (S) genes is $x=0.679$, whereas the frequency of A_1 among the normal (A) genes is 0.034. Therefore, $d=0.645$ and $A_j=0.958$ from equations (7.56) and (7.58), respectively. The latter value indicates that the association between the β globin and restriction fragment polymorphisms is very strong. Livingstone (1962) and Wiesenfeld (1967) suggested that the sickle-cell mutant gene arose about 2,000 to 7,000 years ago, when agriculture was introduced in Africa. If this is the case, the S gene has existed for only about 80 to 280 generations, assuming that one generation in primitive societies was 25 years. If we use these estimates of evolutionary time, the recombination value (r) between the β globin locus and the 7.6 kb restriction site can be estimated by using (7.57). In this case, $d_0=1$ and $d_t=\exp(-rt)$ approximately. Therefore, we have $rt=-\log_e d=0.4385$, so that r is estimated to be 0.002 to 0.005. This value is considerably higher than the recombination value (5×10^{-5}) expected for a DNA region of about 5,000 bases (between the 3' end of the β-globin gene and the 7.6 kb restriction site) from information on the total map length of the human genome (Kurnit 1979). This result suggests either that there might be a hot spot of recombination in the

DNA region concerned or that the sickle-cell mutation occurred much earlier than 7,000 years ago.

Migration

Single Locus

Migration between populations often has a profound effect on the genetic structure of populations. Here, we consider a simple Wright's (1931) island model to see the effect of migration on gene frequency in a population. Consider a population (X) into which gene migration occurs from a neighboring population (Y). Let x and x_I be the frequency of allele A_1 in populations X and Y, respectively, and assume that migration occurs only from population Y to X with a rate of m per generation. The gene frequency in population X in the next generation is then given by

$$x' = (1 - m)x + mx_I, \tag{12.8}$$

the amount of change in gene frequency per generation being

$$\Delta x = m(x_I - x). \tag{12.9}$$

This is mathematically similar to equation (12.2) and indicates that x eventually reaches the equilibrium frequency x_I. The migration rate is usually much higher than the mutation rate, so that the approximation similar to (12.4) may not be justified. However, if we note the following relation between the gene frequencies (x_t and x_{t-1}) in generations t and $t - 1$,

$$x_t - x_I = (1 - m)(x_{t-1} - x_I),$$

we can easily obtain

$$x_t - x_I = (1 - m)^t(x_0 - x_I), \tag{12.10}$$

where x_0 is the initial gene frequency.

We note that to obtain the relationship among x_0, x_I, and x_t, m need not be the same for all generations. If the migration rate in the ith generation is m_i, equation (12.10) may be written as

$$x_t - x_I = \prod_{i=1}^{t}(1 - m_i)(x_0 - x_I).$$ (12.11)

We also note that the total proportion of genes that entered into population X during t generations is given by $M = 1 - \prod_{i=1}^{t}(1 - m_i)$. If we know x_0, x_I, and x_t, this M can be estimated by

$$M = (x_t - x_0)/(x_I - x_0).$$ (12.12)

This formula has been used by human geneticists to estimate the proportion of genes introduced into one population from another. For example, Reed (1969) studied the proportion of Caucasian genes that entered into the American black population by using gene frequency data for several blood group loci. His estimate of the proportion is about 20 percent. By contrast, the proportion of Negroid genes in the gene pool of American Caucasians seems to be less than 1 percent.

Two Loci

The effect of migration on the chromosome frequencies or linkage disequilibrium for two loci is somewhat complicated (Cavalli-Sforza and Bodmer 1971; Prout 1973; Nei and Li 1973). To obtain a rough idea about this effect, we consider a simple model of two populations between which migration occurs. Let N_1 and N_2 be the sizes of populations 1 and 2, respectively. We assume that N_1 and N_2 are fairly large and remain constant for all generations. We further assume that the two populations exchange N_m individuals per generation, so that the migration rates in populations 1 and 2 are $m_1 = N_m/N_1$ and $m_2 = N_m/N_2$, respectively.

Let X_{1i}, X_{2i}, X_{3i}, and X_{4i} be the frequencies of the four chromosome types A_1B_1, A_1B_2, A_2B_1, and A_2B_2 in population i ($i = 1$ or 2). For simplicity, we assume that migration occurs after mating. Then, the chromosome frequencies after migration in population 1 become $(1 - m_1)X_{11} + m_1X_{12}$, $(1 - m_1)X_{21} + m_1X_{22}$, $(1 - m_1)X_{31} + m_1X_{32}$, and $(1 - m_1)X_{41} + m_1X_{42}$. The chromosome frequencies in the next generation, after meiosis, are therefore given by

$$X'_{11} = (1 - m_1)(X_{11} - rD_1) + m_1(X_{12} - rD_2),$$

$$X'_{21} = (1 - m_1)(X_{21} + rD_1) + m_1(X_{22} + rD_2),$$

$$X'_{31} = (1 - m_1)(X_{31} + rD_1) + m_1(X_{32} + rD_2),$$

$$X'_{41} = (1 - m_1)(X_{41} - rD_1) + m_1(X_{42} - rD_2),$$

where $D_1 = X_{11}X_{41} - X_{21}X_{31}$, $D_2 = X_{12}X_{42} - X_{22}X_{32}$, and r is the recombination value between the A and B loci. Similar recurrence equations may be obtained also for the chromosome frequencies in population 2. If random mating occurs in each population, the linkage disequilibria for populations 1 and 2 in the next generation are given by $D'_1 = X'_{11}X'_{41} - X'_{21}X'_{31}$ and $D'_2 = X'_{12}X'_{42} - X'_{22}X'_{32}$, respectively. Therefore, we obtain the following recurrence equations for linkage disequilibria.

$$D'_1 = (1 - m_1)(1 - m_1 - r)D_1 + m_1(1 - m_1)D_{12} + m_1(m_1 - r)D_2, \tag{12.14a}$$

$$D'_2 = m_2(m_2 - r)D_1 + m_2(1 - m_2)D_{12} + (1 - m_2)(1 - m_2 - r)D_2, \tag{12.14b}$$

$$D'_{12} = [(1 - m_1)(2m_2 - r) - m_2 r]D_1 + (1 - m_1 - m_2 + 2m_1 m_2)D_{12}$$

$$+ [(1 - m_2)(2m_1 - r) - m_1 r]D_2, \tag{12.14c}$$

where $D_{12} = X_{11}X_{42} + X_{12}X_{41} - (X_{21}X_{32} + X_{22}X_{31})$.

Equation (12.14a) indicates that even when $D_1 = D_2 = 0$, D'_1 is not equal to 0 if the chromosome frequencies in populations 1 and 2 are not the same, i.e., $D_{12} \neq 0$. That is, if $D_{12} \neq 0$, linkage disequilibrium may be generated by migration even when the two parental populations are in linkage equilibrium (Cavalli-Sforza and Bodmer 1971).

By using equations (12.14a)–(12.14c), it is possible to derive the formulas for the linkage disequilibria for the tth generation after migration started (Nei and Li 1973), but the general formulas are quite complicated. However, Nei and Li's formulas indicates that when r is relatively small, a considerable amount of linkage disequilibrium may be developed by migration alone. Of course, if there is no selection and the population size is large, the two populations will eventually have the same chromosome frequencies as long as m_1 and m_2 are larger than 0. However, it may take a long time before this situation is attained. Feldman and Christiansen (1975) extended the above mathematical model to the case of n subpopulations.

In the above formulation, we ignored the effect of mutation. If one is interested in long-term evolution, however, the effect of mutation as well as the effect of genetic drift should be considered. If we consider all these factors, the mathematical formulation necessarily becomes very complicated. The reader who is interested in this problem may refer to Ohta (1973a, 1982a, 1982b). Note also that if we consider long-term evolution, the assumption of constant migration pattern and constant population sizes becomes questionable. In practice, the migration pattern in natural populations almost never stays constant for a long time. Nevertheless, Ohta's theoretical study is useful for obtaining a rough idea about the joint effect of mutation, migration, and genetic drift on the dynamics of linkage disequilibrium.

Natural Selection

In population genetics, *natural selection* means the differential rates of reproduction among different genotypes. Thus, when viability and fertility are the same for all genotypes, there is no natural selection. Natural selection is an important factor that causes *adaptive change* of populations. Adaptive change of a population occurs by substitution of selectively advantageous genes for less advantageous ones in the population. In population genetics, the process of gene substitution is described by the change of gene frequency in population. Selective advantage of a gene depends on whether or not the gene increases the *fitness* of the individual carrying the gene. Fitness is measured in terms of the number of offspring an individual produces. Since the size of a natural population is more or less constant in ordinary circumstances, it is convenient to measure fitness in terms of the relative number of offspring among different genotypes.

In the classical theory of natural selection developed by Fisher (1930), Wright (1931), and Haldane (1932), it is customary to assign a constant value of fitness for each genotype throughout the evolutionary process. In practice, the fitness of a genotype would not stay constant for all generations. However, this simple theory is useful for understanding how the genetic structure of a population changes by natural selection. In the following, we consider the basic principles of this theory.

Selection at a Single Locus

Consider a pair of alleles, A_1 and A_2, at a locus in a randomly mating diploid population. We assume that generations are discrete. Let x_1 and x_2 ($= 1 - x_1$) be the relative frequencies of alleles A_1 and A_2 in a generation, respectively, and designate the fitnesses of the three possible gen-

Table 12.1   Frequencies and fitnesses of
genotypes A_1A_1, A_1A_2, and A_2A_2 at a locus.

Genotype	A_1A_1	A_1A_2	A_2A_2
Frequency	x_1^2	$2x_1x_2$	x_2^2
Fitness	W_{11}	W_{12}	W_{22}

otypes A_1A_1, A_1A_2, and A_2A_2 by W_{11}, W_{12}, and W_{22}, respectively.
Under random mating, the frequencies of the three genotypes before
selection follow Hardy-Weinberg proportions and become as given in
table 12.1. The gene frequency in the next generation is therefore given
by

$$x_1' = [x_1^2 W_{11} + (1/2) \times 2x_1x_2 W_{12}]/\bar{W},$$

$$= x_1[x_1 W_{11} + x_2 W_{12}]/\bar{W}, \tag{12.15}$$

where $\bar{W} = x_1^2 W_{11} + 2x_1x_2 W_{12} + x_2^2 W_{22}$ is the mean fitness of the population. The amount of change in gene frequency per generation then be-
comes

$$\Delta x_1 = x_1' - x_1$$

$$= x_1(1-x_1)[x_1(W_{11}-W_{12}) + (1-x_1)(W_{12}-W_{22})]/\bar{W}. \tag{12.16}$$

From (12.16), it is easy to see that Δx_1 depends on the relative values
of W_{11}, W_{12}, and W_{22} and not on the absolute values. Thus, we can
write $W_{11}=1$, $W_{12}=1-h$, and $W_{22}=1-s$ or $W_{11}=1-s_1$, $W_{12}=1$,
and $W_{22}=1-s_2$. The quantities h, s, etc., are called *selection coefficients*.
Let us consider some special cases.

(1) Semidominant gene ($W_{11}=1$, $W_{12}=1-s/2$, $W_{22}=1-s$).

$$\Delta x_1 = sx_1x_2/(2\bar{W}). \tag{12.17}$$

(2) Completely dominant gene ($W_{11}=W_{12}=1$, $W_{22}=1-s$).

$$\Delta x_1 = sx_1x_2^2/\bar{W}. \tag{12.18}$$

(3) Completely recessive gene ($W_{11}=1$, $W_{12}=W_{22}=1-s$).

$$\Delta x_1 = sx_1^2x_2/\bar{W}. \tag{12.19}$$

(4) Overdominant gene ($W_{11}=1-s_1$, $W_{12}=1$, $W_{22}=1-s_2$).

$$\Delta x_1 = x_1 x_2 [s_2 - (s_1 + s_2) x_1] / \bar{W}. \tag{12.20}$$

Formulas (12.16)–(12.20) are nonlinear difference equations, so that it is not easy to solve for the gene frequency for an arbitrary generation. However, if we use a computer, the gene frequency can easily be obtained by recurrence formula (12.15), starting from a given initial value. Thus, the entire process of gene frequency change can be studied. In (12.17)–(12.19), Δx_1 is always positive as long as s remains positive. Therefore, the frequency of A_1 increases until it is fixed in the population. On the other hand, Δx_1 in (12.20) is positive if x_1 is less than $\hat{x}_1 = s_2 / (s_1 + s_2)$ but negative if x_1 is larger than $\hat{x}_1$. Therefore, the frequency of A_1 tends to be $\hat{x}_1$, where $\Delta x_1 = 0$. We shall discuss this problem in more detail later. Note also that the amount of gene frequency change due to natural selection is generally much larger than that due to mutation, since selection coefficients are usually larger than the mutation rate. For example, in the case of semidominant genes with $s = 0.02$ and $x_1 = 0.5$, Δx_1 becomes 0.0025, which is much larger than the value obtained by (12.2) when u and v are of the order of 10^{-5}.

Unless selection coefficients are very large, $\bar{W}$ is close to 1 and Δx_1 is small. In this case, equation (12.16) may be approximated by

$$\frac{dx}{dt} = x(1-x)[x(W_{11} - W_{12}) + (1-x)(W_{12} - W_{22})], \tag{12.21}$$

where $x = x_1$ and t stands for time in generations. It is easy to solve the above differential equation. For semidominant genes, (12.21) becomes

$$\frac{dx}{dt} = \frac{1}{2} s x (1-x). \tag{12.22}$$

Solution of this equation gives

$$t = \frac{2}{s} \log_e \frac{x_t (1 - x_0)}{x_0 (1 - x_t)}, \tag{12.23}$$

or

$$x_t = \frac{1}{1 + \left(\dfrac{1 - x_0}{x_0}\right) e^{-st/2}}, \tag{12.24}$$

where x_0 is the initial frequency of x. Therefore, the gene frequency increases logistically. For the cases of dominant and recessive genes, we can obtain similar equations. In these cases, however, it is more convenient to use equations equivalent to (12.23). They become as follows.

For a dominant gene,

$$t = \frac{1}{s}\left[\log_e \frac{x_t(1-x_0)}{x_0(1-x_t)} + \frac{1}{1-x_t} - \frac{1}{1-x_0}\right]. \tag{12.25}$$

For a recessive gene,

$$t = \frac{1}{s}\left[\log_e \frac{x_t(1-x_0)}{x_0(1-x_t)} - \frac{1}{x_t} + \frac{1}{x_0}\right]. \tag{12.26}$$

These equations are useful when we want to know the number of generations required for gene frequency to change from a given value to

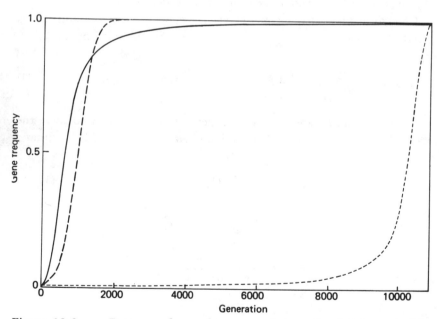

Figure 12.2.   Patterns of gene frequency changes for dominant (solid line), semidominant (broken line), and recessive (dotted line) genes under selection. The initial gene frequency (x_0) is 0.01 and selection coefficient (s) is 0.01 for all cases.

another. They can also be used for estimating s when x_0, x_t, and t are known (e.g., Mukai and Yamazaki 1980).

Figure 12.2 shows the pattern of gene frequency changes for dominant, semidominant, and recessive genes with an initial frequency of $x_0 = 0.01$. For all cases, $s = 0.01$ is assumed. The frequency of a semidominant gene increases logistically and reaches 0.999 in about 2,000 generations. The frequency of a dominant gene increases rapidly in early generations, but the rate of increase gradually declines in later generations. The frequency of a recessive gene increases very slowly when it is small but very rapidly when it is high.

Selection at Two Loci

The above theory of gene frequency change for a single locus applies to any locus in the genome, as long as the effect of the locus on the fitness is independent of the effects of other loci. When there is gene interaction or epistasis among different loci, however, the single-locus theory is no longer applicable.

Consider two loci each with two alleles A_1, A_2 and B_1, B_2, and denote the frequencies of chromosomes A_1B_1, A_1B_2, A_2B_1, and A_2B_2 in a generation by X_1, X_2, X_3, and X_4, respectively. There are nine possible genotypes, and their frequencies are given in table 7.5. This table also includes the fitnesses of the nine genotypes. It is not difficult to obtain the chromosome frequencies in the next generation from this table. The frequency of A_1B_1 is given by

$$X_1' = [X_1^2 W_{11} + X_1 X_2 W_{12} + X_1 X_3 W_{13} + \{X_1 X_4(1-r) + rX_2 X_3\}W_{14}]/\bar{W}$$
$$= [X_1 W_1 - rW_{14}D]/\bar{W}, \tag{12.27a}$$

where $W_1 = X_1 W_{11} + X_2 W_{12} + X_3 W_{13} + X_4 W_{14}$, and

$$\bar{W} = X_1^2 W_{11} + 2X_1 X_2 W_{12} + 2X_1 X_3 W_{13} + 2(X_1 X_4 + X_2 X_3)W_{14}$$
$$+ X_2^2 W_{22} + 2X_2 X_4 W_{24} + X_3^2 W_{33} + 2X_3 X_4 W_{34} + X_4^2 W_{44}.$$

Similarly, the frequencies of A_1B_2, A_2B_1, and A_2B_2 in the next generation are given by

$$X_2' = [X_2 W_2 + rW_{14}D]/\bar{W}, \tag{12.27b}$$

$$X_3' = [X_3 W_3 + r W_{14} D]/\bar{W}, \tag{12.27c}$$

$$X_4' = [X_4 W_4 - r W_{14} D]/\bar{W}, \tag{12.27d}$$

where

$$W_2 = X_1 W_{21} + X_2 W_{22} + X_3 W_{23} + X_4 W_{24},$$

$$W_3 = X_1 W_{31} + X_2 W_{32} + X_3 W_{33} + X_4 W_{34},$$

$$W_4 = X_1 W_{41} + X_2 W_{42} + X_3 W_{43} + X_4 W_{44}.$$

The changes of chromosome frequencies per generation are therefore given by

$$\Delta X_1 = X_1' - X_1$$

$$= [X_1(W_1 - \bar{W}) - r W_{14} D]/\bar{W}, \tag{12.28a}$$

$$\Delta X_2 = [X_2(W_2 - \bar{W}) + r W_{14} D]/\bar{W}, \tag{12.28b}$$

$$\Delta X_3 = [X_3(W_3 - \bar{W}) + r W_{14} D]/\bar{W}, \tag{12.28c}$$

$$\Delta X_4 = [X_4(W_4 - \bar{W}) - r W_{14} D]/\bar{W}. \tag{12.28d}$$

These equations are due to Lewontin and Kojima (1960), but equivalent equations had been obtained earlier by Kimura (1956) using a continuous time model.

The above expressions are simultaneous nonlinear difference equations, and the general solutions are not available. However, if we use a computer, the chromosome frequencies for any generation can be obtained by using equations (12.27a–d). The patterns of chromosome frequency changes by natural selection vary greatly with genotype fitness, recombination value, and initial linkage disequilibrium. If there is no gene interaction between loci and the initial linkage disequilibrium is 0, the chromosome frequencies are approximately given by the products of gene frequencies, and the gene frequency at one locus changes independently of the gene frequency at the other locus.

The departure of chromosome frequencies from linkage equilibrium can be measured by another quantity. That is,

$$Z = X_1 X_4 / (X_2 X_3), \tag{12.29}$$

which is related to D by $Z = 1 + D/(X_2 X_3)$. The natural logarithm of Z,

$$\log_e Z = \log_e X_1 - \log_e X_2 - \log_e X_3 + \log_e X_4,$$

has the same sign as that of D. If D is 0, $\log_e Z$ is also 0. If the amounts of changes in chromosome frequencies per generation are small, we have

$$\Delta \log_e Z = \frac{\Delta Z}{Z} = \frac{\Delta X_1}{X_1} - \frac{\Delta X_2}{X_2} - \frac{\Delta X_3}{X_3} + \frac{\Delta X_4}{X_4} \qquad (12.30)$$

approximately. Mathematically, the above formula does not hold when the effects of the second and higher-order terms of chromosome frequency changes are large, but if the two loci are loosely linked with weak gene interaction, it gives a good approximation (Kimura 1965a; Nagylaki 1974). Substituting ΔX_i ($i = 1, \cdots, 4$) into the above expression, we have

$$\bar{W} \Delta \log_e Z = W_1 - W_2 - W_3 + W_4 - r W_{14} D \left(\frac{1}{X_1} + \frac{1}{X_2} + \frac{1}{X_3} + \frac{1}{X_4} \right)$$

$$= E - r W_{14} D X, \qquad (12.31)$$

where $E = W_1 - W_2 - W_3 + W_4$ and $X = \sum_{i=1}^{4} X_i^{-1}$. Since W_i is the average fitness of the ith gamete, E measures the effect of gene interaction or *epistasis* in fitness. If $E = 0$, there is no epistasis.

In the case of $E = 0$, (12.31) reduces to

$$\bar{W} \Delta \log_e Z = -r W_{14} D X. \qquad (12.32)$$

Since $\bar{W}$, W_{14}, and X are all positive and $Z = 1 + D/(X_2 X_3)$, $\log_e Z$ and D decrease if D is positive, but increase if D is negative unless r is 0. Therefore, D eventually becomes 0. Namely, if there is no epistasis, linkage disequilibrium converges to 0. If there is epistasis, the change in $\log_e Z$ is determined by $E - r W_{14} D X$. If $E > 0$, $\log_e Z$ and D will increase whenever D is negative or zero. If $E < 0$, they will decrease whenever D is positive or zero. Thus, D tends to have the same sign as E (Felsenstein 1965). Note, however, that E is not constant when chromosome frequencies are changing in the presence of epistasis.

An important aspect of linkage disequilibrium is that the gene frequency change at a locus may be affected by selection at a closely linked second locus. In general, it is not known what kind of selection is operating at closely linked loci. If there is linkage disequilibrium between two loci and one of the loci is subject to natural selection, the gene frequency at the other locus may change even if there is no selection at all at this locus (e.g., Maynard Smith and Haigh 1974; Thomson 1977; Hedrick et al. 1978).

One possible example is given in figure 12.3, where the frequency changes of allele F at the esterase 6 locus in laboratory populations of *Drosophila melanogaster* (MacIntyre and Wright 1966) are presented, along with the result from a computer simulation. In this simulation, the esterase locus is assumed to be neutral but linked with a second locus that is subject to overdominant selection. The recombination value between the two loci is 0.15. The esterase 6 locus has two alleles, F and S, whereas the second locus is assumed to have B and b. The fitnesses of BB, Bb, and bb used are 0.6, 1, and 0.9, so that the equilibrium

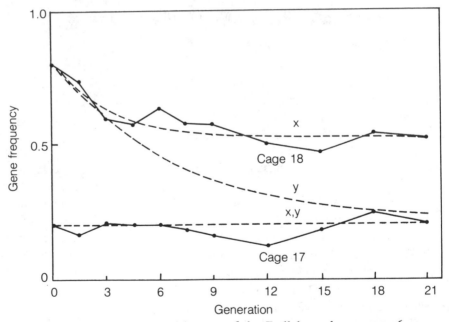

Figure 12.3.   Frequency changes of the *F* allele at the esterase 6 locus in two cage populations of *Drosophila melanogaster* studied by MacIntyre and Wright (1966) and the results of a computer simulation (broken lines). In this computer simulation, the esterase 6 locus was assumed to be neutral but linked with an overdominant locus (*B locus*). *x* is the frequency of the *F* allele, while *y* is the frequency of an allele at the *B* locus. From Nei (1975).

gene frequency of B is 0.2. The initial frequencies of chromosomes FB, Fb, SB, and Sb were 0.2, 0, 0, and 0.8, respectively, in one experiment (cage 17) and 0.8, 0, 0, and 0.2 in the other (cage 18). In the former case, the frequency (y) of allele B was 0.2 from the beginning, so that there was no change. Consequently, the frequency (x) of allele F also did not change. In the latter case, however, the B gene frequency gradually declined with increasing generation, and the frequency of the F allele followed the change of the B gene frequency in early generations because of linkage, even if this locus was subjected to no selection. In both cases, the frequency change of the F allele is close to the experimental result. It should be noted, however, that this is not the only computer simulation that closely mimics the experimental data. Similar results may be obtained by changing the initial chromosome frequencies and the recombination value, and also by adding some more loci. In fact, if we consider a number of linked loci with deleterious alleles, as was considered in figure 8.6, a similar result may be obtained without the aid of any overdominant loci. This sort of linkage effect always makes it difficult to draw a definite conclusion about selection from experimental data.

Equilibrium Frequencies of Genes and Chromosomes

In the foregoing sections, we were mainly concerned with the directional changes of gene frequency in populations. However, if there are some opposing factors such as mutation and selection, gene frequency may reach an *equilibrium gene frequency.* Theoretically, there are many different ways in which gene frequency equilibria may arise (Crow and Kimura 1970; Hedrick 1983). In this book, we discuss only a few important cases.

In the classical theory of population genetics, the equilibrium gene frequency was an important subject of study. Until around 1966, a majority of genetic polymorphisms observed in nature were thought to be *stable polymorphisms,* in the sense that if the gene frequency is perturbed from the equilibrium point by some factor, it comes back to the original point sooner or later. In particular, stable polymorphism due to overdominant selection was regarded as an important source of genetic variation in natural populations (Dobzhansky 1951; Ford 1964). However, there are only a few cases in which authentic overdominance exists at the gene level, and recent studies of the evolutionary changes of proteins and DNA indicate that there must be a large number of transient poly-morphisms that arise in the process of gene substitution in populations. Also, the classical theory of gene frequency equilibrium due to the for-ward and backward mutations between a pair of neutral alleles is now known to be unrealistic, as was mentioned earlier.

Mutation-Selection Balance for Deleterious Genes

Although most new mutations would be different from the alleles extant in the population at the nucleotide level, deleterious mutations often result in the same or similar phenotypic effects. In this case, all deleterious genes can be pooled and treated as a single allele. Since most deleterious mutations are selected against, the gene frequency ultimately reaches an equilibrium point. Let us denote the deleterious allele and its wild-type allele by A_2 and A_1, respectively, and let x_2 be the frequency of A_2, so that the frequency of A_1 is $x_1 = 1 - x_2$. If the fitnesses of gen-otypes A_1A_1, A_1A_2, and A_2A_2 are 1, $1 - h$ and $1 - s$, respectively, the amount of change in x_2 per generation is

$$\Delta x_2 = -x_1 x_2 [h + (s - 2h)x_2]/\bar{W} \tag{12.33}$$

from (12.16), where $\bar{W} = 1 - 2hx_1x_2 - sx_2^2$. On the other hand, the amount of change in gene frequency due to mutation is $\Delta x_2 = ux_1$, where u is the mutation rate from A_1 to A_2. Therefore, combining these two ef-fects, we have

$$\Delta x_2 = ux_1 - x_1 x_2 [h + (s - 2h)x_2]/\bar{W}. \tag{12.34}$$

At equilibrium, Δx_2 should be 0, so that

$$u = x_2 [h + (s - 2h)x_2], \tag{12.35}$$

approximately, since $\bar{W}$ is close to 1 for a deleterious gene at equilib-rium.

The equilibrium gene frequency ($\hat{x}_2$) can be obtained by solving (12.35) for x_2. It becomes

$$\hat{x}_2 = \frac{-h + \sqrt{h^2 + 4u(s - 2h)}}{2(s - 2h)}. \tag{12.36}$$

In the case of completely recessive genes $h = 0$, so that

$$\hat{x}_2 = \sqrt{u/s}. \tag{12.37}$$

If h is much larger than $\sqrt{(su)}$, the square root term in (12.36) can be written as

$$h\left[1 + \frac{4u(s - 2h)}{h^2}\right]^{1/2} = h\left[1 + \frac{2u(s - 2h)}{h^2} + \cdots\right]$$

by using Taylor's series. Therefore, if the degree of dominance of the deleterious gene is sufficiently large, we have

$$\hat{x}_2 = u/h, \tag{12.38}$$

approximately.

Equations (12.37) and (12.38) have been used by many authors particularly in man and *Drosophila*. When equation (12.37) is used, however, some caution should be exercised. First, (12.37) is correct only in very large populations. If population size is smaller than the reciprocal of the mutation rate, the actual gene frequency is expected to be smaller than the value given by this formula (Wright 1969). Second, the equilibrium gene frequency of a recessive deleterious gene is affected considerably by a small positive or negative selection in heterozygotes. In most cases, such a small heterozygous effect on fitness cannot be detected experimentally. Third, it takes a long time for the frequency of a recessive gene to reach the equilibrium value once it is disturbed. Particularly in human populations, mating and migration patterns have changed considerably in the last few centuries. Therefore, the frequencies of many recessive deleterious genes may not be at equilibrium.

Equation (12.38) is, however, applicable for a variety of situations, if h is large. As an example, let us consider achondroplastic dwarfism in man, which is caused by a single dominant gene. The fitness of heterozygotes for this gene has been estimated to be $1 - h = 0.196$ (cf. Stern 1973). In a survey conducted in Denmark, ten heterozygotes were found in a sample of 94,075 newborns. Eight out of these ten heterozygotes were fresh mutations. Thus, the mutation rate is $8/(2 \times 94,075) = 4.25 \times 10^{-5}$ per generation. On the other hand, the gene frequency ($\hat{x}_2$) in newborns is estimated to be $10/(2 \times 94,075) = 0.0000531$. Using this value and the estimate of fitness, the mutation rate is computed to be $u = h\hat{x}_2 = 0.0000426$ per generation. This estimate agrees quite well with the direct estimate of mutation rate.

Overdominant Selection

If there are two opposing forces of selection, gene frequency equilibria may arise. The simplest case is overdominant selection first studied by Fisher (1922). Let the fitnesses of A_1A_1, A_1A_2, and A_2A_2 be $1-s_1$, 1, and $1-s_2$. Then, the amount of change in the frequency of A_1 per generation is given by (12.20), where $\bar{W} = 1 - s_1x_1^2 - s_2x_2^2$. At equilibrium, $\Delta x_1 = 0$, so that the equilibrium gene frequency is

$$\hat{x}_1 = s_2/(s_1 + s_2). \tag{12.39}$$

Using this equilibrium gene frequency, (12.20) may be written as

$$\Delta x_1 = (s_1 + s_2)x_1x_2(\hat{x}_1 - x_1)/\bar{W}. \tag{12.40}$$

x_1 increases if it is smaller than $\hat{x}_1$, but decreases if it is larger than $\hat{x}_1$. Thus, if there is any deviation of x_1 from the equilibrium gene frequency, the deviation is reduced in each generation, and the gene frequency eventually reaches the equilibrium value. Therefore, this equilibrium is stable. Once the gene frequency reaches the stable equilibrium, it will stay there forever unless the selection coefficients change. Note also that, unlike the case of the mutation-selection balance, the equilibrium gene frequency can be high. Therefore, a relatively small number of overdominant loci may create a large amount of genetic variability.

Because of its simplicity, the overdominance model has been used by many authors to explain genetic polymorphisms in natural populations. As mentioned earlier, however, there are not many cases in which overdominance has been proven. An oft-cited example of overdominance is the polymorphism of chromosome inversions in *Drosophila pseudoobscura*. In the third chromosome of this species, there are many different gene arrangements in natural populations. Since there is virtually no recombination within the inverted segment in heterozygotes, each gene arrangement behaves just like a single allele. Wright and Dobzhansky (1946) studied the changes in frequency of gene arrangements Standard (ST) and Chiricahua (CH) in a laboratory population and showed that the ST chromosome eventually reaches an equilibrium frequency of about 70 percent. From data on chromosome frequency changes, they estimated the genotype fitnesses as follows:

Genotype	ST/ST	ST/CH	CH/CH
Relative fitness	1–0.3	1	1–0.7

The expected equilibrium frequency of the ST chromosome is therefore $0.7/(0.3+0.7)=0.7$, which agrees quite well with the observed value.

However, this sort of overdominance at the chromosome level does not necessarily mean overdominance at the gene level, since the inverted segment of a chromosome generally includes a large number of genes and the genes in this segment are almost completely isolated from those of other chromosomes. Therefore, as explained in chapter 8, associative overdominance is expected to occur for a pair of gene arrangements. The extent of associative overdominance would be particularly high when a cage population is started from a small number of individuals (Ohta 1971; Jones and Yamazaki 1974).

A well-established case of overdominance is the polymorphism at the sickle cell anemia gene locus in African black populations. This anemia is caused by the abnormal hemoglobin S. The β chain of the normal hemoglobin A has glutamic acid at position 6, whereas hemoglobin S has valine, as was mentioned in chapter 10. The homozygotes for the S gene are almost lethal in Africa, but the gene frequency is as high as 10 to 20 percent in some areas. The prevalence of this gene is associated with a high endemic incidence of malaria. Allison (1955) showed that the heterozygotes for the S gene are more resistant to malaria than normal homozygotes and thus have a higher fitness than both homozygotes. This was later confirmed by studies on mortality due to malaria (Allison 1964; Motulsky 1964). It seems that in malaria-endemic areas the sickle cell heterozygotes have a selective advantage of about 10 to 20 percent over normal homozygotes.

There are several other mutant genes which apparently show heterozygote advantage due to increased resistance to malaria. The genes for hemoglobin variants C (Glu $\rightarrow$ Lys at position 6 of the β chain), E (Glu $\rightarrow$ Lys at position 26 of the β chain), and thalassemia (reduced production of hemoglobins), which also cause anemia in homozygous condition, all show a high frequency in malaria-endemic areas (Livingstone 1967). Furthermore, a mutant gene which causes the deficiency of the enzyme glucose-6-phosphate dehydrogenase (G6PD) is also frequent in malarial areas.

Overdominance with Epistasis

Overdominance is an interaction between two alleles at a locus, while epistasis is an interaction between alleles of two different loci. Thus, one might suspect that epistasis itself is sufficient to maintain stable polymorphism without overdominance. Actually, this is not the case. To maintain polymorphisms, there must be overdominance at least at one locus but not necessarily at both loci (Kimura 1956).

If there is overdominant selection for both loci, there may be several stable or unstable equilibria for a given set of genotype fitnesses. This problem has been studied by Wright (1952), Lewontin and Kojima (1960), Karlin and Feldman (1969), and many others. Let us consider the following simple fitness model:

	A_1A_1	A_1A_2	A_2A_2
B_1B_1	$(1-s)(1-t)$	$1-t$	$(1-s)(1-t)$
B_1B_2	$1-s$	1	$1-s$
B_2B_2	$(1-s)(1-t)$	$1-t$	$(1-s)(1-t)$

Clearly, the fitnesses at the two loci are multiplicative and symmetric about heterozygotes; s and t are the selection coefficients for either homozygotes at the A and B loci, respectively. Multiplicative fitness is expected to occur if selection occurs independently at the two loci. It involves epistatic interaction since there are deviations in genotype fitnesses from additivity between two loci. By using (12.28), it can be shown that there are three equilibria (Bodmer and Felsenstein 1967; Kimura and Ohta 1971a):

$$\hat{X}_1 = \hat{X}_4 = \frac{1}{4}\left[1 + \left(1 - \frac{4r}{st}\right)^{1/2}\right], \qquad (12.41a)$$

$$\hat{X}_1 = \hat{X}_4 = \frac{1}{4}\left[1 - \left(1 - \frac{4r}{st}\right)^{1/2}\right], \qquad (12.41b)$$

$$\hat{X}_1 = \hat{X}_4 = 1/4, \qquad (12.41c)$$

whereas $\hat{X}_2 = \hat{X}_3 = 1/2 - \hat{X}_1$ for each of the above equilibria. Note that the gene frequencies of A_1 and B_1 are both 0.5 in all cases. The first two equilibria with $\hat{D} = \pm(1/4)\sqrt{\{1 - (4r/st)\}}$ exist only when $r < st/4$. Otherwise, the system will move to the third equilibrium. In practice, s and t would rarely exceed 0.1. If $s = t = 0.1$, r must be smaller than 0.0025 for the first two equilibria to exist. Therefore, only when the recombination value is extremely small do the equilibria with linkage disequilibria become important.

However, when there are many loci showing multiplicative overdominance, strong linkage disequilibria may be developed even if selection coefficients are relatively small and the recombination values among loci are rather large (Franklin and Lewontin 1970; Slatkin 1972; Feldman et al. 1974). This is caused by higher-order interaction among loci, and in the presence of random genetic drift only a few types of chromosomes become predominant in the population. When this status is attained, many loci show strong linkage disequilibria.

Following this discovery, Franklin and Lewontin (1970) and Lewontin (1974) proposed the theory of "chromosome as a unit of selection." According to this theory, a chromosome with many linked loci behaves just like a single locus with multiple alleles. Consequently, the genetic load or the amount of selective deaths in the population is kept at a relatively low level. Lewontin (1974) took Prakash and Lewontin's (1968) data on strong linkage disequilibria between inversion chromosomes and electromorphs in *Drosophila* as support for this theory. As discussed earlier, however, this type of linkage disequilibrium can be explained by neutral alleles that happened to exist on a new inversion chromosome. Therefore, it cannot be support for the theory. Furthermore, extensive study of linkage disequilibria among enzyme loci, conducted by Mukai et al. (1971), Langley et al. (1974), and others, has shown that alleles at different enzyme loci are generally in linkage equilibrium. Therefore, as far as enzyme loci are concerned, overdominant selection with epistasis seems to be rare in nature.

In general, however, there arise stable equilibria with $D \neq 0$ if the two interacting loci are closely linked and there is overdominance at both loci. The human HLA loci are highly polymorphic and show strong linkage disequilibria among them. This polymorphism may be maintained by overdominant selection (Hedrick and Thomson 1983). Nevertheless, it should be noted that linkage disequilibrium is generated by a number of factors, i.e., unique mutation, migration, epistatic selection, and random genetic drift (Hill and Robertson 1968; Ohta and Kimura 1969, 1971b). Selection without epistasis may also cause a stable linkage disequilibrium in the presence of migration (Karlin and McGregor 1972; Li and Nei 1974; Christiansen and Feldman 1975). It is, therefore, difficult to determine the presence or absence of selection or its pattern by the study of linkage disequilibrium alone.

Other Types of Balancing Selection

Theoretically, there are several other types of balancing selection which may produce stable polymorphism with intermediate gene frequency. Wright and Dobzhansky (1946) showed that their experimental data on the frequency changes of inversion chromosomes can also be explained by frequency-dependent selection. Their model of frequency-dependent selection is as follows:

Genotype	Frequency	Fitness
A_1A_1	x_1^2	$1 + a - bx_1$
A_1A_2	$2x_1(1-x_1)$	1
A_2A_2	$(1-x_1)^2$	$1 - a + bx_1$

That is, the fitness of A_1A_1 decreases as the gene frequency (x_1) of A_1 increases, whereas that of A_2A_2 increases with increasing x_1. Therefore, the gene frequency, x_1, reaches a stable equilibrium. The amount of change of gene frequency per generation is given by

$$\Delta x_1 = x_1(1-x_1)(a-bx_1)/\bar{W}, \qquad (12.42)$$

where $\bar{W} = 1 - (a - bx_1)(1 - 2x_1)$. Therefore,

$$\hat{x}_1 = a/b. \qquad (12.43)$$

Wright and Dobzhansky's estimates of a and b for their experimental data are 0.902 and 1.288, respectively, so that $\hat{x}_1 = 0.7$, as obtained earlier.

In recent years, many other models of frequency-dependent selection have been developed (see Wright 1969; Hedrick 1983). Experimental data which support frequency-dependent selection have also increased (Wallace 1981; Hedrick 1983). Yet the biological mechanism of frequency-dependent selection is not well understood (Lewontin 1974). It is possible that some seemingly frequency-dependent selection is actually caused by loci closely linked to a marker gene or by subtle environmental changes during the process of population changes. In some cases, earlier claims of frequency-dependent selection were not supported by later studies (e.g., Dolan and Robertson 1975; Yoshimaru and Mukai 1979). More studies on the biological mechanism of frequency-dependent selection are required.

Levene (1953) showed that stable polymorphism may occur when a population occupies a wide variety of niches among which the selection coefficient for an allele varies. Several similar models are reviewed by Hedrick et al. (1976) and Felsenstein (1976). In these models, however, rather severe conditions are required for the equilibrium to be stable. Under certain circumstances, stable polymorphism may also occur when selection coefficients vary in different generations (Haldane and Jayakar 1963; Gillespie and Langley 1974; Gillespie 1978). Here again, however, a severe condition is required. Particularly in finite populations, the "power of holding polymorphisms" seems to be very weak (Hedrick 1974; Takahata 1981).

CHAPTER THIRTEEN

POPULATION GENETICS THEORY: STOCHASTIC MODELS

The deterministic theory of population genetics discussed in the preceding chapter is based on the assumption that the population size is so large that there is no sampling error in the process of gene frequency change from one generation to the next. In reality, however, the number of breeding individuals in natural populations is often quite small, so that sampling error cannot be neglected. In this case, gene frequency does not change uniquely from one value to another, but any particular change occurs only with a certain probability. This sort of probabilistic change is called *stochastic change*. In population genetics, this stochastic change is referred to as *genetic drift*. The stochastic change of gene frequency may also be generated by random fluctuation of selection intensity among different generations. In general, a stochastic model is more realistic than a deterministic one; the latter is merely a special case of the former. However, the mathematics of stochastic models is more complicated, and exact solutions are often difficult to obtain. Nevertheless, after the pioneering work of Fisher and Wright, many important problems have been solved in terms of stochastic models. The stochastic theory of population genetics has been particularly important in the analysis and interpretation of data on molecular polymorphism and evolution (Kimura 1983a).

Stochastic Change of Gene Frequency

Mathematical Models

If a mutation occurs in a population, the initial survival of the mutant gene depends largely on chance regardless of its selective advantage or the population size. This can be seen in the following way. Let A_1 and A_2 be the mutant and the wild-type genes in a population, respectively. In a diploid organism, the mutant gene appears first in heterozygous condition (A_1A_2). In a bisexual organism, this individual will mate with a wild-type homozygote (A_2A_2). The mating $A_1A_2 \times A_2A_2$, however, may not produce any offspring for some biological reason other than the effect of the A_1 gene. The mutant gene will then disappear from the population in the next generation. Survival of the mutant gene is not assured even if $A_1A_2 \times A_2A_2$ produces some offspring. This is because the A_1A_2 genotype will appear only with probability 1/2 in the offspring. Thus, if two offspring are born from this mating, the chance that no A_1A_2 will appear is 0.25.

Let us now study this problem in more detail. Consider a randomly mating population of a monoecious diploid organism. We assume that each individual produces a large number of offspring and that exactly N of all offspring survive to maturity. Let x be the frequency of mutant gene A_1 among the gametes produced in a generation. The expected frequencies of genotypes A_1A_1, A_1A_2, and A_2A_2 after fertilization are then given by x^2, $2x(1-x)$, and $(1-x)^2$, respectively. We now consider selection with constant fitness, and let the fitnesses of A_1A_1, A_1A_2, and A_2A_2 be $1+s$, $1+h$, and 1, respectively. After selection, the gene frequency of A_1 changes from x to

$$\xi = \frac{x\{1+sx+h(1-x)\}}{1+2hx(1-x)+sx^2} \tag{13.1}$$

[see equation (12.15)]. The number of individuals that survive to maturity is N by definition. We assume that the 2N genes carried by these N individuals are a random sample from the gene pool after selection, neglecting the fact that the actual survivors are genotypes rather than genes. It is known that this assumption does not affect the result appreciably unless the population size is extremely small. Since the frequency of A_1 after selection is ξ and since 2N genes are chosen at random from the gene pool, the number of A_1 genes among the adults may vary from 0 to 2N. The probability that the number of A_1 genes becomes j is given by the jth term of the binomial expansion of $[\xi+(1-\xi)]^{2N}$. That is,

$$p(j) = \binom{2N}{j} \xi^j (1-\xi)^{2N-j}. \tag{13.2}$$

The gene frequency is given by $x'=j/2N$, and the mean [$M(x')$] and variance [$V(x')$] of x' are

$$M(x') = \xi, \qquad V(x') = \frac{\xi(1-\xi)}{2N}. \tag{13.3}$$

Note that the mean gene frequency is x if there is no selection, since $\xi = x$ in this case.

If $x' = 0$, there are no longer A_1 genes in the population, and no gene frequency change occurs in the subsequent generations. Similarly, if $x' = 1$, A_1 genes are fixed in the population, and there is again no change in gene frequency in the subsequent generations. However, if $0 < x' < 1$, selection and random sampling of genes again operate in the next generation. This process continues until the A_1 gene is lost or fixed in the population.

Mathematically, this process is called a Markov chain. If there are N individuals in a population, there are $2N + 1$ possible gene frequency classes, i.e., 0, 1/2N, 2/2N, $\cdots$, 2N/2N. These classes are called *states* in probability theory. We call the gene frequency class $i/2N$ state i and denote by $f_t(x)$ the probability that the gene frequency is in state i at the tth generation, where $x = i/2N$. We have already seen that when the gene frequency at a generation is x, the probability that the gene frequency becomes x' in the next generation is given by (13.2). That is, this is the probability that the number of A_1 genes in the population changes from $i = 2Nx$ to $j = 2Nx'$. This is called the *transition probability* from state i to state j, and we now denote this by $p_{i,j}$. If we know $p_{i,j}$'s and $f_t(x)$ is given, we can easily obtain $f_{t+1}(x)$ by the following equations.

$$f_{t+1}(0) = p_{0,0} f_t(0) + p_{1,0} f_t\left(\frac{1}{2N}\right) + \cdots + p_{2N,0} f_t(1),$$

$$f_{t+1}\left(\frac{1}{2N}\right) = p_{0,1} f_t(0) + p_{1,1} f_t\left(\frac{1}{2N}\right) + \cdots + p_{2N,1} f_t(1), \tag{13.4}$$

$$\cdots$$

$$\cdots$$

$$f_{t+1}(1) = p_{0,2N} f_t(0) + p_{1,2N} f_t\left(\frac{1}{2N}\right) + \cdots + p_{2N,2N} f_t(1).$$

If we use matrix notation, the above simultaneous equations may be expressed in a simpler form. Let $\mathbf{f}_t$ be the column vector of state probabilities $f_t(0), f_t(1/2N), \cdots, f_t(1)$, and $\mathbf{P}$ be the following matrix.

$$\mathbf{P} = \begin{bmatrix} p_{0,0} & p_{1,0} & \cdots & p_{2N,0} \\ p_{0,1} & p_{1,1} & \cdots & p_{2N,1} \\ \cdot & \cdot & \cdots & \cdot \\ \cdot & \cdot & \cdots & \cdot \\ p_{0,2N} & p_{1,2N} & \cdots & p_{2N,2N} \end{bmatrix}.$$

The equation (13.4) can then be written as

$$\mathbf{f}_{t+1} = \mathbf{P}\mathbf{f}_t. \tag{13.5}$$

Therefore, the probability distribution of gene frequencies at the tth generation is given by

$$\mathbf{f}_t = \mathbf{P}^t\mathbf{f}_0, \tag{13.6}$$

where $\mathbf{f}_0$ is the initial probability distribution.

For a relatively small population, it is possible to obtain $\mathbf{f}_t$ by using a computer. In this case, either (13.4) or (13.6) may be used. One such example is given in figure 13.1, where $N = 10$ and no selection ($h = 0$ and $s = 0$) are assumed. The initial gene frequency was 0.5, so that $f_0(x) = 1$ for $x = 0.5$ but $f_0(x) = 0$ for all other states.

In the first generation, gene frequency is distributed as a binomial variate with mean 0.5 and variance $(0.5)^2/20 = 0.0125$. In subsequent generations, the distribution becomes flatter and flatter, and by the twentieth generation it becomes virtually uniform except for the terminal ($x = 0$ and $x = 1$) and a few subterminal classes. By this time, gene A_1 is lost from or fixed in the population with a probability of about 0.5. After this generation, the shape of the probability distribution of gene frequency among unfixed classes remains virtually the same, though the absolute probability of each gene frequency class is reduced at a rate of $1/(2N) = 0.05$ in every generation. The probabilities of classes $x = 0$ and $x = 1$ gradually increase and eventually become 0.5 when gene A_1 is completely lost or fixed. In the present case, there is no selection, so that the mean gene frequency is 0.5 throughout the process of gene frequency change.

In the study of evolution, it is important to know the probability of fixation of an advantageous mutation. This can also be studied by using equation (13.6). An example is given in table 13.1, where the fitnesses of A_1A_1, A_1A_2, and A_2A_2 are assumed to be 1, 0.9, and 0.8, respec-

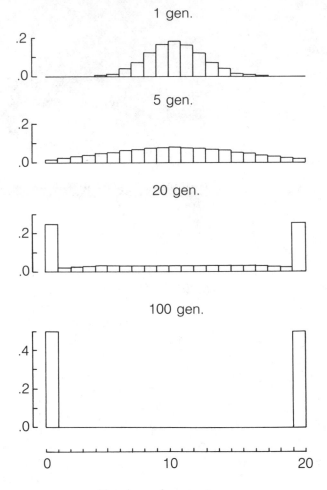

Figure 13.1.   Probability distributions of gene frequencies in randomly mating populations of $N = 10$. The initial gene frequency is 0.5. No selection is assumed.

tively. The population size is again 10, but the initial frequency is now $1/(2N) = 0.05$. The probability of fixation is very low in early generations, but gradually increases and eventually reaches 0.1755. If there were no selection, the gene would have been fixed with probability $1/(2N) = 0.05$. So, selection has increased the probability of fixation by

Table 13.1   Probabilities of fixation $[f(1)]$ and loss $[f(0)]$ of a
mutant gene (A_1) in a population of size $N = 10$. The fitnesses of
A_1A_1, A_1A_2, and A_2A_2 are assumed to be 1, 0.9, and 0.8,
respectively. The initial gene frequency is assumed to
be $1/2N = 0.05$.

Generation	1	2	3	10	50	∞
$f(1)$	6×10^{-26}	6×10^{-13}	5×10^{-9}	3×10^{-3}	.1540	.1755
$f(0)$	.3246	.4694	.5528	.7390	.8211	.8245

0.1255, but the gene has still been lost from the population with probability 0.8245.

In the above treatment of the stochastic change of gene frequencies, we used the Markov chain method. Although this method gives the exact distribution of gene frequencies for small populations, it cannot be used for large populations. Kimura (1955a,b) has shown that the stochastic change of gene frequencies can be studied more easily by using the diffusion method in probability theory. Indeed, he (Kimura 1955a) obtained a general formula for the gene frequency distribution for neutral alleles by using this method. The diffusion method requires a high level of mathematics, and the reader who is interested in this problem may refer to his original papers or to his 1964 review article.

So far we have considered the stochastic change of gene frequencies due to genetic drift. As mentioned earlier, however, the stochastic change may also occur by random fluctuation of selection intensity in different generations. This problem has been studied by Wright (1948), Kimura (1954, 1962), Jensen and Pollak (1969), Gillespie (1978), Takahata (1981), and others. The general effect of this factor is to spread the gene frequency distribution, but under certain circumstances an effect to retard the fixation of genes may be generated.

Variance of Gene Frequencies and Heterozygosity

We have seen that one of the properties of genetic drift is to spread the gene frequency distribution as time goes on. In the absence of selection, this property can be studied by the variance of gene frequencies. Consider a large number of populations of equal size N, in each of which random mating occurs. We assume that the initial gene frequency, p, is

the same for all populations. If there is no selection, the probability that
the gene frequency of A_1 in the first generation becomes $x = i/(2N)$ is

$$f(x) = \binom{2N}{i} p^i (1-p)^{2N-i},$$

from (13.2). This probability is equal to the relative frequency of populations that have gene frequency x. Therefore, the mean and variance
of x among all populations are

$$\bar{x} = E(x) = p, \tag{13.7a}$$

$$V_x = E(x-p)^2 = \frac{p(1-p)}{2N}, \tag{13.7b}$$

respectively.

In the next generation, the same random process operates for each
gene frequency class x in the first generation. Therefore, denoting by x'
the gene frequency in the second generation, we have

$$\bar{x}' = E(x') = E_1[E_2(x')] = E_1(x) = p,$$

where E_1 and E_2 denote expected value operators in the first and second
generations, respectively. Clearly, the mean of x' is the same as that of
x. The variance of x' is computed in the following way.

$$
\begin{aligned}
V_{x'} &= E(x'-p)^2 \\
&= E_1 E_2[(x'-x)+(x-p)]^2 \\
&= E_1 E_2[(x'-x)^2 + 2(x'-x)(x-p) + (x-p)^2] \\
&= E_1\left[\frac{x(1-x)}{2N} + (x-p)^2\right],
\end{aligned}
$$

since $E_2(x'-x)^2 = x(1-x)/(2N)$ and $E_2(x'-x) = 0$. Noting that

$$E_1[x(1-x)] = E_1(x) - E_1(x-p)^2 - p^2$$

$$= p - p^2 - \frac{p(1-p)}{2N},$$

we have

$$V_{x'} = \frac{1}{2N}\left[p - p^2 - \frac{p(1-p)}{2N}\right] + \frac{p(1-p)}{2N}$$

$$= \frac{p(1-p)}{2N}\left[\left(1 - \frac{1}{2N}\right) + 1\right].$$

It is now obvious that if the same process continues for t generations, the mean $(\bar{x}_t)$ and variance (V_t) of the gene frequency in the tth generation are given by

$$\bar{x}_t = p, \tag{13.8a}$$

$$V_t = \frac{p(1-p)}{2N}\left\{\left(1 - \frac{1}{2N}\right)^{t-1} + \cdots + \left(1 - \frac{1}{2N}\right) + 1\right\},$$

$$= p(1-p)\left\{1 - \left(1 - \frac{1}{2N}\right)^{t}\right\}. \tag{13.8b}$$

Therefore, the mean gene frequency remains constant for all generations, while the variance gradually increases as t increases. At $t = \infty$, the variance becomes $p(1-p)$. This corresponds to the case of complete fixation of alleles. Since we have assumed no selection and no mutation in the present case, alleles A_1 and A_2 are eventually fixed in the population with probabilities p and $1-p$, respectively.

Wright (1951, 1965) has called the ratio (F_{ST}) of V_t to $p(1-p)$ the fixation index. Clearly,

$$F_{ST} = V_t/[p(1-p)]$$

$$= 1 - \left(1 - \frac{1}{2N}\right)^{t}$$

$$\simeq 1 - e^{-t/2N}, \tag{13.9}$$

when N is large. Therefore, the fixation index is independent of the initial gene frequency and increases from 0 to 1 as t increases.

We have seen that genetic drift gradually increases the interpopulational variation of gene frequency. However, the genetic variability within populations gradually declines. This can be studied by considering the average frequency of heterozygotes within populations (H_t). The frequency of heterozygotes in a population having gene frequency x_t in the tth generation is given by $2x_t(1-x_t)$. Taking the average of $2x_t(1-x_t)$ over all populations, we have

$$H_t = 2E[x_t(1-x_t)] = 2E[x_t - (x_t^2 - p^2) - p^2]$$
$$= 2(p - V_t - p^2)$$
$$= 2p(1-p)\left(1 - \frac{1}{2N}\right)^t. \tag{13.10}$$

This is related to V_t by $H_t = 2p(1-p) - 2V_t$.

In the derivation of the above equation, we considered a single locus in a large group of populations of equal size. The above theory, however, can also be applied to a large number of independent neutral loci in a single population, if the initial gene frequency is the same for all loci. In this case, H_t stands for the average frequency of heterozygotes per locus in the population or the average frequency of heterozygous loci for an individual. This quantity is generally called *average heterozygosity* (chapter 8). In practice, of course, the assumption of an equal initial gene frequency is unrealistic except in artificial populations. However, if we replace $2p(1-p)$ by the average heterozygosity over all loci at the 0th generation, i.e., by $\overline{2p(1-p)}$, then equation (13.10) holds.

Equation (13.10) was derived for the case of two alleles at a locus, but it holds true for any number of alleles. Suppose that there are m alleles at a locus, and let x_i be the frequency of the ith allele in generation t. The heterozygosity is therefore given by $H_t = 2 \sum_{i<j} x_i x_j$. The next generation is formed by sampling $2N$ genes at random from this population, so that the gene frequencies (x_i') in generation $t+1$ follow a multinomial distribution. Thus, the expected heterozygosity in generation $t+1$ is

$$H_{t+1} = 2E\left(\sum_{i<j} x_i' x_j'\right) = 2 \sum_{i<j} E(x_i' x_j')$$
$$= 2\left(1 - \frac{1}{2N}\right) \sum_{i<j} x_i x_j$$
$$= \left(1 - \frac{1}{2N}\right) H_t. \tag{13.11}$$

Therefore, if we denote by H_0 the heterozygosity in generation 0, we have

$$H_t = H_0\left(1 - \frac{1}{2N}\right)^t$$
$$\simeq H_0 e^{-t/(2N)}. \tag{13.12}$$

This indicates that the average heterozygosity per locus (chapter 8) will decline at a rate of $1/(2N)$ per generation in the absence of mutation and selection.

Equation (13.11) can be used to derive the recurrence formula for homozygosity ($J_t = \Sigma x_i^2$) between two generations. Since $H = 1 - J$, we have

$$J_{t+1} = \frac{1}{2N} + \left(1 - \frac{1}{2N}\right)J_t. \tag{13.13}$$

This formula will be used in a later section. Also, from (13.12),

$$J_t = 1 - (1 - J_0)\left(1 - \frac{1}{2N}\right)^t. \tag{13.14}$$

Note that if $J_0 = 0$, J_t becomes identical to F_{ST}. For this reason, the two quantities are often confused. In practice, however, J_0 never becomes 0. Furthermore, if we take into account mutation and migration, J_t and F_{ST} take different forms, as will be seen later.

Effective Population Size

In the above formulation, we have assumed that the organism in question is monoecious and that all individuals in the population contribute gametes to the next generation with equal probability, though there may be chance variation. In practice, however, many organisms have separate sexes, and there are almost always some deviations from this idealized reproduction even in a monoecious organism. These deviations introduce many complications in mathematical formulation, but they can be avoided if we consider an idealized population that would have the same gene frequency distribution as that of the actual population.

The size of this idealized population is called the *effective population size*. This concept is due to Wright (1931) and simplifies the mathematical treatment enormously.

Crow (1954) has distinguished between the inbreeding effective size and the variance effective size. The former is defined as the reciprocal of the probability that two uniting gametes come from the same parent, whereas the latter is the population size that would give the same variance of gene frequency change due to sampling error as that in an idealized population (13.7b). Namely, the variance effective size is

$$N_e = x(1-x)/(2V_{\delta x}), \tag{13.15}$$

where $V_{\delta x}$ is the variance of gene frequency change for a particular case. In practice, there is not much difference between the two effective sizes except in some special cases. In the following, I shall list the mathematical formulas for computing effective size for various cases without going into detail.

Separate Sexes.   If a population consists of N_m males and N_f females, the effective size (N_e) is given by

$$N_e = 4N_m N_f/(N_m + N_f) \tag{13.16}$$

(Wright 1931). Unless $N_m = N_f$, this is always smaller than the actual size $(N_m + N_f)$.

Cyclic Change of Population Size.   If population size changes cyclically with a relatively short period of n generations and N_i is the population size in the ith generation in the cycle, then N_e is given by the following harmonic mean of N_i,

$$N_e = n/\sum_{i=1}^{n} N_i^{-1}. \tag{13.17}$$

Therefore, N_e is close to a smaller size rather than to a larger size in the cycle (Wright 1938a).

Variation in Progeny Size.

$$N_e = 2N/(1 + V_k/\bar{k}), \tag{13.18}$$

where $\bar{k}$ and V_k are the mean and variance of progeny number per individual. If progeny number follows the Poisson distribution, then $V_k = \bar{k}$, and $N_e = N$ (Wright 1938a; Crow 1954). In general, however, $V_k > \bar{k}$, so that $N_e < N$. Crow and Morton (1955) estimate that the ratio N_e/N is about 0.75 for many organisms.

Overlapping Generations. If N_a is the number of individuals born per year who survive up to reproductive age and τ is the mean age of reproduction, then

$$N_e = \tau N_a \tag{13.19}$$

(Nei and Imaizumi 1966b; Crow and Kimura 1972; Hill 1979). In human populations, the value of N_e computed from the above formula is 30–40 percent of the total population, including nonreproductive individuals.

Extinction and Recolonization of Subpopulations. A species usually consists of many subpopulations, and these subpopulations are often subject to extinction and recolonization. Suppose that a diploid species consists of n subpopulations, each of which is subject to random extinction and subsequent replacement by one of the remaining subpopulations. Let $\tilde{N}$ and n be the harmonic mean of the effective size for a subpopulation and the number of subpopulations, respectively. The effective size of the entire species is then given by

$$N_e = \tilde{N} + n/[4(\lambda + v + m)] + n\tilde{N}(v+m)/(\lambda + v + m), \tag{13.20}$$

where λ, v, and m are the rate of extinction of a subpopulation, the mutation rate, and the migration rate per generation, respectively (Maruyama and Kimura 1980).

Consider a species (similar to a *Drosophila* species) in which there are $n = 10,000$ subpopulations and each subpopulation grows rapidly from a small number N_0 but becomes extinct on the average in six generations. In this case, although the average size of a subpopulation may be very large, the average effective number ($\tilde{N}$) of a subpopulation is very small. For example, if a subpopulation increases tenfold in each of five consecutive generations starting from $N_0 = 10$, the harmonic mean of population size for this period of time is $\tilde{N} = 6/0.111 = 54$, although the arithmetic mean is 185,185. Let us now assume that $\lambda = 1/6 = 0.17$ and $v + m = 10^{-4}$. Then the effective size for the entire species becomes 15,077 from equation (13.20). This is a small fraction of the actual size, which is expected to be about $10,000 \times 185,000 = 1.85 \times 10^9$. Although this example is certainly artificial, a similar situation is likely to occur in many organisms, including *Escherichia coli* and *Drosophila* (Jones et al. 1981).

It is clear from the above considerations that the effective size of a population is generally much smaller than the actual size. Particularly in lower organisms, where extinction and recolonization of subpopulations occur very frequently, the effective size is expected to be drastically smaller than the actual size.

Gene Frequency Distribution

As we have seen in the previous section, genetic variability in a finite population will eventually disappear unless there is mutation or migration from an outside population. This is true even if there is overdominant selection, though it might take a long time. In reality, however, the reduction in genetic variability by genetic drift is counteracted by mutation or migration, and in a population of fixed size the gene frequency distribution reaches a certain stable form. Let us now discuss this problem.

Gene Frequency Distribution for a Pair of Alleles

GENERAL FORMULA

Consider alleles A_1 and A_2 at a locus, and let x be the frequency of allele A_1. We denote the fitnesses of A_1A_1, A_1A_2, and A_2A_2 by $1 - s$, $1 - h$, and 1, respectively, and assume that mutation occurs from A_2 to A_1 with a rate of u per generation and from A_1 to A_2 with a rate of v. Under these assumptions, Wright (1937) showed that the equilibrium distribution of allele frequency x is given by

$$\phi(x) = Ce^{-4Nhx - 2N(s - 2h)x^2} x^{4Nu - 1}(1 - x)^{4Nv - 1}, \qquad (13.21)$$

where N is the effective population size, and C is a constant such that

$$\int_0^1 \phi(x)dx = 1.$$

This distribution takes various forms, depending on the values of N, h, s, u, and v (Wright 1931, 1969). In the following, I consider two special cases.

NEUTRAL GENES WITH MUTATION

When there is no selection, $h = s = 0$. In this case, $\phi(x)$ takes the following form:

$$\phi(x) = \frac{\Gamma\{4N(u+v)\}}{\Gamma(4Nu)\Gamma(4Nv)} x^{4Nu-1}(1-x)^{4Nv-1}, \tag{13.22}$$

where $\Gamma(\cdot)$ is the gamma function. The mean of this distribution is given by

$$x = u/(u+v). \tag{13.23}$$

When $4Nu = 4Nv = 1$, $\phi(x)$ becomes a uniform distribution. When both $4Nu$ and $4Nv$ are smaller than 1, it takes a U-shaped distribution, whereas when both parameters are larger than 1, it takes a bell-shaped form. However, this two-allele model is now obsolete, since at the nucleotide or amino acid level most loci produce a large number of different alleles. Nevertheless, equation (12.22) is useful for deriving the infinite-allele model of mutation, as will be seen later.

NEUTRAL ALLELES WITH MIGRATION

In the preceding chapter, we noted that the deterministic change of gene frequency due to migration is very similar to that due to mutation. Therefore, the gene frequency distribution under the effects of migration and genetic drift is also expected to be similar to that of (13.22). Consider a group of partially isolated island populations, each of which exchanges genes with a nearby large population at a rate of m per generation. We assume that the size of the nearby population is so large that the gene frequency (x_I) of A_1 remains constant over generations. This type of model is called the *island model* (Wright 1931, 1943). In this case, the distribution of allele frequencies (x) among the island populations is given by

$$\phi(x) = \frac{\Gamma(4Nm)}{\Gamma(4Nmx_I)\Gamma[4Nm(1-x_I)]}x^{4Nmx_I-1}(1-x)^{4Nm(1-x_I)-1}, \quad (13.24)$$

where N is the effective size of an island population (Wright 1943).

Although the above model was originally developed for partially isolated island populations, it is also applicable to a large population that is subdivided into many subpopulations if the migrant genes entering into each subpopulation are a random sample from the entire population. The actual pattern of gene migration in natural populations is usually much more complicated than the above island model, but this model is known to be useful for obtaining a rough idea about the effect of migration on gene frequency distributions (Maruyama 1977).

The mean and variance of x in (13.24) among subpopulations are given by x_I and $V_x = x_I(1-x_I)/[4Nm+1]$. Therefore, the fixation index defined by (7.13) becomes

$$F_{ST} = 1/(4Nm+1). \quad (13.25)$$

This indicates that the degree of differentiation of gene frequencies among subpopulations is high when the product of N and m is small.

Nei and Imaizumi (1966b) studied the variances (and also the covariances) of the ABO blood group gene frequencies among small, isolated (mostly island) populations in Japan. It is believed that a very small amount of migration has occurred between these isolated populations and the general Japanese population for many generations. Their estimate of F_{ST} was 0.00191, which is significantly different from 0. From the demographic data of these populations, the average effective size of the populations was estimated to be 1,993. Therefore, the migration rate (m) can be estimated from equation (13.25), if we assume that the stationary distribution has been reached. From $F_{ST} = 1/[4 \times 1,993 \times m + 1] = 0.00191$, we obtain $m = 0.06$. Thus, a substantial amount of migration must have occurred between the isolated populations and the general Japanese population.

One problem that arises in the application of the island model to a subdivided population is that it does not take into account the relationship between migration rate and geographic distance. In general, migration rate decreases with increasing geographical distance, and there are several models in which this property is taken into account. The reader who is interested in this problem may refer to Malécot (1948), Wright (1969), Crow and Kimura (1970), Maruyama (1977), and Endler (1977).

Gene Frequency Distribution for Multiple Alleles

At the nucleotide level, the number of possible alleles at a locus is enormous. For example, a gene of 1,000 nucleotide pairs may produce $4^{1,000} \simeq 10^{602}$ alleles. Even if the majority of these alleles are deleterious, there must be a huge number of alleles that are functional. Therefore, most new mutations occurring in a population are expected to be different from the extant alleles in the populations. Considering this possibility, Kimura and Crow (1964) proposed the so-called infinite-allele model of mutations. Let us now consider the extent of genetic variability using this model.

NEUTRAL MUTATIONS

Following Kimura (1968b), let us first assume that there are k possible alleles at a locus, and each allele mutates with a frequency of $v_1 = v/(k-1)$ to one of $k-1$ remaining alleles, so that v is the mutation rate per gene per generation. Denote by x the frequency of a particular allele in a population. Under the assumption that all alleles are selectively neutral, the mean change of gene frequency per generation is given by $\Delta x = -vx + (1-x)v_1$, which is of the same form as that of the two-allele case (12.2). Therefore, the stationary distribution of gene frequency x may be expressed by (13.22), replacing u by v_1. Namely,

$$\phi(x) = \frac{\Gamma(M + M')}{\Gamma(M)\Gamma(M')} (1-x)^{M-1} x^{M'-1}, \qquad (13.26)$$

where $M = 4Nv$ and $M' = M/(k-1)$. Clearly, the mean of x is $\bar{x} = v_1/(v_1 + v) = 1/k$.

Since the total number of possible alleles is k and each allele behaves in the same way, the expected number of alleles whose frequency is from x to $x + dx$ is given by $\Phi(x)dx = k\phi(x)dx$. In practice, k is very large, so that the distribution for the expected number of alleles is given by

$$\Phi(x) = \lim_{k \to \infty} \frac{k\Gamma(M + M')}{\Gamma(M)\Gamma(M')} (1-x)^{M-1} x^{M'-1}$$

$$= M(1-x)^{M-1} x^{-1} \qquad (13.27)$$

(Kimura and Crow 1964). Note that $\Gamma(M') \to 1/M'$ as $M' \to 0$.

This distribution is different from the ordinary distribution in statistics, and $\int_{1/2N} \Phi(x)dx$ gives the total number of alleles that are expected to exist in the population. For this reason, Ewens (1979) called this the *frequency spectrum*. Note also that the number of alleles existing at a locus greatly varies from locus to locus even if M is the same. That is, even if M is the same, a locus may be monomorphic or highly polymorphic, depending on chance. However, if we consider a large number of loci with the same M value, the average number of alleles (per locus) with frequencies between x and $x + dx$ is given by $\Phi(x)dx$. In practice, of course, the mutation rate (v) is expected to vary greatly from locus to locus, even if N remains constant. If we assume that v follows the gamma distribution given by equation (9.57), $\Phi(x)$ becomes

$$\Phi(x) = \frac{\bar{M}x^{-1}(1-x)^{-1}}{[1 - (\bar{M}/\alpha)\log_e(1-x)]^{\alpha+1}}, \tag{13.28}$$

where $\bar{M}$ and $\alpha = \bar{M}^2/V_M$ are the mean and the reciprocal of the squared coefficient of variation of M over loci, V_M being the variance of M (Nei et al. 1976a). In practice, however, there is not much difference between (13.27) and (13.28) for reasonable values of $\bar{M}$ and α, i.e., $\bar{M} \leq 1$ and $\alpha \geq 1$.

   Chakraborty et al. (1980) (see also Latter 1975 and Ohta 1976) compared the observed distributions of allele frequencies for protein loci with the expected distribution in 138 different species or subspecies. Their results have shown that the observed distribution is U-shaped in all species examined and agrees fairly well with the expected distribution from equation (13.28), though there was a significant excess of rare alleles in about a quarter of the species examined (see four examples in figure 13.2). This excess of rare alleles was interpreted as a reflection of either the existence of many slightly deleterious alleles or the effect of population size reduction at the time of the last glaciation. [See Nei et al. (1975), Watterson (1984a), and Maruyama and Fuerst (1984) for the mathematical study of bottleneck effect.]

   As discussed in chapter 8, one of the most important measures of genetic variability of a population is heterozygosity or gene diversity defined by $h = 1 - \Sigma x_i^2$. This quantity is expected to vary greatly from locus to locus even if M is the same. The analytical theory of the distribution of h has not been obtained, but the distribution can easily be obtained with a computer (Fuerst et al. 1977; Griffiths and Li 1983). One example is presented in figure 8.1.

   The mathematical formulas for the mean and variance of h can be obtained relatively easily. The expectation of Σx_i^2 over loci is

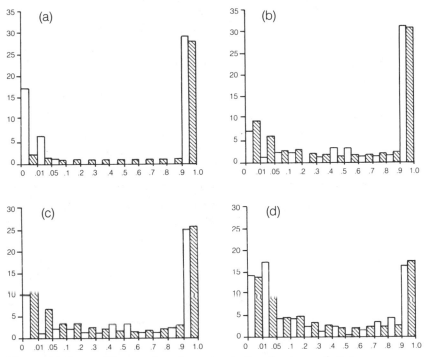

Figure 13.2.   Observed and expected distributions of allele frequencies in four species representing different average heterozygosity values. (a) Japanese macaque (20 loci, average heterozygosity $\hat{H} = 0.018$, $n = 1976$. (b) *Taricha rivularis* (salamanders) (37 loci, $\hat{H} = 0.077$, $n = 784$). (c) *Zoarces viviparus* (eelpouts) (32 loci, $\hat{H} = 0.102$, $n = 757$). (d) *Drosophila heteroneura* (25 loci, $\hat{H} = 0.162$, $n = 605$). The observed distributions are represented by open columns, and the expected distributions by hatched columns. The abscissa gives allele frequency classes and the ordinate the number of alleles. From Chakraborty et al. (1980).

$$J = \int_0^1 x^2 M(1-x)^{M-1}x^{-1}dx$$
$$= 1/(M+1) \tag{13.29}$$

Therefore, the expected heterozygosity is

$$H = 1 - J = M/(M+1). \tag{13.30}$$

(Kimura and Crow 1964). Similarly, the variance $[V(h)]$ of h has been obtained by Watterson (1974) and Stewart (1976). It becomes

$$V(h) = \frac{2M}{(1+M)^2(2+M)(3+M)}.$$ (13.31)

[For an elementary derivation of equations (13.29) and (13.30), see Tajima (1983a).]

Nei (1975), Fuerst et al. (1977), Yamazaki (1977), and others have used the relationship between the mean and variance of observed heterozygosity among different loci to test the "null hypothesis" of the neutral mutation theory, but they could not reject it. The variance of h for the cases of slightly deleterious and overdominant mutations have been studied by Li (1978), Maruyama and Nei (1981), and Kimura (1983a).

The genetic variability of a population can also be measured by the number of alleles in the population. The expected number of alleles that exist in a population is given by

$$n_a = \int_{1/2N}^{1} \Phi(x)dx.$$ (13.32)

Note that n_a is different from the effective number of alleles (n_e) defined by Kimura and Crow (1964), i.e.,

$$n_e = 1/E(\Sigma x_i^2) = M + 1.$$ (13.33)

The effective number is usually much smaller than the actual number. It should also be noted that $E(1/\Sigma x_i^2) \neq M + 1$. An approximate formula for $E(1/\Sigma x_i^2)$ is given by Zouros (1979).

In practice, however, the total number of alleles existing in a population is not usually observable. Instead, the number of alleles is examined by sampling a certain number of genes from the population. The expected number of alleles in a sample of $2n$ genes is given by equation (8.16). As discussed in chapter 8, this number is highly dependent on sample size.

Another parameter which is often used in the study of polymorphism is the proportion of polymorphic loci. We define a locus as *polymorphic* if the frequency of the most common allele is equal to or less than $1 - q$, where q is a small quantity. The most commonly used value of q is 0.01 (chapter 8). If all loci have the same mutation rate, then the expected proportion of polymorphic loci is given by

$$P = 1 - q^M$$ (13.34)

(Kimura 1971). Since M can be estimated by $M = H/(1 - H)$ from average heterozygosity H when a large number of loci are examined, one can examine the relationship between H and P to test the neutral mutation hypothesis. Such tests have shown that the agreement between theory and data is quite good with respect to protein polymorphism (Kimura and Ohta 1971b; Fuerst et al. 1977).

HOMOZYGOSITY TEST

Using the relationship among the number of genes sampled ($2n$), number of alleles observed (k), and sample homozygosity (j_1), Watterson (1978) developed a method for testing neutral mutations. This test (homozygosity test) is based on Ewens' (1972) sampling theory of neutral alleles. Studying the probability of observing a set of sample allele frequencies ($\hat{x}_1, \hat{x}_2, \cdots, \hat{x}_k$) for given values of $2n$ and k, Ewens showed that the probability is independent of population size and mutation rate. It is, therefore, possible to compute the distribution of sample homozygosity $j_1 = \Sigma \hat{x}_i^2$ for any given values of $2n$ and k (table 13.2), and this distribution can be used for testing whether or not j_1 is significantly higher or lower than the neutral expectation. Suppose that one obtains $k = 10$ and $j_1 = 0.16$ examining $2n = 200$ genes. Table 13.2 indicates that $j_1 = 0.16$ is significant at the 2.5 percent level, suggesting that the homozygosity is too low for neutral alleles. If j_1 is proven to be too low, it suggests that there is some type of balancing selection, whereas if it is too high, it suggests that there is purifying selection. Evidence of

Table 13.2 Values of j_1 corresponding to various cumulative probabilities (%) for given values of n and k under the assumption of neutral mutations. From Watterson (1978).

k	$2n$	$j_{1(min)}$	1	2.5	5	10	50	90	95	97.5	99	$j_{1(max)}$
3	100	0.33	0.34	0.37	0.40	0.44	0.67	0.96	0.96	—	—	0.96
	200	0.33	0.34	0.37	0.40	0.46	0.68	0.95	0.98	—	—	0.98
5	100	0.20	0.24	0.26	0.28	0.30	0.44	0.73	0.80	0.84	0.89	0.92
	200	0.20	0.25	0.28	0.30	0.32	0.49	0.80	0.85	0.90	0.92	0.96
7	100	0.14	0.19	0.20	0.22	0.24	0.34	0.57	0.66	0.72	0.81	0.88
	200	0.14	0.20	0.22	0.23	0.25	0.39	0.65	0.74	0.79	0.83	0.94
10	100	0.10	0.14	0.15	0.16	0.17	0.25	0.40	0.45	0.50	0.54	0.83
	200	0.10	0.15	0.16	0.18	0.19	0.28	0.49	0.55	0.62	0.68	0.91

— denotes significance level not possible.

purifying selection does not disprove Kimura's neutral mutation hypothesis, but evidence of balancing selection will unless there are some other disturbing factors. A simple way to obtain the distribution of j_1 for given values of k and $2n$ is to use the computer program developed by F. M. Stewart (see Fuerst et al. 1977) or by Griffiths and Li (1983).

A number of authors have applied this test to protein polymorphism data. In most cases, either no selection or purifying selection was indicated (e.g., Watterson 1978; Keith et al. 1985). In the case of human HLA (MHC) loci, however, Hedrick and Thomson (1983) found a significant reduction in homozygosity. It is, therefore, likely that the high polymorphism at these loci is maintained by some sort of balancing selection.

One troublesome problem with Watterson's test is that the population is not always in equilibrium. Recently, Watterson (1986) studied the effect of population size change on his test. His results show that in an expanding population there is an excess homozygosity relative to the number of observed alleles. The human population has certainly expanded in the past 10,000 years. Yet Hedrick and Thomson found a deficiency of homozygosity in the HLA loci. This strengthens their view that the polymorphism at the HLA locus is maintained by balancing selection. Another factor that would affect Watterson's homozygosity test is intragenic recombination. However, this factor is also known to increase homozygosity relative to the number of alleles (Morgan and Strobeck 1979).

MUTATIONS AND SELECTION

Suppose that there are k possible alleles at a locus and mutation occurs with equal frequency among them, as in the case of neutral mutation. We assume that the fitness of genotype $A_i A_j$ is W_{ij} so that the mean fitness is $\bar{W} = \Sigma x_i x_j W_{ij}$. It can then be shown that the joint distribution of allele frequencies $x_1, x_2, \cdots, x_k$ at equilibrium is

$$\phi(x_1, x_2, \cdots, x_k) = C \exp(2N\bar{W}) \prod_{i=1}^{k} x_i^{M'-1}, \tag{13.35}$$

where $M' = 4Nv/(k-1)$ and C is a constant (Wright 1949, 1969).

When all alleles are subject to the same type of selection or a few different types of selection, the one-dimensional allele frequency distribution (frequency spectrum) similar to equation (13.27) can be obtained by letting $k \rightarrow \infty$. However, the mathematical expression for the distribution is generally quite complicated (Watterson 1977; Li 1977a, 1978). Because of this complication, a number of authors have derived various approximate formulas for several different cases (e.g., Kimura and Crow 1964; Ewens 1964b; Watterson 1977). For example, in the case of symmetric overdominant selection, where the fitnesses of all heterozygotes (A_iA_j) are 1 and the fitnesses of all homozygotes are $1-s$, Ewens (1964b) and Yokoyama and Nei (1979) derived a relatively simple formula for $\Phi(x)$ under the assumption that $M \equiv 4Nv$ and $S \equiv 4Ns$ are sufficiently large so that the expected homozygosity (J) is lower than 0.25. It is given by

$$\Phi(x) = Me^{Sx}(1-x)^{B-1}x^{-1}, \tag{13.36}$$

where $B = M + S(1-J)$, and

$$J = \frac{\Gamma(1+B)_1F_1(2, 2+B, S)}{\Gamma(2+B)_1F_1(1, 1+B, S)}. \tag{13.37}$$

Here, $\Gamma(\cdot)$ is the gamma function, and $_1F_1(\cdot,\cdot,\cdot)$ is the confluent hypergeometric function (see Abramowitz and Stegun 1964).

Another approach to this complicated problem is to use numerical methods to find the distribution. Indeed, even if an analytical formula for a certain type of selection is derived, extensive numerical computations are required to evaluate the properties of the distribution (Li 1977a, 1978). For this reason, some authors (e.g., Maruyama and Nei 1981; Takahata 1981) have used direct numerical methods to study the properties of $\Phi(x)$.

The types of selection so far studied include (1) overdominant selection (Kimura and Crow 1964; Ewens 1964b; Watterson 1977; Li 1978; Yokoyama and Nei 1979; Maruyama and Nei 1981), (2) slightly disadvantageous selection (Li 1977a, 1978; Ohta 1977; Kimura 1979; Ewens and Li 1980), and (3) fluctuating selection (Nei and Yokoyama 1976; Takahata and Kimura 1979; Takahata 1981; Gillespie 1983, 1985). Compared with the case of neutral mutations, overdominant selection usually increases the number of intermediate frequency alleles, whereas slightly disadvantageous selection decreases this number and increases rare alleles (figure 13.3). In the case of random fluctuation of selection intensity, the pattern of allele frequency distribution is highly depen-

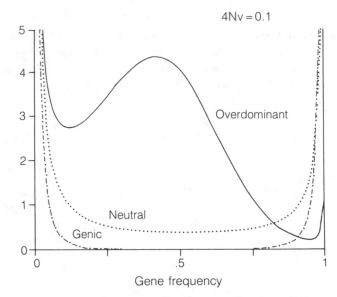

Figure 13.3.   Allele frequency distributions for three different types
of selection [overdominant, neutral, and slightly deleterious (genic)].
The ordinate stands for $\Phi(x)$. From Li (1978).

dent on population size and on whether or not a stabilizing factor is
introduced. If no stabilizing factor is introduced, fluctuating selection
always reduces intermediate frequency alleles. Even in the presence of a
stabilizing factor, fluctuating selection generally reduces the intermediate frequency alleles unless the population size is very large.

Once $\Phi(x)$ is obtained, the expected heterozygosity (H) can be obtained from $1 - J$, where $J = \int_0^1 x^2 \Phi(x)dx$. Many authors have investigated the relationship between effective population size (N) and expected
heterozygosity for different types of selection. In the case of overdominant selection, H increases with increasing N much more rapidly than
in the case of neutral mutations, unless selection coefficient s is very
small (figure 13.4). That is, overdominant selection is very powerful in
maintaining polymorphism. By contrast, disadvantageous selection (purifying selection) substantially reduces genetic variability (Ohta 1977; Li
1978; Kimura 1979). Yet, if we use Kimura's (1979) model of slightly
deleterious mutations, the expected heterozygosity slowly increases with
increasing N and eventually reaches 1 (see figure 8.3).

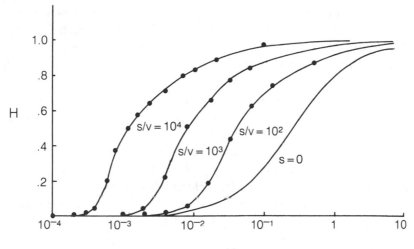

Figure 13.4.   Relationships between Nv and mean heterozygosity (H) for overdominant alleles, where N and v are the population size and mutation rate, respectively. The fitness of heterozygotes and homozygotes are assumed to be 1 and $1-s$, respectively. When s and v remain constant, each curve represents the relationship between N and H. The curve for $s=0$ was obtained by $H = 4Nv/(1 + 4Nv)$. From Maruyama and Nei (1981).

Infinite-Site Model of Mutation

In the foregoing section, we regarded each nucleotide (or codon) sequence as an allele and assumed that there are virtually infinitely many possible alleles at a locus. Genetic variation of a population can also be studied by looking at each nucleotide site separately and examining whether or not the site is polymorphic. In most organisms, the mutation rate at the nucleotide level is so low that most nucleotide sites are monomorphic for one of the four possible nucleotides, and when a site is polymorphic, there are only two different nucleotides segregating (Kreitman 1983). (In RNA viruses, the mutation rate at the nucleotide level is so high that this is not the case; chapter 10.) Therefore, we can use a model in which a gene consists of n nucleotides and any of the n nucleotides mutates to a new nucleotide with a rate of μ per generation. However, the mutation rate is so low that a new mutation occurs almost

POPULATION THEORY: STOCHASTIC MODELS

always at a site which was previously monomorphic. In practice, n is usually very large, so that it may be assumed to be practically infinite. This is particularly so if we consider all nucleotides in the genome, as was originally done by Kimura (1969, 1971). Kimura (1971) called this model the infinite-site model. Mathematically, however, this is equivalent to Fisher's (1930) or Wright's (1931, 1938b) irreversible mutation model in which a collection of loci in the genome rather than a collection of nucleotides was considered (see also Haldane 1939).

In Kimura's (1969) original infinite-site model, each nucleotide was assumed to be inherited independently of others. Later, Watterson (1975) considered a model in which all nucleotides in a gene are completely linked. This version of the infinite-site model seems to be more realistic, but many of Kimura's (1969) results still hold. Note that Watterson's version is also applicable to restriction-site data for mitochondrial DNA (chapter 10).

Neutral Mutations

If we use the infinite-site model, genetic variation can be studied with several different measures. However, let me first present the nucleotide (or gene) frequency distribution that can be obtained under the assumption that the input of variation is balanced by the loss of variation by genetic drift. Here, we consider a sequence of n nucleotides (a locus) and let x be the frequency of a mutant nucleotide (a gene in the Fisher-Wright version) at a given site in a population. We then count the number of nucleotide sites in which the frequency of the mutant nucleotide (A') is between x and $x + dx$. If we exclude all monomorphic sites $(x = 0$ or $1)$, the expected number of these nucleotide sites per locus is given by

$$\Phi_1(x)dx = \frac{4Nv}{x}dx,$$ (13.38)

where v is the total mutation rate per locus, i.e., $v = n\mu$ (Wright 1938b). Note that $\Phi_1(x)$ becomes identical with $\Phi(x)$ in (13.27) only when $4Nv = 1$.

One quantity that can be used for measuring genetic variability is the number of nucleotide differences between two randomly chosen genes. The expectation of this number is given by

$$H_1 = \int_{1/2N}^{1-1/2N} 2x(1-x)\Phi_1(x)dx \simeq 4Nv, \qquad (13.39)$$

if the population size N is large (Kimura 1969). Under the assumption of complete linkage of nucleotides, Watterson (1975) showed that the probability of two randomly chosen sequences having m nucleotide differences is given by

$$P(m) = \frac{1}{M+1}\left(\frac{M}{M+1}\right)^m, \qquad (13.40)$$

where $M - 4Nv$. Therefore, the mean ($\bar{m}$) and variance [$V(m)$] of m are

$$\bar{m} \simeq M, \qquad V(m) \simeq M + M^2. \qquad (13.41)$$

The mean agrees with equation (13.39). We have used these equations in the analysis of DNA polymorphism in chapter 10.

Equation (13.40) can also be used for computing the expected heterozygosity per locus. That is, the probability that two randomly chosen sequences are identical is $P(0) = 1/(M+1)$. Therefore, the expected heterozygosity is given by

$$H = 1 - P(0) = M/(M+1), \qquad (13.42)$$

which is identical with the expected heterozygosity obtained from the infinite-allele model.

Genetic variability of a population can also be measured by the number of segregating (polymorphic) sites per locus. The expectation of this number is given by

$$I = \int_{1/2N}^{1-1/2N} \Phi_1(x)dx \simeq M \log_e(2N) \qquad (13.43)$$

(Kimura 1969). In practice, however, the number of segregating sites is computed from the genes sampled. The expected number of segregating sites in a sample of n genes ($n/2$ individuals) becomes

$$I_p = M[1 + 1/2 + 1/3 + \cdots + (n-1)^{-1}] \qquad (13.44)$$

(Watterson 1975). It is clear that I_p increases with increasing sample size.

MUTATIONS AND SELECTION

When mutant nucleotides (or genes) are subject to selection, the general formula for $\Phi_1(x)$ is quite complicated (Fisher 1930; Wright 1938b, 1942, 1969; Ewens 1964a; Kimura 1964, 1979, 1983a). In the case of genic selection, however, a relatively simple formula has been obtained. Let A' be the mutant nucleotide and A be its parental nucleotide, and denote the fitnesses of AA, AA', and $A'A'$ by 1, $1+s$, and $1+2s$, respectively. Then, $\Phi_1(x)$ is given by

$$\Phi_1(x) = \frac{M}{x(1-x)} \frac{1 - e^{S(1-x)}}{1 - e^{-S}}, \tag{13.45}$$

where $S = 4Ns$ (Fisher 1930; Wright 1938a). Yamazaki and Maruyama (1974) used (13.38) and (13.45) for testing the neutral mutation hypothesis.

Kimura (1969) derived the formulas for H_1 and I for the case of genic selection with $4Ns \gg 1$. They are

$$H_1 \simeq 8Nv, \tag{13.46}$$

$$I = 4Nv[\log_e(4N^2s) + 1.577]. \tag{13.47}$$

Equation (13.46) indicates that the expected number of nucleotide differences per locus for advantageous mutations is twice that for neutral mutations if the mutation rate is the same.

Stepwise Mutation Model

This model was proposed by Ohta and Kimura (1973) to explain the extent of protein polymorphism. The electrophoretic mobility of a protein is primarily determined by the total net charge of the protein. If a positively charged amino acid is replaced by a neutral one, the mobility of the protein changes (chapter 3). Therefore, if we assume that the total charge is solely responsible for electrophoretic mobility and express the allelic state of a gene in terms of the total charge, mutation will occur stepwise as shown in figure 13.5. This is called the *stepwise mutation model*. In practice, the mutational change is not always stepwise, since a

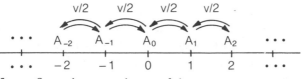

Figure 13.5. Stepwise mutation model.

positively charged amino acid may change to a negatively charged amino acid or vice versa. However, the probability of occurrence of these two-step changes is so small (table 3.3) that they can be neglected.

The allele frequency distribution (spectrum) for the stepwise mutation model is too complicated to be presented here (Kimura and Ohta 1978). However, the mean heterozygosity is given by a simple formula. That is,

$$H = 1 - 1/\sqrt{1 + 2M}, \tag{13.48}$$

where $M = 4Nv$ (Ohta and Kimura 1973; Moran 1975). For a given value of M, the above formula gives a smaller value than (13.30), but when $M < 0.25$, the difference is small. The exact formula for the variance of single-locus heterozygosity (h) is more complicated than the mean (Moran 1975), but Moran (1976) produced the following approximate formula,

$$V(h) = M/(3 + 11.25M + 13M^2 + 1.7M^3). \tag{13.49}$$

After the proposal of the stepwise mutation model, a number of authors examined the applicability of the model to actual data. Analyzing Ayala and Tracey's (1974) data on electromorphs, Weir et al. (1976) concluded that at about three-quarters of the polymorphic loci the mobility changes of proteins follow the stepwise mutation model approximately. Ramshaw et al. (1979) and Fuerst and Ferrell (1980) examined the electrophoretic mobilities of hemoglobins of which the amino acid sequences were known. They concluded that although the net electric charge is the principal determinant of electrophoretic mobility, a significant deviation from strict stepwise change occurs for 30 to 40 percent of the variant electromorphs studied. On the other hand, McCommas (1983) showed that in sea anemone the stepwise mutation model is applicable to most of the variation existing as common polymorphisms within species, but it cannot account for the variation between species.

These studies indicate that the stepwise mutation model is not as realistic as was originally thought. However, it is useful for determining the lower bound of genetic variability for a given value of M. The expected heterozygosity for electromorphs must be somewhere between the values obtained from the infinite-allele model and the stepwise mutation model. Li (1976c) has shown that the expected heterozygosity quickly approaches that of the infinite-allele model as the proportion of nonstepwise mutations increases.

Gene Substitution

Rate of Gene Substitution

As mentioned earlier, the basic process of evolution is gene substitution in populations. Gene substitution consists of two events, i.e., (1) occurrence of mutations and (2) fixation of mutations in the population. At the nucleotide level a new mutation occurring in a population is almost always different from the extant alleles in the population. If the mutation rate per generation is v at a locus, then there occur $2Nv$ mutations at this locus in the entire population. Therefore, if we consider long-term evolution, assuming that the effects of mutation, selection, and genetic drift are balanced, the rate of gene substitution per generation is given by

$$\alpha = 2Nvu, \tag{13.50}$$

where u is the probability of fixation of a mutant allele (Kimura and Ohta 1971a).

In chapters 4 and 5, we considered the rate of amino acid or nucleotide substitution per site rather than the rate of substitution per locus. The rate of amino acid or nucleotide substitution per site per year (λ) can be expressed in the same way as (13.50), if we replace v by the mutation rate per site per year (μ_y). That is,

$$\lambda = 2N\mu_y u. \tag{13.51}$$

Here, μ_y may be expressed as $\mu_y = v/(ng)$, where n is the total number of amino acid or nucleotide sites per locus and g is the generation time measured in years.

The probability of fixation of a mutant allele in a finite population has been studied by Haldane (1927), Fisher (1930), Wright (1931, 1942), Malécot (1952), Kimura (1957, 1962), and others (see Crow and Kimura 1970; Nei 1975; Ewens 1979). It depends on the type of selection involved. In the case of neutral alleles, it is equal to the initial gene frequency (p). When a new mutation appears in a population of size $2N$, the initial gene frequency is $p = 1/(2N)$. Therefore, $u = 1/(2N)$, and the rate of gene substitution is given by

$$\alpha = 2Nvu - v. \tag{13.52}$$

That is, the rate of gene substitution is equal to the mutation rate per locus. This simple rule was first noted by Kimura (1968a).

In the case of genic selection, the fitnesses of the three genotypes AA, AA', and $A'A'$ are given by 1, $1+s$ and $1+2s$, respectively, where A and A' represent the original and mutant alleles, respectively. The probability of fixation of a mutant allele is then given by

$$u = \frac{1 - e^{-4Nsp}}{1 - e^{-4Ns}} \tag{13.53}$$

(Malécot 1952; Kimura 1957). When $p = 1/(2N)$ and $4Ns \gg 1$, u becomes $2s$ approximately (Haldane 1927). In this case, therefore, we have

$$\alpha = 2Nv \cdot 2s = 4Nsv. \tag{13.54}$$

This equation also applies to a dominant mutant gene, where the fitnesses of AA, AA', and $A'A'$ are 1, $1+s$, and $1+s$, respectively.

Comparison of equations (13.52) and (13.54) indicates that natural selection is very effective in speeding up gene substitution when $4Ns$ is large. For example, if $N = 10^5$ and $s = 10^{-3}$, α is 400 times greater for advantageous genes than for neutral genes, as long as v remains the same.

Another feature of equation (13.54) is its dependence on both N and s in addition to v. Therefore, the rate of gene substitution for advantageous mutations is expected to be higher in organisms with large population size than in organisms with small population size. In practice, however, the rate of amino acid substitution per year seems to be roughly the same for all organisms. This observation can be easily accommodated with the neutral mutation theory, if we assume that the rate of neutral mutations is constant per year rather than per generation. In the case of advantageous mutations, however, it is very difficult to explain rate constancy, since different organisms have vastly different population sizes and each mutant gene would have a different selection coefficient. Kimura and Ohta (1971b) therefore took the observation of rate constancy as support of the neutral mutation theory.

The probabilities of fixation of a mutant gene for other types of selection have also been studied. In the case of a recessive mutant gene with fitnesses 1, 1, and $1+s$ for AA, AA', and $A'A'$, respectively, the probability is given by

$$u = \sqrt{2s/(\pi N)} \tag{13.55}$$

approximately (Kimura 1957). For an underdominant gene with fitnesses of 1, $1-s$, and 1, we have

$$u = \frac{1}{2N}e^{-s(N-2)} \tag{13.56}$$

approximately for a small value of Ns (M. Nei in Wilson et al. 1977b), and

$$u = \frac{1}{N}e^{-Ns}(Ns/\pi)^{1/2} \tag{13.57}$$

approximately for $Ns \geq 2$ (Lande 1979). The above two equations seem to be useful for studying the probability of fixation of chromosomal mutations (Bush et al. 1977; Hedrick 1981). The probability of fixation of an overdominant gene with fitnesses $1-s_1$, 1, and $1-s_2$ is a complicated function of s_1, s_2, and N, and it may be higher or lower than that of a neutral mutation (Nei and Roychoudhury 1973a). The rate of gene substitution is also a quite complicated function of these parameters, but in a large population the rate is usually lower than that for neutral alleles (Maruyama and Nei 1981).

Unit Substitution Time

In chapter 4, we considered the unit evolutionary time for amino acid substitution. Mathematically, a more convenient quantity is the average time required for one gene substitution (figure 13.6). It is given by

$$\bar{T}_s = 1/\alpha. \tag{13.58}$$

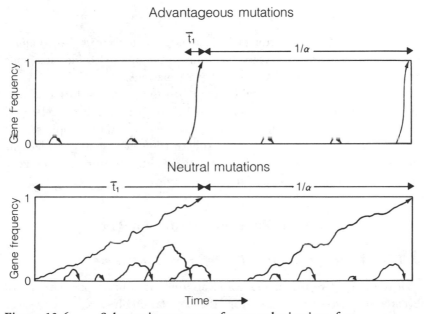

Figure 13.6.   Schematic patterns of gene substitutions for advantageous and neutral mutations. Advantageous mutations are either quickly lost from or quickly fixed in the population, so that their contribution to polymorphism for a given rate of gene substitution α is small. On the other hand, the frequency changes of neutral alleles are very slow in large populations and generate a large amount of transient polymorphism. The expected time ($\bar{t}_1$) for a mutation to be fixed is approximately $(2/s)\log_e(2N)$ generations for advantageous mutations and $4N$ generations for neutral mutations, where N and s are the effective population size and selection coefficient. The expected time interval between two consecutive gene substitutions is $1/\alpha$, where $\alpha = 4Nsv$ for advantageous mutations and $\alpha = 1/v$ for neutral mutations. v is the mutation rate per generation.

In the case of the Poisson process considered in chapter 4, the substitution time (T_s), i.e. the time between two gene substitutions, follows the exponential distribution. That is,

$$P(T_s) = \alpha e^{-\alpha T_s} \qquad (13.59)$$

The mean of this distribution is given by equation (13.58), whereas the variance is

$$V(T_s) = 1/\alpha^2. \qquad (13.60)$$

Therefore, the standard error of T_s is as large as the mean. To compute the substitution time in years for amino acid or nucleotide sites, one may simply replace α by λ in the above equations.

The above formulation is valid only when long-term evolution is considered for a single amino acid or nucleotide sequence. When a relatively short period of evolutionary time is considered, we must sample several genes to determine the rate of gene substitution. We shall consider this problem in a later section, where we discuss the measurement of gene substitutions between two isolated populations.

Fixation Time and Extinction Time

Time Until Fixation or Loss of a Mutant Gene

If there is no recurrent mutation, an allele in a finite population will eventually be fixed in or lost from the population. In a reasonably large population, the time required for the fixation or loss is quite long unless there is strong selection. The mean time to either fixation or loss of a mutant gene with an initial frequency of p has been studied by Ewens (1963). In evolutionary studies, however, it is often necessary to consider the event of fixation of a mutant gene separately from the event of loss. This is because in a retrospective study of evolution we consider only those mutations which are fixed in the population. Considering this situation, Kimura and Ohta (1969), Kimura (1970), and Narain (1970) studied the time until fixation of a new mutation in a population of size N, given that the mutant gene is fixed. In the case of neutral genes, the mean $(\bar{t}_1)$ and standard deviation $[s(t_1)]$ of this *conditional* fixation time become

$$\bar{t}_1 = 4N, \qquad s(t_1) = 2.14N, \tag{13.61}$$

where time is measured in generations. This indicates that it takes a long time for a mutant gene to be incorporated into the genome unless N is very small. Note that the standard deviation of the fixation time is also very large.

The mean fixation time can be substantially shortened if the mutant gene is selectively advantageous (see figure 13.6). However, except for the case of a large Ns, the mathematical formula is quite complicated (Kimura and Ohta 1969). The mean fixation time for an overdominant gene has also been studied (Nei and Roychoudhury 1973a).

Time Until Extinction of an Allele Under Recurrent Mutation and Selection

NEUTRAL GENES OR UNUSED GENES

In the infinite-allele model of neutral mutations, an allele existing in a population will eventually be lost from the population because of mutation and genetic drift. However, it usually takes a long time for an allele to be lost. To study this problem, let us consider an allele (A) that exists in a population of size N with an initial frequency of p. We assume that this allele mutates to new alleles with a rate of v per generation. Once allele A becomes extinct, it will never be produced again. Because of the stochastic process involved, the time required for extinction greatly varies from case to case. Crow and Kimura (1970) have worked out a mathematical formula for the probability that allele A becomes lost by the tth generation. Using this formula, one can show that the mean ($\bar{t}$) and variance [$V(t)$] of the time (t) at which the allele is lost are

$$\bar{t} = 4N + 1/v, \qquad V(t) = (1 - M^2)/v^2, \tag{13.62}$$

where $p = 1$ is assumed (Nei 1976). Therefore, when the mutation rate is low and N is large, it takes a long time for allele A to disappear from the population.

Figure 13.7 shows the distribution of t for the case of $p = 1$ and $4Nv = 0.1$. If $v = 10^{-6}$ and $N = 25,000$, the mean extinction time be-

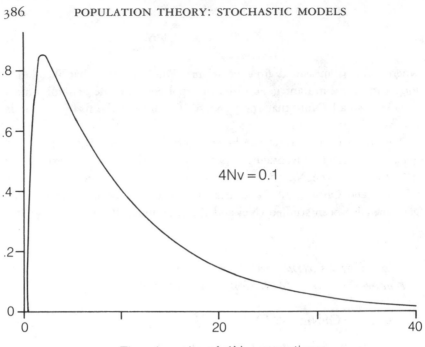

Figure 13.7.   Probability distribution of extinction time of a neutral allele under irreversible mutation pressure. The initial allele frequency is 1. From Nei (1976).

comes $1.1 \times (4N)$ or 1.1×10^6 generations. However, figure 13.7 shows that the actual extinction time may vary greatly from case to case.

It is well known that many cave-dwelling animals lack pigmentation and eyesight. These animals apparently colonized caves during the Pleistocene, some of them as late as about 300,000 years ago (Barr 1968). It has been controversial, however, whether the loss of pigmentation and eyesight occurred simply because of mutation pressure or because the individuals lacking these characters had a selective advantage in cave conditions. Wright (1964) and Barr (1968) are of the opinion that the loss occurred as a by-product of selection for other advantageous, pleiotropic characters, because the time involved is too short to be explained by mutation pressure alone. By using the deterministic theory [equation (12.5)], Kosswig (see Barr 1968) estimated that it would require one million or more generations for unused characters to degenerate by mutation pressure alone.

To study the problem of degeneration of unused characters, however, the stochastic theory presented above is more appropriate, because unused characters would not be subject to strong selection pressure and the population size is usually very small. As an example, let us consider the cave population (eyeless and depigmented individuals) of the characid fish *Astyanax mexicanus* in Pachon, Mexico. Using Avise and Selander's (1972) data on protein polymorphism, Chakraborty and Nei (1974) estimated that the cave population was formed about 710,000 years ago. This corresponds to about 140,000 generations, since the generation time of this fish is about five years (P. Sadoglu, personal communication). On the other hand, if we assume that the rate of mutations to *nonfunctional alleles* is 10^{-5} per locus per generation, the mean time for the original allele to become lost from the population is about 100,000 generations, since the effective size of this population is known to be very small (less than 100). Therefore, it is possible to explain the degeneration of eyes and pigmentation in this cave fish in terms of mutation pressure without assuming any selective advantage (Li and Nei 1977).

In the absence of selection with $p \geq 0.4$ and $M \ll 1$, the probability that the original allele becomes lost from the population by generation t is given by

$$f(p,t) = 1 - pe^{-vt} \tag{13.63}$$

approximately (Crow and Kimura 1970). In the above example of *Astyanax mexicanus,* this probability becomes 0.75 if we assume $v = 10^{-5}$, $p = 1$, and $t = 140,000$. Thus, the probability of loss of gene function is very high. On the other hand, if t is 14,000, the probability becomes 0.13. This indicates that even in such a short evolutionary time as 14,000 generations, the loss of eyes and pigmentation may occur with an appreciable probability.

To prove the above contention, however, it is necessary to identify the gene or genes responsible for unused characters and examine whether the genes are really nonfunctional. Quax-Jeuken et al. (1985) have examined the function of the genes for lens-specific proteins (crystallins) in blind species of moles *(Talpa europaea)* and mole rats *(Spalax ehrenbergi)* by using the techniques of DNA hybridization and immunological tests of gene expression. They have shown that the crystallin genes in these blind animals show a rather high similarity to the genes from the rat and that they still produce proteins. If we note that these burrowing animals apparently evolved many millions of years ago, it seems that the crystallin genes have maintained some function other than vision and that this has prevented the genes from being silenced. It is hoped that a similar study will be conducted with the Mexican cave fish in the near future.

MUTATIONS AND SELECTION

The above story changes considerably if new mutant genes (A') have selective advantage or disadvantage over the original allele A. Li and Nei (1977) studied this problem, and some of their results for the case of genic selection are presented in table 13.3. In this table, $M = 4Nv$, $S = 4Ns$, and time is measured in $4N$ generations. It is clear that positive selection (positive S) substantially speeds up the extinction of allele A or fixation of allele A' compared with the case of neutral genes. For example, in the case of $N = 2,500$, $s = 0.01$, and $v = 10^{-6}$, we have $M = 0.01$ and $S = 100$. In this case, it takes $0.052 \times 4N = 520$ generations for A to become extinct. This is about 1/20 of the time (1.05×10^4 generations) for neutral genes with $S = 0$. This indicates that a relatively small amount of selection substantially accelerates the extinction of allele A or the fixation of allele A'. Note also that in the absence of selection the

Table 13.3 Mean extinction time (measured in units of $4N$ generations) of the original allele (A) when mutant genes (A') are selectively advantageous or disadvantageous. The fitnesses of AA, AA', and $A'A'$ are assumed to be 1, $1 + s$, and $1 + 2s$, respectively. $M = 4Nv$, $S = 4Ns$, and p is the initial frequency of allele A. From Li and Nei (1977).

Selection		M			
intensity	p	1	0.1	0.01	0.001
$S = 0$	1	1.65	10.94	101.0	1001.0
	0.5	1.06	5.80	50.8	500.8
$S = 10$	1	0.54	1.53	10.4	99.1
	0.5	0.29	0.32	0.39	0.99
$S = 100$	1	0.104	0.197	1.05	9.61
	0.5	0.052	0.052	0.052	0.052
$S = -5$	1	9.93	251	—	—
$S = -50$	1	2×10^{18}	7×10^{20}	—	—

extinction time of A is approximately inversely proportional to M as long as $M \leq 0.1$, and that the value for $p = 0.5$ is about half that for $p = 1$. When there is selection, however, the effect of M becomes smaller. Particularly when $S = 100$ and $p = 0.5$, the extinction time is the same for all values of M examined. This is because gene frequency change occurs almost deterministically in this case. When $p = 1$, however, the mean extinction time always depends on M, as expected.

Nei (1985) applied this theory to evaluate the time required for the differentiation of skin pigmentation between the Caucasian and Negroid populations of man. Considering various possible values of N, s, and v, he concluded that it probably took more than 100,000 years for the skin color differentiation to occur.

When the mutant genes are selectively disadvantageous, the extinction time of the original allele is very long, even if selection coefficient is extremely small. For example, in the case of $N = 25,000$, $s = -0.0005$, and $v = 10^{-5}$, we have $M = 1$ and $S - -50$. In this case, it takes $2 \times 10^{18} \times 4N = 2 \times 10^{23}$ generations. That is, A would practically never be lost from the population. This indicates that in the presence of slight negative selection the rate of gene substitution is slowed down tremendously.

Time for Gene Silencing at Duplicate Loci

As mentioned in chapter 6, there are many pseudogenes in the genome of higher organisms. A substantial proportion of these pseudogenes seem to have occurred by silencing of duplicate genes. A duplicate gene may lose its function by null mutation and become a pseudogene, as long as the other duplicate gene is functioning normally. Fixation of pseudogenes in a population occurs mainly by mutation and genetic drift. In a very large population, a pseudogene may never be fixed in the population because of selection against homozygotes for null genes at both loci (Fisher 1935). In a relatively small population, however, a null allele may be fixed with a rather high probability, owing to the effect of genetic drift (Nei and Roychoudhury 1973b). In recent years, a number of authors have studied the expected time for a pseudogene to be fixed at one of the two duplicate loci (Bailey et al. 1978; Takahata and Maruyama 1979; Kimura and King 1979; Li 1980; Watterson 1983). Because of the mathematical difficulty, most of these authors used either computer simulation or numerical evaluation of differential equations. For large N and S, however, Watterson produced a useful analytical formula for the mean fixation time, which is given by

$$\bar{t} = N[\log_e(2N) - \psi(2Nv)], \tag{13.64}$$

where $\psi(\cdot)$ is the digamma function (see Abramowitz and Stegun 1964).

Numerical computations have shown that the mean fixation time of a null allele is usually very long unless $M \equiv 4Nv$ is small. For example, if $v = 10^{-5}$ and $N = 10^5$, it takes about one million generations (Li 1980). This fixation time is much longer than the time given by equation (13.62) for strictly neutral mutations. When $M < 0.04$, however, $\bar{t}$ is nearly the same as that for neutral alleles (for two loci) and is approximately given by $1/(2v)$ (Li 1980; Watterson 1983).

Among the teleost fish, all salmonids, catostomids, and some loaches are thought to be of tetraploid origin (Bailey et al. 1978). Tetraploidization seems to have occurred about 100 million years ago in salmonids, about 50 million years ago in catostomids, and 15 to 40 million years ago in loaches (see Li 1980, 1982 for review). A number of authors (e.g., Allendorf et al. 1975; Ferris and Whitt 1977, 1979; Buth 1979) have studied the proportions of enzyme loci in which one of the two duplicate loci has been silenced. The proportions for salmonids, catostomids, and loaches have been estimated to be 50, 53, and 75 percent, respectively. Since the generation time for these species of fish is about three years and the long-term effective population size seems to be much less than 10^6 (Li 1980), these observations suggest that gene silencing of duplicate loci did not occur as fast as theoretically expected.

A number of authors proposed hypotheses to explain this discrepancy. Some of them are as follows. (1) Null mutations are not completely recessive, and heterozygotes for these mutations are subject to a small amount of negative selection (Takahata and Maruyama 1979). (2) Some duplicate genes have already diverged in their function or regulatory system (Ferris and Whitt 1979; Allendorf 1979). (3) Meiosis continued to be of the tetraploid form for a long time after the number of chromosomes was doubled, and the diploidization of meiosis occured relatively recently (Li 1980). The reader who is interested in this problem may refer to Ferris and Whitt (1979) and Li (1980, 1982).

Gene Genealogy in a Finite Population

In recent years, a number of authors (e.g., Griffiths 1980b; Kingman 1982; Tajima 1983a; Watterson 1984b; Tavaré 1984) have studied the genealogical relationship of genes in a finite population. These studies are useful for finding the expected genealogical relationship of sampled genes. In the absence of recombination, all genes sampled from a population trace back to a single common ancestral gene, and thus all genes are genealogically related. In the following, we examine the expected genealogical relationship of n genes sampled from a randomly mating population without selection. The topology and branch lengths of a genealogical tree will be considered separately. Since we are interested only in the branching pattern of the genes sampled, we shall not consider the effect of mutation.

Topological Relationships

We start from the simplest case of two genes sampled from a population of effective size N. In this case, we obviously have a single branching pattern as shown in figure 13.8a. When three genes are sampled, we have two different branching patterns. One is the case where a common ancestral gene bifurcates at one time and later one of the two resulting branches again bifurcates (figure 13.8b). The other branching pattern obtains when the common ancestral gene trifurcates. However, the probability of occurrence of the latter event is very small when N is large. This is true for any n as long as n is much smaller than N. We, therefore, assume that branching always occurs by bifurcation. When there are four genes sampled, there are two different branching patterns (figure 13.8c and d). The probabilities of obtaining these two branching patterns can be computed by using figure 13.8b. If bifurcation occurs at point C or D in this diagram, pattern c is obtained, whereas if bifurcation takes place at point E, pattern d is obtained. Since these three bifurcation events occur with the same probability, the probabilities of obtaining patterns c and d are 2/3 and 1/3, respectively. When five genes are sampled, five different branching patterns occur, and the probability of obtaining each pattern is given in figure 13.9.

In general, the probability of obtaining a particular branching pattern for n sampled genes is given by

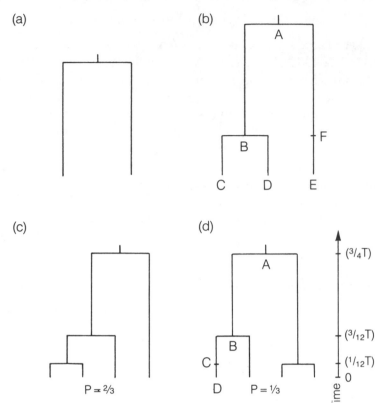

Figure 13.8.   Expected evolutionary relationships of genes sampled from a population. (a) Two genes. (b) Three genes. (c, d) Four genes. From Tajima (1983a).

$$P = 2^{n-1-s}/(n-1)!, \qquad (13.65)$$

where s is the number of branching points that lead to exactly two descendant genes (Tajima 1983a). For example, in the case of figure 13.9a, there is only one such branching point. Therefore, we have $P = 2^{5-1-1}/4! = 1/3$. In all other geneaologies in figure 13.9, $s = 2$. Therefore, $P = 2^{5-1-2}/4! = 1/6$.

We are often interested only in topological relationships. For example, patterns c, d, and e in figure 13.9 can be regarded as the same topology. In this case, the probability of obtaining this topology is com-

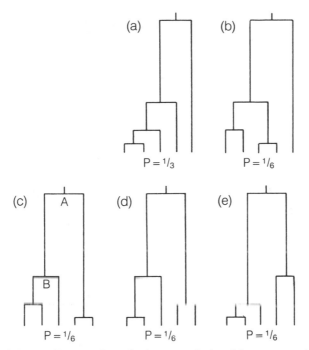

Figure 13.9. Expected evolutionary relationships among five genes sampled from a population. From Tajima (1983a).

puted by summing the probabilities of patterns c, d, and e, i.e., $1/6 + 1/6 + 1/6 = 1/2$. In general, the probability of obtaining a particular topology for n genes sampled can be obtained by using the following method. The probability that a certain branching point divides n genes into two groups of n_1 and n_2 genes (order not considered) is

$$P(n_1, n_2) = 2/(n-1) \qquad \text{if } n_1 \neq n_2,$$
$$P(n_1, n_2) = 1/(n-1) \qquad \text{if } n_1 = n_2, \qquad (13.66)$$

where $n = n_1 + n_2$ (Tajima 1983a). To show how to use the above equation, let us again consider the topology given in figure 13.9c. Point A divides the five genes into groups of two genes and three genes. Therefore, this probability is $2/(5-1) = 1/2$. Branching point B divides the three genes into the two groups of one gene and two genes, and the probability of occurrence of this event is $2/2 = 1$. Thus, the probability of obtaining the topology is $1/2 \times 1 = 1/2$, which agrees with the above computation.

Branch Lengths

Let us now consider the branch lengths of a genealogical tree. It is convenient to measure the branch length in terms of the number of generations. Let $f_n(t)$ be the probability that n genes randomly sampled from a population are derived from $n-1$ genes $t+1$ generations ago and the divergence took place t generations ago. Here, t is a random variable, and n is a fixed number.

We first consider the case in which two genes are sampled. The topological relationship is shown in figure 13.8a. The probability that the two genes are derived from a common ancestral gene in the immediately previous generation is $1/(2N)$, whereas the probability that they are derived from two different genes is $1-1/(2N)$. Here, N is the number of diploid individuals in a population. Therefore, we have

$$f_2(t) = f_2(0)\{1 - f_2(1)\}^t = \{1/(2N)\}\{1 - 1/(2N)\}^t$$
$$\approx \{1/(2N)\}\exp\{-t/(2N)\}. \tag{13.67}$$

This formula gives the probability distribution of branch lengths in terms of the number of generations (t). The mean $(\bar{t})$ and variance $[V(t)]$ of the distribution are

$$\bar{t} = 2N, \qquad V(t) = 4N^2. \tag{13.68}$$

Thus, the standard deviation of t is as large as the mean.

When three genes are sampled, there is only one topological relationship as shown in figure 13.8b. The branch length between A and B in this figure has the same probability distribution as $f_2(t)$ by definition. The probability that three genes were derived from two genes in the immediately previous generation is

$$f_3(1) = \binom{3}{2}[1/(2N)][1 - 1/(2N)] \approx 3/(2N),$$

whereas the probability that three genes are derived from three different genes in the immediately previous generation is

$$[1 - 1/(2N)][1 - 2/(2N)] \approx 1 - 3/(2N).$$

Here, we have ignored the probability that the three genes are derived from a common ancestral gene in the immediately previous generation because the probability $[1/(2N)^2]$ is very small. From the above probabilities, we obtain

$$f_3(t) = [3/(2N)]\{1 - 3/(2N)\}^t \simeq [3/(2N)]\exp[-3t/(2N)].\qquad(13.69)$$

This gives the probability distribution of the branch length between B and C in figure 13.8b. The mean and variance of the branch length (t) are given by

$$\bar{t} = 2N/3, \qquad V(t) = 4N^2/9.\qquad(13.70)$$

This indicates that the mean branch length between B and C is three times shorter than that between A and B.

Similarly, we can obtain $f_n(t)$, which becomes

$$f_n(t) \simeq [\tbinom{n}{2}/(2N)]\exp[-\tbinom{n}{2}t/(2N)]\qquad(13.71)$$

(Kingman 1982; Hudson 1983; Tajima 1983a). The mean and variance of t are given by

$$\bar{t} = 4N/[n(n-1)], \qquad V(t) = 16N^2/[n(n-1)]^2.\qquad(13.72)$$

One important quantity in the study of gene genealogy is the *coalescence time*, i.e., the time at which all of the n genes sampled converge to a single ancestral gene. The mean of this coalescence time is

$$\bar{t} = 2N\sum_{i=2}^{n}[1/\tbinom{i}{2}] = 4N(1 - 1/n)\qquad(13.73)$$

(Kingman 1982; Tajima 1983a), whereas the variance is

$$V(t) = 4N^2\sum_{i=2}^{n}[1/\tbinom{i}{2}^2]\qquad(13.74)$$

(Tajima 1983a). As n increases, $\bar{t}$ and $V(t)$ quickly approach $4N$ and $4.6N^2$, respectively. Note that when $n = 2N$, these values are essentially the same as those of fixation time of a newly arisen neutral mutation (Kimura and Ohta 1969; Burrows and Cockerham 1974).

Let us now compute the expected times of gene splitting in the genealogical tree of figure 13.8d. The expected branch length between A and B is equal to $(1/2)T$ from (13.68) or (13.72), where $T = 4N$ generations. The branch length between B and C is $(1/6)T$, whereas the branch length between C and D is $(1/12)T$. Therefore, the total time is $(1/2 + 1/6 + 1/12)T = (3/4)T$, which agrees with the coalescence time obtained from (13.73). Note, however, that each branch length has a large standard deviation which is equal to the mean.

In actual data analysis, a genealogical tree is constructed by using information on the numbers of nucleotide differences between each pair of genes (e.g., figure 10.5). Under the assumption of neutral mutations, the expectation of such a tree is given by the above theory. In practice, however, both topology and branch lengths will vary greatly from locus to locus because of stochastic error.

Gene Differentiation Between Populations

Differentiation with Migration

Previously, we discussed Wright's island model without mutation. Let us now extend this model to the case of the infinite-allele model of neutral mutations. We shall also remove the assumption of an infinite number of subpopulations. We assume that there are s subpopulations of effective size N and that immigrants into a subpopulation are a random sample of individuals from the entire population. We denote the migration rate by m and the mutation rate by v. Let J_0 be the probability of identity of two randomly chosen genes from a subpopulation and J_1 be the probability of identity of two randomly chosen genes, one from each of two subpopulations. Clearly, J_0 is equal to the expected homozygosity within populations, i.e., $J_0 = E(\Sigma x_i^2)$, where x_i is the frequency of the ith allele in a subpopulation. On the other hand, J_1 is given by $E(\Sigma x_i y_i)$, where x_i and y_i are the frequencies of the ith allele in populations X and Y, respectively. We have seen that when there is no migration and no mutation, the recurrence equation for J_0 is given by equation (13.13). We now assume that sampling of genes, migration, and mutation occur in this order. We can then derive the following recurrence equations for J_0 and J_1.

$$J_0^{(g+1)} = (1-v)^2 \left[a \left\{ \frac{1}{2N} + \left(1 - \frac{1}{2N}\right) J_0^{(g)} \right\} + (1-a) J_1^{(g)} \right], \qquad (13.75a)$$

$$J_1^{(t+1)} = (1-v)^2\left[b\left\{\frac{1}{2N} + \left(1 - \frac{1}{2N}\right)J_0^{(t)}\right\} + (1-b)J_1^{(t)}\right], \qquad (13.75b)$$

where $a = (1-m)^2 + m(2-m)/s$, $b = m(2-m)/s$, and superscripts refer to generations (Maruyama 1970; Nei 1975).

It is not difficult to obtain general equations for $J_0^{(t)}$ and $J_1^{(t)}$ from the above equations, but they are too complicated to be useful (see Latter 1973b and Li 1976b). The equilibrium values of J_0 and J_1 are, however, obtained easily by putting $J_0^{(t+1)} = J_0^{(t)} = J_0^{(\infty)}$ and $J_1^{(t+1)} = J_1^{(t)} = J_1^{(\infty)}$. They become

$$J_0^{(\infty)} = (1-v)^2[a - (1-m)^2(1-v)^2]/(2NG), \qquad (13.76a)$$

$$J_1^{(\infty)} = b(1-v)^2/(2NG), \qquad (13.76b)$$

where

$$G = 1 - (1-v)^2\left[1 + (1-m)^2 - \frac{a}{2N}\right] + (1-m)^2(1-v)^4\left(1 - \frac{1}{2N}\right).$$

In chapter 9, we considered the normalized identity of genes between two populations defined as $I = J_{XY}/\sqrt{(J_X J_Y)}$, where J_X and J_Y are the values of J_0 in populations X and Y, respectively, and J_{XY} is the value of J_1 between X and Y. In the present case, $J_{XY} = J_1$ for any pair of subpopulations and $J_X = J_Y = J_0$. Therefore, we have

$$I = m(2-m)/[s(1-m)^2\{1 - (1-v)^2\} + m(2-m)]$$

$$\approx m(2-m)/[2vs(1-m)^2 + m(2-m)]. \qquad (13.77)$$

Thus, as long as vs is small compared with m, I is close to 1, and the gene differentiation between populations is small. For the gene differentiation to be substantially large, the migration rate must be very small.

Of special interest is the case of $s = 2$ with a small migration rate. In this case, we may redefine the migration rate as the proportion of immigrant genes (m') entering from one population to the other, rather than the proportion of immigrant genes which are a random sample from the entire population (m). In general m' is related to m by $m' = m(s-1)/s$ (Nagylaki 1983). In the present case, $m' = m/2$. If we use this definition, I becomes $m'/(m' + v)$ approximately, ignoring the higher-order terms of m' and v. This is identical with equation (9.65).

In the above island model, the geographic distance between populations is disregarded. Maruyama (1970, 1973, 1977) studied the relationship between J_{XY} and geographic distance, assuming that s is finite. The results obtained indicate that in the case of one-dimensional distribution of populations J_{XY} decreases roughly exponentially as geographic distance increases, but the rate of decrease depends on the total length of the distribution of populations and migration distance. In the case of two-dimensional distribution, J_{XY} does not change so much with increasing distance as in the case of one-dimensional distribution. The value of I can be close to 1, even if the geographic distance between a pair of populations is a thousand times greater than the migration distance in one generation (Maruyama and Kimura 1974).

Another measure of population differentiation (chapter 8) is

$$G_{ST} = D_{ST}/H_T, \tag{13.78}$$

where H_T is the gene diversity in the total population and D_{ST} is the interpopulational gene diversity. In the present case, $D_{ST} = (1 - 1/s)(J_0 - J_1)$ and $H_T = 1 - J_0 + D_{ST} = 1 - J_1 - (J_0 - J_1)/s$. Therefore,

$$G_{ST} = 1/\left[1 + 4N\left(\frac{s}{s-1}\right)(m+v) \right], \tag{13.79}$$

approximately (Takahata and Nei 1984).

Note that if $m \gg v$ and s is sufficiently large, G_{ST} is approximately $1/(4Nm + 1)$, which is equal to F_{ST} given by equation (13.25). Therefore, G_{ST} is determined mainly by the $4Nm$ value. It is known that this equilibrium value of G_{ST} is reached relatively rapidly if m is sufficiently large. This is true even if D_{ST} and H_T deviate substantially from the equilibrium values (Nei et al. 1977; Crow and Aoki 1984).

Gene Differentiation Under Complete Isolation

We have seen that, as far as neutral genes are concerned, substantial differentiation of genes among populations occurs only when there is little or no migration. Let us now consider how gene differentiation proceeds under complete isolation.

With no migration, (13.75a) and (13.75b) reduce to

$$J_0^{(t+1)} = (1-v)^2 \left[\frac{1}{2N} + \left(1 - \frac{1}{2N}\right) J_0^{(t)} \right],$$

$$J_1^{(t+1)} = (1-v)^2 J_1^{(t)}.$$

Therefore,

$$J_0^{(t)} = J_0^{(\infty)} + (J_0^{(0)} - J_0^{(\infty)}) \left[(1-v)^2 \left(1 - \frac{1}{2N}\right) \right]^t$$

$$\simeq J_0^{(\infty)} + (J_0^{(0)} - J_0^{(\infty)}) e^{-(2v + 1/2N)t}, \qquad (13.80a)$$

$$J_1^{(t)} = (1-v)^{2t} J_1^{(0)}$$

$$\simeq J_1^{(0)} e^{-2vt}, \qquad (13.80b)$$

where

$$J_0^{(\infty)} = (1-v)^2 / [2N - (2N-1)(1-v)^2]$$

$$\simeq 1/(4Nv+1). \qquad (13.81)$$

A formula equivalent to (13.80a) was first derived by Malécot (1948). Formula (13.81) is the same as equation (13.29), as expected.

In the presence of mutation, $J_1^{(t)} = J_1^{(0)} e^{-2vt}$, whereas $J_0^{(t)} = J_0^{(\infty)}$ if the populations are in equilibrium. Therefore, if $J_1^{(0)} = J_0^{(0)}$, the genetic identity becomes

$$I = J_1^{(t)} / J_0^{(t)} = e^{-2vt} \qquad (13.82)$$

(Nei and Feldman 1972). This is identical with equation (9.53) when $I_0 = 1$ and $\alpha = v$.

Nucleotide Differences Between Genes from Different Populations

In chapter 10, we discussed the estimation of the number of net nucleotide substitutions (d_A) between populations. If we consider the infinite-site model of neutral mutation, equation (10.22) can easily be justified. Previously, we showed that in an equilibrium population the mean number of nucleotide differences per locus between two randomly chosen alleles is $M = 4Nv$, where N is the effective population size and v is the mutation rate per locus per generation [equation (13.51)]. Therefore, the mean number of nucleotide differences per nucleotide site is $L = 4N\mu$, where $\mu = v/n$ and n is the total number of nucleotides per locus. We assume that the equilibrium population splits into two identical populations (X and Y) in a generation and thereafter no migration occurs between the two populations. We also assume that the mutation-drift balance is maintained throughout the evolutionary process. Suppose that two genes, one from each of populations X and Y, are randomly chosen at the tth generation after population splitting. The expected number of mutations that occurred in the two genes after population splitting is $2\mu t$ per site. However, the two genes are expected to have diverged and had $L \equiv 4N\mu$ nucleotide differences at the time of population splitting. Therefore, the expected number of nucleotide substitutions (d_{XY}) per site between two randomly chosen genes from X and Y is given by

$$E(d_{XY}) = 2\mu t + L. \tag{13.83}$$

Mathematically, the above equation can be derived more rigorously if we use the method of probability-generating function (Li 1977b; Gillespie and Langley 1979). Furthermore, the variance of d_{XY} can be shown to be

$$V(d_{XY}) = 2\mu t + L + L^2. \tag{13.84}$$

In the present case, we have assumed that both populations X and Y are in mutation-drift balance throughout the evolutionary process. Therefore, the expectations of the numbers of nucleotide differences between two randomly chosen genes within X and Y (d_X and d_Y, respectively) are both L. Thus, the expectation of equation (10.21), i.e., $d_A = d_{XY} - (d_X + d_Y)/2$, is equal to $2\mu t$, which is the same as (10.22) if μ and t are replaced by λ and T, respectively.

When there is selection, the rate of nucleotide substitution (λ) is not necessarily equal to the mutation rate. However, as long as the expectations of d_X and d_Y remain constant, equation (10.22) holds true. For example, in the case of advantageous mutations with genic selection, $\lambda = 4Ns\mu$ and $L = 8N\mu$ where time is measured in generations [see equations (13.54) and (13.46)].

The stochastic variance of d_A has been studied by Takahata and Nei (1985) under the assumption of no selection. Their study indicates that the coefficient of variation of d_A is very large when d_A is small and that this cannot be reduced substantially by increasing the number of genes sampled. To reduce the coefficient of variation, one must examine genes from many independent loci.

Genealogical Relationship of Genes from Different Populations

In a previous section, we considered the genealogy of genes sampled from a single population. A similar study can be made for the genealogy of genes sampled from closely related species or populations. However, the genealogical relationship of genes is much more complicated in this case (Takahata and Nei 1985). In the following, I shall consider the simplest case and discuss its relationship with the phylogenetic tree of populations.

Let us consider the genealogical relationship of three genes sampled from three different populations that diverged following the evolutionary tree given in figure 11.1. There are three possible relationships of genes in the present case (a, b, and c in figure 11.1). In (a), the three genes are derived from two genes in the ancestral species A, whereas in the other two cases they are derived from three genes. In cases (a) and (b) the topology of the gene tree is the same as that of the species tree, but in case (c) it is different. Therefore, if we estimate the topology of a species tree by using single gene data, we are expected to make an erroneous inference with a certain probability.

Let us now consider the probability of making an erroneous topology under the assumption of no selection. This probability is expected to be high when the time interval between the first and second species splittings is short. It is also expected to be high when the extent of polymorphism is high. The probability depends on (1) whether the three genes from populations X, Y, and Z are derived from two genes or three genes in A and, if they are derived from three genes, (2) whether or not the splitting of genes from X and Y occurred earlier than the other splitting. Here, we neglect the event of trifurcation, since the probability of this event is negligible. The probability that the three genes in X, Y, and Z are derived from two genes in A is obviously the same as the probability that the two genes in the ancestral species B are derived from one gene in A. This probability is given by $1 - \exp(-T/2N)$, where T is the number of generations between the first and second speciation events $(t_1 - t_0)$ and N is the effective population size. This is because the probability that the two genes in B are derived from two genes in A is

$$\int_0^T f_2(t)dt = 1 - e^{-T/2N},$$

where $f_2(t)$ is given by (13.67). Thus, the probability that the three genes from X, Y, and Z are derived from three genes in A is $\exp(-T/2N)$. In this case, however, the topology of the gene tree is correct with a probability of 1/3 [case (b) in figure 11.1]. Therefore, the probability of obtaining an erroneous tree is

$$P = \frac{2}{3} e^{-T/2N}. \qquad (13.85)$$

This probability is high when T is small and N is large, as expected. If $T/(2N)$ is very small, the topology of a gene tree will be different from that of the species tree with a probability of 2/3. On the other hand, if $T/(2N)$ is large, the topology of the gene tree almost always agrees with that of the species tree. How large must $T/(2N)$ be to obtain the correct topology with a probability of $1 - E$? This is given by

$$T/(2N) = -\log_e(3E/2). \qquad (13.86)$$

If $E = 0.05$, we have $T/(2N) = 2.6$ or $T = 5.2N$. If $E = 0.01$, $T = 8.4N$.

In chapter 11, we constructed a gene tree for humans and apes from mitochondrial DNA data. In this tree, the human and chimpanzee were genetically closer to each other than to the gorilla. However, this branching order could be different from that of the species tree if the time interval between the first and second speciation events among the three species is short. In the ancestors of the human, chimpanzee, and gorilla, the long-term effective size was probably of the order of 10^4 (Nei and Graur 1984). Then, T must be 5.2×10^4 generations for $E = 0.05$. If the generation time for these organisms is 15 years, it corresponds to 780,000 years. If we use $E = 0.01$, this becomes 1.26 million years. Thus, if the first and second speciation events in the human, chimpanzee, and gorilla triad had a time interval of one million years or less, it would be very difficult to know the actual branching pattern of the species tree by using information from single genes. One might think that if several alleles are sampled from each species, gene trees would be closer to the species trees. Unfortunately, this is not the case (P. Pamilo and M. Nei, unpublished). Furthermore, when there are more than three species, the probability of making an erroneous topology increases unless the time interval between two consecutive speciation events is very large. To make a reliable tree, we must use many different genes which are independently evolving.

Recently, Ueda et al. (1985) used a qualitative approach to resolve the order of branching of the human, chimpanzee, and gorilla. Studying the genes for immunoglobulin C_ϵ chain, they found that the human and gorilla share a young truncated pseudogene whereas the old world monkeys and chimpanzee lack it. From this observation, they suggested that the gorilla is closer to man than to the chimpanzee unless the chimpanzee has lost the pseudogene after its divergence from the other two hominoid species. It should be noted, however, that in the presence of polymorphism the gene tree does not necessarily agree with the species tree, as discussed above, and thus Ueda et al.'s approach may not give the correct answer. Indeed, Hixson and Brown (1986) recently identified a one-base deletion shared by the chimpanzee and the gorilla but not by the human in their study of mitochondrial DNA, whereas Koop et al. (1986) reported two deletions and one insertion shared by the human and the chimpanzee in the η globin gene. These observations are contradictory with Ueda et al.'s and suggests that the common ancestor of the three species was quite polymorphic at the time of the first species splitting.

CHAPTER FOURTEEN

IMPLICATIONS FOR EVOLUTIONARY THEORY

While the purpose of this book is to discuss the theoretical framework of molecular evolutionary genetics, I have presented a fair amount of data to illustrate various points in discussion. I have also provided relevant references to which the reader may turn for further information. Yet my discussion of experimental data in the previous chapters was not intended to be systematic. In this chapter, I will discuss some of the major findings in the study of molecular evolution and their implications for general evolutionary theory. The discussion begins with some historical background, as viewed by a geneticist.

Historical Background

Darwinism

The key elements of Darwin's theory of evolution are chance (fluctuating) variation and natural selection. Darwin did not know how genetic variation arises, but he clearly visualized how natural selection operates on genetic variation once it is produced. In his book *The Origin of Species,* Darwin accepted a weak form of the Lamarckian theory of use and disuse to explain the origin of variation. In the late nineteenth century, there was considerable controversy over the mechanism of evolution (see Allen 1978; Mayr 1982; Provine 1986), mainly because the origin of genetic variation was unknown. Some authors attributed great significance to natural selection, arguing that natural selection is responsible for creating new adaptive characters. Other authors (e.g., St. George Mivart and William Bateson) questioned the importance of natural selection and advocated the so-called saltational change of species. The latter authors argued that natural selection acting on continuous variation within species is not sufficient to transform one species into another. There was also a group of biologists who advocated Lamarckism.

Mutationism and Its Extension

It was at the time of this controversy that de Vries (1903) proposed his theory of the formation of new species by single mutations. It was welcomed by many biologists because his theory was based on seemingly solid experimental data. At the time when genetics was just born, his theory was not scrutinized. De Vries' mutation theory was based on his research on an American species of evening primrose, *Oenothera lamarckiana,* which was growing in the Netherlands. He first observed that there were several aberrant forms of this species in wild populations. Conducting breeding experiments for many generations, he showed that *O. lamarckiana* continually produced small proportions of aberrant forms and that they either bred true or segregated into *O. lamarckiana* and the new type. Some of the aberrant forms (e.g., *O. gigas*) were morphologically quite different from *O. lamarckiana,* so new species names were assigned to them (see Cleland 1972). However, de Vries' contention that new species arise by single mutational changes was later disproved. Renner (1914) and Cleland (1923) showed that *O. lamarckiana* is a permanent heterozygote for two chromosomal complexes and that de Vries' mutants were merely segregants from this unusual genetic form. This was a serious blow to de Vries' mutation theory. [Of course, this theory still applies to the special case of formation of new species by polyploidization, particularly in plants.]

However, the idea that new species arise by a special type of mutation did not disappear even after disproof of de Vries' contention. Goldschmidt (1940) examined the biological nature of local differentiation of polymorphic characters such as the color pattern of the moth *Lymantria* and concluded that microdifferentiation of polymorphic characters is qualitatively different from interspecific differences of characters and that to form a new species a special type of mutation called *systemic mutation* is necessary. He made a clear distinction between *microevolution* and *macroevolution* and contended that microevolution does not lead beyond the confines of the species. He believed that the development of an organism is determined by the total composition of chromosomes and a systemic mutation (macromutation) occurs when this composition changes. An example of systemic mutation he presented is the *position effect,* which arises when the gene arrangement of a chromosome is changed by inversion or translocation. However, Goldschmidt's theory of systemic mutation was not widely accepted because of lack of evidence. Most macromutations or "hopeful monsters" observed were deleterious and did not seem to be useful at all for evolution. Recently, this theory has been revived by some authors, and I shall return to this problem later.

While de Vries' and Goldschmidt's views on the origin of species were extreme and unrealistic, from our current knowledge of genetics, de Vries' idea led Morgan (1925, 1932) to propose a more reasonable theory of evolution. Conducting an extensive breeding experiment with *Drosophila melanogaster*, Morgan and his colleagues (Morgan et al. 1915) obtained many mutations and showed that most of them are controlled by Mendelian genes. They also showed that these mutations occur with a very low but measurable frequency. From these observations, Morgan advocated the importance of mutation in evolution. He was aware that most new mutations are deleterious and are quickly eliminated from the population but realized that a small proportion of mutations are advantageous and may spread through the population. He visualized that evolution occurs by replacement of extant genes by more advantageous mutations in the population. In his view, selection plays a less important role than mutation, its chief role being to preserve useful mutations and eliminate unfit genotypes. That is, natural selection is regarded merely as a sieve to choose beneficial mutations. For this reason, his theory is often called mutationism, but a better terminology would be the mutation-selection theory or classical theory in Dobzhansky's (1955) sense, since he did not neglect natural selection.

It is often said that Morgan was a typologist and did not have a population concept (e.g., Dobzhansky 1980; Allen 1978). In his 1932 book *The Scientific Basis of Evolution,* however, he presents a rather incisive discussion of population and quantitative genetics. On page 132, he states, "If the mutant is dominant . . . and is bred to the wild type, the mutant character will appear again in half of the progeny, and if advantageous, i.e., if one that increases the chance of survival of the individual and the race, it will gradually spread through the race. If the new mutant is neither more advantageous than the old character, nor less so, it may or may not replace the old character, depending partly on chance; but if the same mutation recurs again and again, it will most probably replace the original character. If the new dominant character is a disadvantageous one, it will soon be eliminated." This statement is as modern as that of Wright (1969) or Kimura (1983a), and indicates that Morgan had a fairly good grasp of population genetics concepts.

Morgan's mutation-selection theory was quite popular in the 1920s and 1930s, but its popularity gradually declined as neo-Darwinism gained force. There seem to be two main reasons for the decline of Morgan's theory. First, most mutations observed in laboratory experiments were deleterious and did not appear to be useful for adaptive evolution. Second, Morgan's theory appeared too simplistic to explain the evolution of complex organisms, where gene interaction exists among many different loci.

Neo-Darwinism

The term *neo-Darwinism* has been used to represent various versions of modified Darwinism since the late nineteenth century, but at the present time it usually means the evolutionary theory formulated by Fisher (1930), Wright (1931), and Haldane (1932). In the 1920s and 1930s, these three authors conducted extensive mathematical studies on the change in gene frequencies due to mutation, selection, and genetic drift and reached the conclusion that selection is much more effective in changing gene frequencies than mutation. This theoretical work was soon accepted by a number of leading experimental geneticists, notably Dobzhansky (1937), who wrote an influential book on evolution entitled *Genetics and the Origin of Species*. Through this book, neo-Darwinism was gradually disseminated among biologists, and various authors (Dobzhansky 1937, 1952; Muller 1940; Simpson 1944, 1949, 1953; Huxley 1942; Mayr 1942, 1963; Stebbins 1950; Ford 1964) made further refinements of the theory. Because of these works, neo-Darwinism was accepted by most biologists by 1960.

Following King (1972), we can characterize neo-Darwinism by the following statements [see Dobzhansky (1951, 1970), Simpson (1953), and Mayr (1963) for the details].

1. Mutation is random with respect to gene function and recurs with a reasonably high frequency.
2. Mutation is the primary source of variation, but its effect on gene frequency change is so small that it plays a minor role in evolution.
3. Because of mutations that have occurred in the past, natural populations contain a sufficient amount of genetic variability to respond to almost any kind of selection.
4. Evolution is determined mainly by environmental changes and

natural selection. Since there is enough genetic variability, no new mutations are required for a population to evolve in response to an environmental change. There is no relationship between the rate of mutation and the rate of evolutionary change.

5. Because mutations tend to recur at reasonably high frequencies, the majority of clearly advantageous mutations should have been fixed or should have reached their optimum frequencies in the population. Therefore, the genetic structure of a population is almost always at or near its optimum for a given environment.

6. Evolutionary change of a species occurs gradually by means of natural selection. Thus, macroevolution is nothing but the accumulation of the effects of microevolution.

Although the above statements characterize the general properties of neo-Darwinism, there are some differences among the views of different authors. The views of Fisher, Haldane, and Mayr are more or less the same as the one mentioned above. Ford held the view that most evolutionary changes of organisms occur through strong natural selection, whereas Simpson left substantial room for the role of mutation in creating new adaptive characters. Wright's view is also somewhat different. In his 1931 and 1932 papers, he developed an elaborate theory of shifting-balance evolution in which the mean fitness increases owing to a combined effect of selection and genetic drift in subdivided populations. Dobzhansky (1937, 1951) and Simpson (1944, 1953) accepted and used the theory to explain various levels of evolution, including evolution above the species level. However, British geneticists such as Fisher and Ford did not accept the importance of genetic drift even for this case.

Another subject over which considerable disagreement existed among neo-Darwinians is the maintenance of genetic polymorphism. One school, led by Dobzhansky (1955), Ford (1964), and Wallace (1968), believed that a large proportion of genetic polymorphisms affecting viability and fertility are maintained by some types of balancing selection. The other school, led by Muller (1950) and Crow (Morton et al. 1956), believed that they are maintained mainly by mutation-selection balance. Dobzhansky termed the former view the balance theory and the latter the classical theory. (The classical theory is essentially the same as Morgan's mutation-selection theory.) There was rather keen debate between the two schools in the 1950s and 1960s, particularly about the heterozygous effects of radiation-induced mutations that are deleterious in homozygous condition. This debate was later extended into the controversy over the maintenance of protein polymorphisms (see Lewontin 1974; Nei 1975; Wallace 1981; Wills 1981; Kimura 1983a).

Theory of Molecular Evolution

Experimental Observations

Let us now consider the general pattern of molecular evolution, taking into account the above historical background. In the previous chapters, I have discussed various findings on molecular evolution and polymorphism. They can be summarized by the following statements. [The first four statements are included in Kimura and Ohta's (1974) five principles of molecular evolution.]

1. For each protein or gene, the rate of evolution as measured by amino acid or nucleotide substitution is approximately constant per site per year for various evolutionary lineages, as long as the function of the gene remains the same.
2. Functionally less important genes or parts of genes evolve faster than functionally more important ones.
3. Nucleotide or amino acid substitutions that impair the function of a protein or an RNA product occur less frequently than those that do not affect the function.
4. A gene with a new function generally evolves from a duplicate gene, but duplicate genes often become nonfunctional by deleterious mutations.
5. The genomes of higher organisms are highly flexible, and the number of copies of a given DNA sequence may increase or decrease rapidly under certain conditions.
6. The extent of genetic variability within species is generally higher in large populations than in small populations.
7. Functionally less important parts of genes (DNA) are generally more polymorphic than functionally more important parts.

The first two properties of molecular evolution were recognized as soon as Zuckerkandl and Pauling (1965) and Margoliash and Smith (1965) started extensive studies on evolutionary changes of hemoglobins and cytochrome c. They tried to explain these observations in terms of neo-Darwinism. For example, Margoliash and Smith thought that a constant rate of amino acid substitution per site per year is possible if various types of selection are averaged out. For these biochemists and even evolutionists such as Simpson (1964) and Mayr (1965), the idea of fixation of a mutant gene in a large population without the aid of natural selection was unthinkable at that time.

A careful examination of the first two properties of molecular evolution, however, indicates that they contradict most of the principles of neo-Darwinism mentioned earlier. In neo-Darwinism, the rate of evolution should depend on how often and how fast the environment changes. Thus, we would expect that the rate of evolution in living fossils such as the lamprey is much slower than that in rapidly evolving groups such as primates. In reality, however, the hemoglobins in living fossils seem to have evolved nearly at the same rate as those of primates (Jukes 1971). According to neo-Darwinism, the rate of evolution should also depend on population size and generation time (chapter 13). As we have already discussed, this prediction does not hold for molecular evolution. Similarly, the observation that functionally less important genes or parts of genes evolve faster than functionally more important ones is incompatible with neo-Darwinism. In neo-Darwinism, most gene substitutions are supposed to occur by positive Darwinian selection, and thus functionally important genes or parts of genes are expected to show a higher rate of evolutionary change than functionally less important ones (chapters 4 and 13).

It is also difficult to explain both the level of protein or DNA polymorphism and the rate of nucleotide substitution with a mathematical model based on neo-Darwinism (chapters 8 and 10). Furthermore, the genomes of higher organisms are quite flexible and subject to rather rapid structural change without much phenotypic effect (chapter 6).

Neutral Theory and Its Alternatives

Realization of the difficulty of explaining molecular evolution by means of neo-Darwinism led Kimura (1968a) to propose the neutral theory of molecular evolution. This theory has then been refined by a number of authors. The current neutral theory may be characterized as follows.

1. Most amino acid and nucleotide substitutions are caused by random fixation of neutral or nearly neutral mutations. Therefore, the rate of substitution is equal to the rate of occurrence of nondeleterious mutations [equation (13.52)].

2. Genetic polymorphism in a population represents a phase of the process of gene substitution. If the effects of mutation and genetic drift are balanced, the expected heterozygosity per locus is given by $4Nv/(1 + 4Nv)$, where N and v are the effective population size and the mutation rate per locus, respectively (equation 13.30).

3. The neutral theory is concerned with the behavior of a majority of the mutations that are incorporated into the population during evolution and does not preclude the existence of a small proportion of advantageous or overdominant mutations.

4. It is assumed that most new mutations are deleterious and are quickly eliminated from the population. Therefore, they do not contribute very much to genetic variation or gene substitution in populations. Thus, if the total mutation rate per locus is v_T and the fraction of deleterious mutations is f, the rate of gene substitution is given by $(1-f)v_T$, neglecting the effect of advantageous mutation. Here, f is related in some way to the extent of functional constraints of the gene product.

5. Neutral alleles are not functionless genes but are generally of vital importance to the organism. A pair of alleles are called *neutral* if they are functionally equivalent and do not differentially affect the fitness of an organism. In population genetics, however, the definition of neutrality of a pair of alleles depends on the degree to which their behavior in a population is dictated by genetic drift. Suppose that there are two alleles at a locus, and let s be the selective advantage of one allele over the other. Then if $Ns \leq 1$, the pair of alleles are called neutral (Kimura 1968a). Therefore, a mutant gene which is advantageous in a large population may become neutral in a small population.

The neutral mutation theory is capable of explaining most observations on molecular evolution and variation. Thus, the approximate constancy of the rate of amino acid or nucleotide substitution can be explained by assuming that the rate of mutation is constant per year rather than per generation. As discussed in chapter 3, we do not have hard evidence either for or against this assumption, but it is a useful working hypothesis. Variation in substitution rate among different genes or different parts of genes can also be explained by assuming that each gene or gene part has different functional constraints. This view is supported by a substantial amount of data. As we have seen in chapter 5, the rate of nucleotide substitution is highest in functionless pseudogenes and lowest in such important functional genes as those for histones and cytochrome c. The rate of amino acid substitution is also negatively correlated with the degree of functional constraints of proteins (chapter 4).

It should be noted that the constancy of substitution rate in the neutral theory is an approximate one. Strictly speaking, the rate cannot be constant even in the statistical sense. This is because different regions of a gene are generally subject to different intensities of purifying selection, and the extent of purifying selection depends on the type of mutation (amino acid substitution) that occurs. Furthermore, functional constraints seem to change occasionally in the evolutionary process, as in the cases of guinea pig insulin and opossum hemoglobins (see Kimura 1983a). As mentioned above, the genomes of higher organisms are highly flexible and undergo a rather rapid change due to gene duplication, unequal crossover, gene conversion, etc. It seems that a large proportion of these genomic changes is caused by mere mutational changes and random genetic drift (Doolittle and Sapienza 1980; Orgel and Crick 1980; Ohta and Kimura 1981; chapter 6).

There has been much controversy over the neutral theory with respect to protein polymorphism in the past two decades. Initially, most arguments were speculative, but later statistical studies were conducted to examine the agreement between data and predictions from the neutral theory. These studies indicate that the general pattern of protein polymorphism can be explained by the neutral theory reasonably well, particularly when bottleneck effects are taken into account (chapter 8). The pattern of DNA polymorphism so far studied is also compatible with the neutral theory (chapter 10; Nei 1983).

There is still some controversy over the relationship between the number of heterozygous loci and growth rate. As discussed earlier, however, this could largely be due to associative overdominance caused by deleterious mutations at closely linked loci (chapter 8). It should also be noted that a certain proportion of protein polymorphisms must have adaptive significance, since the fundamental process of morphological or physiological adaptation occurs at the protein level. Indeed, there is an increasing amount of data that indicate a selective difference between polymorphic alleles. Nevertheless, this type of polymorphism seems to be a minority among all protein polymorphisms, as discussed in chapter 8.

Role of Mutation in Molecular Evolution

While no definite conclusion can be drawn about the type and intensity of selection at the present time, it is obvious that mutation plays an important role in molecular evolution. The rate of gene substitution and the level of genetic polymorphism are determined primarily by the rate of (nondeleterious) mutation (chapters 4 and 8). In the previous chapters, we have also seen that the most prevalent form of natural selection is purifying selection that eliminates deleterious mutations and maintains the function of a gene already established. That is, selection is mostly conservative.

Of course, occasionally there are mutations that give rise to a new adaptive character. For example, the normal strain of bacteria *Pseudomonas aeruginosa* uses acetamide and propianamide as sources of nitrogen but does not use valeramide and phenylacetamide. However, there are a number of mutant strains that can utilize valeramide or phenylacetamide (Betz et al. 1974) These strains have a higher fitness than the normal strains when they are grown in an environment with valeramide or phenylacetamide. Studies of the biochemical properties of the new enzymes produced by the mutant strains have suggested that only a few steps of mutational changes were involved in the formation of the new genes. At the DNA level, most new genes seem to have been produced by gene duplication and subsequent nucleotide changes (chapters 4 and 6). In these cases, the mutational change of DNA (duplication and nucleotide substitution) is clearly responsible for creating a new gene or character. Natural selection plays no such role. The role of natural selection is to eliminate less fit genotypes and save a beneficial one when there are many different genotypes in the same environment. Therefore, it seems clear that at the molecular level evolution occurs primarily by mutation pressure, though positive selection certainly speeds up gene substitution in populations.

This view also holds for various types of genomic evolution discussed in chapter 6. In a broad sense, gene duplication, gene conversion, and gene transposition are all mutational changes of DNA, and the above principle still applies. Previously, we mentioned that multigene families are subject to concerted evolution mediated by nucleotide substitution, unequal crossover, gene conversion, genetic drift, and selection and that member genes of each gene family evolve in concert rather than independently. However, even in this case the mutational change of DNA, including unequal crossover and gene conversion, seems to be the primary force of genomic evolution.

Adaptive and Nonadaptive Evolution

Relationship Between Molecular Evolution and Phenotypic Evolution

Although mutation and genetic drift are very important in molecular evolution, natural selection plays a significant role in phenotypic evolution. This is obvious if we note that most morphological and physiological characters are well-adapted to the environment in which the organism lives. How can we then reconcile these two types of observations? There are two ways of reconciliation. One is to assume that although most molecular mutations behave just like neutral genes in populations, they are not strictly neutral and their cumulative effects for many loci are responsible for the adaptive change of organisms. For convenience, I call this the minute gene effect hypothesis. The other way of reconciliation is to assume that most nondeleterious mutations are more or less neutral but there are a small proportion of adaptive mutations and these mutations are responsible for adaptation. I call this the major gene effect hypothesis.

The minute gene effect hypothesis was originally proposed by Latter (1972), but it has been formalized by Kimura (1981c, 1983a,b) and Milkman (1982). According to this hypothesis, most important morphological or physiological characters show a continuous variation and are subject to centripetal (stabilizing) selection. It is assumed that a character is normally distributed and controlled by a large number of loci each of which has a very small additive contribution to the character and that the fitness of an individual is determined by its phenotypic value, x, and as x deviates from some optimum value, the fitness declines disproportionately.

One might think that this type of selection leads to a stable polymorphism at each locus, but actually it results in homozygosity in the absence of mutation, as shown by Wright (1935) and Robertson (1956). That is, this is a kind of homogenizing selection. In the presence of mutation, however, a large amount of genetic variability may be maintained (Kimura 1965b). In this case, one can show that although the total amount of selection (genetic load) for a quantitative character is substantial, the selection coefficient for each locus is very small if the number of loci involved is large (Kimura 1981c; Milkman 1982; Lynch 1984). For example, if we consider an abstract character that determines the Darwinian fitness of an individual and assume that all nucleotides in the mammalian genome (3.5×10^9) are concerned with fitness, the selection coefficient(s) for a nucleotide site can be as small as 3.3×10^{-7} if the total amount of selection is 0.5 (Kimura 1983a). From this type of computation, Kimura concluded that "neutral molecular evolution is an inevitable process under stabilizing phenotypic selection when a large number of nucleotide sites are involved."

In this view, "neutral alleles" are not really completely neutral but have very small phenotypic effects, the accumulation of which is significant for adaptive evolution. Of course, this view does not preclude the possibility that a certain fraction of nucleotide changes are completely neutral. Neither Kimura nor Milkman has elaborated how the evolutionary change of morphological characters occurs, but they seem to assume that it is caused by environmental changes, which would shift the optimum phenotypic value.

The major gene effect hypothesis was originally presented by Kimura (1968a) and has been supported by a number of authors (e.g., King and Jukes 1969; Nei 1975; Wilson 1975). In this hypothesis, the majority of nucleotide substitutions are neutral or nearly neutral, and adaptive evolution occurs by a relatively small proportion of gene substitutions. We have seen that at the DNA level most loci are polymorphic and that there are many alleles at each locus. Thus, even if only a small percentage of gene substitutions is adaptive, there are still a large number of substitutions that can result in adaptive evolution. Therefore, there is no problem in explaining adaptive evolution. In this hypothesis, the selective advantage conferred by a mutation is assumed to be generally quite large, so that a rather small number of gene substitutions are required for a particular event of adaptive evolution. Essentially the same view has been expressed by Stebbins and Ayala (1985). Of course, this hypothesis does not preclude the existence of many modifier genes that have been identified in classical genetics but claims that even if these modifier genes are included, advantageous mutations are a minority of the total mutations.

At the present time, however, it is difficult to decide which of the above two hypotheses is correct. I myself prefer the second hypothesis for the following reasons. (1) It is not clear whether there is any character to which a very large number of "nearly neutral loci" contribute additively. For Kimura's model of stabilizing selection to be realistic, the phenotypic character must be completely developed before selection operates (Robertson 1968). This is because a selection coefficient cannot be assigned to each individual unless the phenotypic character is completed. In reality, a large proportion of natural selection seems to occur independently in various stages of development before maturity, as was emphasized by Nei (1975). Furthermore, if there are many quantitative characters each of which is controlled by a different set of relatively small numbers of loci, the overall pattern of selection would be close to the case of independent selection rather than to stabilizing selection. (2) As was discussed earlier, adaptation to new environments often occurs by a few nucleotide or amino acid changes (chapters 4, 8, and 10). Furthermore, there is an increasing amount of data indicating that genetic change of morphological and physiological characters is caused by a few major genes, as will be discussed below.

Evolution by Regulatory Mutation

As mentioned in the introduction, Wilson (1975) (see also King and Wilson 1975) proposed that adaptive evolution occurs mainly by mutational changes of regulatory genes. This proposal was made partly because most amino acid substitutions in a protein do not seem to affect the function of the protein. He also noted that in microorganisms the first stage of adaptation to new nutrients such as carbon sources occurs mainly by regulatory mutations. Indeed, the most common way for a microorganism to acquire a new metabolic capability is to constitutively synthesize an enzyme which already has a specificity for the novel source but for which the novel source is not the primary substrate (Hall 1983). This is accomplished by mutation of a gene that regulates the production of the enzyme. This enzyme is usually quite inefficient in metabolizing the novel source, so that a large quantity of the enzyme is produced. The next step of adaptation is to modify the specificity of the enzyme by further mutations. In this step, the mutations occurring in the structural gene region are important.

Wilson (1975) also argued that the rate of morphological evolution in vertebrates is not correlated with the rate of molecular evolution and that this is because morphological evolution occurs mainly by regulatory mutations. One example he used for this argument is the conspicuous morphological difference between humans and chimpanzees despite the fact that the genetic distance ($D = 0.62$) between them is of the order of interspecific distances in other organisms (chapter 9). He also suggested that the faster development of interspecific hybrid inviability in mammals than in frogs reflects a higher rate of regulatory mutations in mammals.

This hypothesis is attractive, but it is not clear how to distinguish between regulatory genes and structural genes. Presumably, many regulatory genes exert their function by producing proteins or RNAs, and thus regulatory genes themselves are structural genes. It is also quite likely that a large part of the genome of higher organisms functions both as structural genes and regulatory genes. This is because morphogenesis is so well coordinated that the production of a protein or enzyme would often stimulate or restrict the transcription or translation of other genes, as is probably the case with protein hormones. If this is the case, it would not be fundamental to distinguish between "regulatory genes" and "structural genes." Nevertheless, the term "regulatory gene" is useful if it is defined as any gene that affects the transcription or translation of other genes whether or not the gene is structural.

The study of the molecular basis of morphogenesis has already begun. Ambros and Horvitz (1984) identified several mutations that affect the development of the nematode *Caenorhabditis elegans*. *C. elegans* is a tiny organism that consists of only about 1,000 cells. The complete developmental lineage of each of these cells is known, and any deviation from the normal cell lineage can be detected. The mutations Ambros and Horvitz identified are all "heterochronic"; they affect the timing of division of certain types of cells and produce a different type of morphology. For example, two X-linked semidominant mutations at the *lin*-14 locus retard development by repeating certain developmental stages such as molting and larval cuticle synthesis. This results in so-called paedomorphosis. By contrast, some recessive mutations at this locus result in precocious phenotypes. The molecular basis of the expression of these heterochronic mutations is now under investigation. Since heterochrony is an important way of changing morphological characters in evolution (Gould 1977), elucidation of the molecular basis of heterochrony is expected to contribute greatly to our understanding of morphological evolution. It should be noted that heterochrony usually has a profound phenotypic effect even when it is controlled by a single pair of alleles (see Raff and Kaufman 1983 for examples).

Heterochronic mutations are not the only regulatory mutations that affect morphological characters significantly. Mutations of homeotic genes in the *bithorax* and *Antennapedia* gene complexes in *Drosophila melanogaster* also have similar effects. The term "homeosis" refers to the replacement of one structure of the body by a homologous structure from other body segments. For example, a mutation (*bx*) at the *bithorax* gene complex affects the third thoracic segment and produces a fly with four wings instead of the usual two. A mutant allele (*Antp*) at the *Antennapedia* gene complex changes antennae into legs. Existence of these mutations suggests that wild-type homeotic genes control the proper spatial distribution of organs in development. The *bithorax* gene complex is composed of about 400 kb of DNA, and mutations at this complex are known to be due to insertions or deletions (Bender et al. 1983).

Homeotic genes contain a highly conserved sequence of about 180 nucleotides, which is called the *homeo* box (McGinnis et al. 1984). The similarity between the homeo boxes of the *bithorax* and *Antennapedia* complex genes is more than 70 percent. Interestingly, the homeo box occurs not only in insects but also in vertebrates (frogs, mice, and man), and the similarity in the homeo box between insects and vertebrates is very high (Gehring 1985). The homeo box is known to transcribe RNAs with open (unidentified) reading frames which are believed to control the spatial organization of the embryo.

The *Notch* locus in *Drosophila* is also known to be a homeotic gene that switches the developmental fates of cells between alternative pathways of development. Homozygotes for *Notch* mutations show an unusual pattern of development. In wild-type embryos, shortly after gastrulation, cells move up from the ventral ectoderm to become neuroblasts. There are three waves of recruitments, and about one-fourth of the ectodermal cells finally become neuronal. In *Notch* mutant homozygotes, however, most of the ventral ectoderm becomes neuronal. The embryos with all brain and no skin will then die young. Wharton et al. (1985) have shown that this locus produces an RNA transcript that contains an 8,109 bp open reading frame for a 2,703 amino acid polypeptide. This polypeptide does not have the homeo box but contains a repeat structure composed of 36 tandemly arranged 40 amino acid long repeats. Interestingly, these repeats have high homology with the epidermal growth factor (*EGF*) in mammals. They also have homology with the repeats observed in the open reading frame encoded by another homeotic gene, *lin*-12, in the nematode *C. elegans* (Greenwald 1985). The *lin*-12 locus is known to control certain binary decisions concerning the fates of cells in development. Thus, the *EGF, Notch,* and *lin*-12 genes make up a group of homeotic genes that are evolutionarily different from those containing the homeo box.

The regulatory mutations discussed above were all obtained from laboratory strains. Allendorf et al. (1983), however, reported a regulatory mutation obtained from a natural population of rainbow trout. The primary function of this mutant allele [*Pgm-t(b)*] is to control the expression of the enzyme phosphoglucomutase in liver tissue, but the presence of this enzyme in liver tissue makes the embryo hatch earlier and the fish grow faster. This mutant gene, therefore, seems to confer significant selective advantage upon its carriers. This allele exists in low frequency in rainbow trout populations and seems to be absent in its related species. For these reasons, Allendorf et al. speculate that this mutant allele appeared very recently.

Another example of regulatory mutation is the PLA^+ allele at a locus that regulates the production of intestine lactase in man. In most orientals and most mammalian species, the level of intestine lactase rises just before birth and declines at weaning. Adult individuals (genotype PLA^-/PLA^-), therefore, do not have this enzyme and consequently cannot absorb lactose from milk. In many northern Europeans, however, a high level of lactase is maintained even at the adult stage. This is because the mutant allele PLA^+ they carry does not turn off the production of lactase at the time of weaning (Flatz and Rotthauwe 1980). (The lactase enzymes produced by infant and adult Europeans are identical; Potter et al. 1985). In societies with dairy industries, therefore, the PLA^+ allele is expected to affect morphological characters. This mutation seems to have occurred in the early stage of human evolution (Nei and Saitou 1986).

Shifting Balance Theory of Evolution

Wright's (1931, 1932) shifting balance theory was originally proposed to explain morphological evolution. More recently, however, Wright (1978) stated that the discovery of extensive protein polymorphism in natural populations supports this theory, implying that his theory is applicable to molecular evolution. Extensive polymorphism itself is not proof of the shifting balance theory, however. Furthermore, as discussed by Nei (1980), Wright's argument for this theory is largely speculative. Therefore, I would like to discuss it in some detail.

Wright's theory is based on three premises. (1) Populations are subdivided into a large number of subpopulations or demes, among which gene migration occurs with a small probability. (2) Most loci are polymorphic, and evolution occurs mainly by gene frequency shift rather than by complete gene substitution. Genetic polymorphism is maintained by some sort of balancing selection (overdominance or frequency dependent selection). [Wright (1977:455–60) has considered an example of homoallelic (monomorphic) multiple peaks, but in this case the evolutionary rate depends primarily on mutation rate.] (3) There are epistatic gene interaction and pleiotropy among many loci. In general, there is no selective advantage or disadvantage for a specific allele, but a particular combination of alleles at different loci produces a selective advantage or disadvantage.

With these three premises, Wright argues that the rate of evolution in large substructured populations is substantially higher than that in small populations or large nonstructured populations. His argument is as follows. Because of the assumption of gene interaction and pleiotropy for many loci, the mean fitness ($\bar{W}$), expressed as a function of gene or chromosome frequencies, is expected to show numerous peaks in a hyperdimensional space, and the evolutionary change of a population can be described as a process of the population mean, $\bar{W}_p$, moving from one peak on the adaptive ($\bar{W}$) surface to another higher peak [see Wright (1970, 1977) for the detail]. In a small population, $\bar{W}_p$ would not easily reach a higher peak because of genetic drift, or even if it reached the peak temporarily, it would not stay there for a long time. On the other hand, in a large population $\bar{W}_p$ may reach a local peak which is relatively low and stay there indefinitely. It is difficult for this population to reach a higher peak, because it usually has to cross a "valley" or a "saddle" on the adaptive surface. In large subdivided populations, however, the story is different. The effective size of a subpopulation is so small that genetic drift operates in each subpopulation even after the entire population has reached a local peak, and it is possible for a particular subpopulation to cross a "saddle" of the $\bar{W}$ surface and reach a higher peak. Once this happens, the allelic combination of this subpopulation may migrate into its neighboring subpopulations and eventually spread through the entire population. In this way, the mean fitness of a large subdivided population may increase gradually. [This part of Wright's argument is purely speculative, and there is no mathematical proof for it. In a small subpopulation, $\bar{W}_p$ may go down as well as go up owing to genetic drift.]

The first premise for this theory seems to be satisfied for many species, though the migration rate between subpopulations may not be as low as required by the theory (see Simpson 1953). The second premise is based on Wright's (1932) conviction that most loci are polymorphic in natural populations. This conviction was derived from the observation that whenever artificial selection is applied to a quantitative character, there is almost always a response to it (Wright 1977). At the DNA level, this premise is certainly supported, although it is not clear whether most of these polymorphisms are related to the genetic variation of a quantitative character or not. The third premise is based on his early experimental study with guinea pigs, which showed that there is extensive gene interaction and pleiotropy among different loci controlling coat and eye color (see Wright 1982). He also thought that since the development of an organism occurs through complicated biochemical pathways, there must be extensive gene interaction.

The shifting balance theory, however, does not seem to be supported by data on molecular evolution for the following reasons. (1) Comparison of amino acid or nucleotide sequences among different organisms indicates that the genetic change of populations occurs not by gene frequency shift but by complete gene substitution. Even closely related species often have different alleles fixed at many loci. (2) The amino acid substitution in one polypeptide seems to occur independently of that in another polypeptide (Dayhoff 1972). This seems to be true even with polypeptides composing the same protein, such as hemoglobin α and β chains (Langley and Fitch 1974). (3) If gene interaction is prevalent among protein loci, we would expect a substantial amount of linkage disequilibrium among the loci (Wright 1952). Studies on this problem do not support this prediction, as mentioned in chapter 8. (4) Gene substitution occurs roughly at a constant rate irrespective of the organism, whether the organism is rapidly evolving or a living fossil at the morphological level or whether or not it is highly polymorphic. Note that a highly polymorphic and highly structured species such as *Escherichia coli* does not necessarily show a high rate of amino acid or nucleotide substitution (Dayhoff 1972; Hori and Osawa 1980). (5) In the shifting balance theory, the extent of polymorphism for a given rate of gene substitution is expected to be much higher than that for the neutral theory. As discussed in chapter 8, the observed level of protein poly-morphism is even lower than that expected from the pure neutral theory.

Of course, Wright's shifting balance theory is primarily intended to be applied to morphological evolution, and thus molecular data do not really refute his theory. Unfortunately, we know very little about the evolutionary change of genes controlling morphological characters. There is certainly a large component of genetic variation in most quantitative characters in natural populations, and many of the loci controlling these characters probably have some epistatic gene interaction. Indeed, there are data indicating that quantitative characters are subject to centripetal or stabilizing selection, a form of epistatic selection (e.g., Karn and Penrose 1951). However, stabilizing selection is not the kind of selec-tion that is needed for the shifting balance theory, since it is essentially homogenizing selection as mentioned earlier. It should also be noted that for Wright's theory to be applicable natural populations must main-tain the same breeding structure (subdivided population) for a long time with the same migration and selection schemes. In practice, however, the breeding structure and ecological condition of a population often change very rapidly, and this would reduce the applicability of Wright's theory to a great extent, as emphasized by Ford (1964).

The shifting balance theory has been used as a guiding principle for neo-Darwinism by Dobzhansky (1951, 1970) and Simpson (1949, 1952), but there is little empirical evidence for it. Rather, empirical data sug-gest that the evolutionary change of morphological characters occurs faster in small populations than in large populations (Mayr 1970). Many highly evolved species such as primates and carnivores have small population size compared with other organisms such as fish and amphibians.

Nonadaptive Change of Morphological Characters

In the strict form of neo-Darwinism, very little room is given to neutral variation, and virtually all morphological and physiological char-acters are thought to be products of natural selection (Mayr 1963; Ford 1964). Certainly, most morphological and physiological characters are well-adapted to the environment in which the organism lives, and there is no question about the importance of natural selection in the formation of intricate morphological characters. However, are all individual differ-ences in morphological and physiological characters adaptive as claimed by extreme neo-Darwinians? More than 4 billion people live on this planet, and all of them except identical twins are different with respect to various morphological and physiological characters. Are all these differences adaptive? Is random genetic drift unimportant for generating morphological and physiological diversity among organisms? I doubt it. It seems to me that in some morphological characters a substantial part of genetic variation is nonadaptive. For example, the sternopleural or abdominal bristle number of *Drosophila melanogaster* does not seem to be directly related to fitness except the extreme values in both directions (Clayton and Robertson 1955). Robertson (1967) conducted upward artificial selection for six generations with respect to sternopleural bristle number and obtained a rapid response to the selection. He then relaxed the selection and examined the change in mean bristle number in the following generations for nearly two and a half years (about 40 generations). In the first few months, about one-fifth of the selection gain was lost, but thereafter no visible change was observed. However, when downward selection was applied to this line, there was a rapid response. This indicates that this line had enough genetic variability when upward selection was relaxed, but natural selection ceased to operate a few months after relaxation of artificial selection. A similar result was obtained in a different experiment in which downward artificial selection was conducted. From these experiments, he concluded that the genes at a high proportion of the loci responsible for the bristle number must be neutral in their effects on fitness.

There are many other such characters which do not appear to be under strong selection pressure. For example, variation in the dermal ridge count in man seems to be as trivial as that of *Drosophila* bristle number. This character is highly heritable (nearly 100 percent heritability; Holt 1961), and there is a large amount of genetic variation within and between human populations. Human stature also does not seem to be directly related to fitness, except extremely short or tall stature, which is caused mainly by hormonal abnormality. It is possible that individuals whose stature is not far from the mean are virtually identical in fitness and that the genetic variation within this range is maintained largely by the balance among mutation, genetic drift, and weak selection, if any. Even the genetic variability between populations may not always be adaptive. The Ainu living in northern Japan are morphologically different from the Japanese, the male body being quite hairy. However, it is not clear what kind of selective advantage their hairiness confers upon the Ainu. It is possible that this character was brought about by mutation in their ancestral population and that individuals with this mutant gene or genes later formed a new group, which further later became the Ainu population. In the past few decades, many anthropologists, evolutionists, and sociobiologists have tried to explain every detail of morphological, physiological, and behavioral differences between populations in terms of natural selection. However, some of the morphological differences may well be due to nonadaptive genetic change. We should be cautious about the adaptationist program, as was emphasized by Gould and Lewontin (1979) and Gould (1982a).

Note that the possibility of nonadaptive change of morphological characters was recognized even by Charles Darwin. In the sixth edition of *The Origin of Species,* he stated: "Variation neither useful nor injurious would not be affected by natural selection, and would be left either a fluctuating element, as perhaps we see in certain polymorphic species, or would ultimately become fixed, owing to the nature of the organism and the nature of the conditions." Later, many authors (e.g., J. T. Gulick, D. S. Jordan, F. B. Sumner), including such neo-Darwinian pioneers as Wright (1932) and Dobzhansky (1937), discussed the possibility of neutral evolution or variation [see Provine (1986) for an extensive discussion of this subject]. [In his later writings (e.g., Wright 1970), Wright disregarded the importance of genetic drift except as an aid for attaining a better gene combination in his shifting balance theory.] However, with the rise of neo-Darwinism, this view was almost completely obliterated. Thus, Mayr (1963:207) wrote: "Entirely neutral genes are improbable for physiological reasons. Every gene elaborates a 'gene product', a chemical that enters the developmental stream. It seems unrealistic to me to assume that the nature of the particular chemical (enzyme or other product) should be without any effect whatsoever on the fitness of the ultimate phenotype." "Selective neutrality can be excluded almost automatically wherever polymorphism or character clines are found in natural populations." This view has been shared by a large number of authors. However, as emphasized by Lewontin (1974) and Nei (1975), the activity difference between two forms of enzymes would not always result in fitness differences. Many higher organisms have a high degree of phenotypic plasticity and are tolerant of a certain degree of environmental variation. In Waddington's (1957) terminology, many characters are well canalized despite the fact that natural populations contain a large amount of genetic variability.

Role of Mutation in Evolution

I have emphasized the possibility that even the genes controlling morphological characters may be subject to random evolutionary changes. However, no one would deny that the evolution of these characters is controlled by natural selection to a much greater extent than molecular evolution. Is it then unchallengeable that *adaptive* evolution of morphological characters has occurred in the way described by neo-Darwinians? In my view, mutation seems to have played a vital role even in adaptive evolution (Nei 1975, 1983).

In neo-Darwinism, only a minor role in evolution is given to mutation, as mentioned earlier (see also Wright 1982). This view was formed apparently because whenever artificial selection was applied to a quantitative character, a quick response was observed in many organisms. This gave the early population geneticists the impression that a randomly mating population contains almost all kinds of genetic variation and that the only force necessary to achieve a particular evolutionary change is natural selection. However, artificial selection is quite different from natural selection. The response to artificial selection is usually large in the early generations but gradually diminishes as generations proceed. Without further input of new genetic variability, the mean value of a character under selection usually reaches a plateau within a few dozen generations. Note also that the initial response to selection or the heritability of a quantitative character is usually greater in those characters that are remotely related to fitness than in those that are closely related (e.g., Falconer 1981). This suggests that the former characters have not been subjected to strong natural selection, so that the amount of genetic variability accumulated in the population is large. Therefore, if artificial selection is applied to these characters, a rapid response occurs. In the latter characters, however, natural selection seems to have reduced the genetic variation, and thus artificial selection is less effective. If this interpretation is correct, artificial selection does not provide an accurate picture of long-term evolution by natural selection.

It is often said that genetic polymorphism is beneficial to the population because in the presence of genetic variability the population can adapt easily to new environments (Dobzhansky 1970). Thus, any mechanism that increases genetic variability is selected for (Ford 1964). [This view is not shared by all neo-Darwinians; see Wallace (1981), and Stebbins and Ayala (1985).] I am skeptical about this teleological explanation. In my view, the present genetic variability of a population is simply a product of evolution in the past. It may happen to be useful for future evolution, as in the case of industrial melanism (Kettlewell 1955), but I doubt that genetic variability is stored for future use. In many cases, a population may not have the genetic variability that is needed for new adaptation. In this case, the population may stay unchanged until new mutations occur or may simply become extinct. Note that over 99 percent of all species have become extinct in geological time. Note also that rapidly evolving species do not necessarily have a high level of genetic variability. For example, many carnivore and primate species (e.g., cheetahs, weasels, chimpanzees, and macaques) are depauperate in protein polymorphism, despite the fact that they are among the most highly evolved animals (see Nei and Graur 1984; O'Brien et al. 1985b).

In my view, natural selection is a consequence of the existence of two or more functionally different genotypes in the same environment. For example, if one genotype is more efficient in obtaining food or more resistant to a certain disease compared with others, it will have a higher survival value or reproductive success. The functional efficiency of a genotype is determined by the genes possessed by the individual, and natural selection automatically follows when there is a difference in functional efficiency between two different genotypes. Therefore, the most fundamental process of adaptive evolution is the production of better (functionally more efficient) genotypes by mutation or gene recombination. [Here, mutation includes all sorts of genetic changes such as nucleotide substitution and gene duplication.] Simpson (1953), Mayr (1963), and Dobzhansky (1970) have argued that natural selection has creative power just as a sculptor or a composer does (see also Gould 1982a; Stebbins and Ayala 1985). Their argument seems to be based on Muller's (1929) or Haldane's (1932) computation that the chance of combining many favorable mutations into one individual is enhanced tremendously if there is selection for all of them. Furthermore, in the presence of gene interaction a new phenotype may appear when different alleles are combined at different loci. However, it should be noted that natural selection operates only when there are different genotypes that are produced by mutation or recombination. It does not create any new genotypes.

Wright (1931, 1970, 1977) and Mayr (1963) have stated that the fitness of a genotype is determined by the entire set of genes in the individual, and it is difficult to isolate the effect of a single gene because of gene interaction among loci. As mentioned earlier, Wright viewed adaptive evolution as the process of a population climbing up an adaptive surface in a multidimensional space. In randomly mating populations, however, evolution by interactive genes does not occur easily, because random mating quickly breaks down a good combination of alleles at different loci unless these alleles are very tightly linked (Kimura 1974). This suggests that a large part of evolutionary change of genes occurs almost independently, as was conceived by Fisher (1930).

This does not mean that there is no gene interaction in the development and physiology of an organism. Rather, it implies that adaptive evolution occurs by using favorable mutations available at each evolutionary time. Thus, an allele fixed in a population at a locus may not be the best one in combination with the possible alleles at another locus. If there were another allele segregating at the second locus, the allele fixed at the first locus might have been different. Environmental change and genetic drift would also affect the outcome of evolution substantially. In my view, adaptive evolution is quite opportunistic and occurs mainly by mutation, recombination, and mass selection (according to the average gene effect) at each evolutionary time. Thus, the product of evolution could be much inferior to the theoretically possible best result. However, it does not matter if every population evolves in the same way. Furthermore, an allele whose effect on fitness was trifling when it was fixed in the population may become essential at a later time when another mutation is fixed at a different locus, as is often argued. In this scheme of evolution at interactive loci, polymorphism is not required for all loci simultaneously. It should also be noted that selective advantage of an allele depends on the environment and even apparently neutral alleles in one environment may change to selectively advantageous alleles in another environment (Simpson 1953; Dykhuizen and Hartl 1980).

The above view of evolution is similar to Morgan and Muller's classical theory but includes some new elements. It may be characterized as follows (Nei 1983).

1. At the nucleotide level, many mutations are deleterious, but a substantial proportion of them are neutral or nearly neutral. Only a small proportion of mutations are advantageous, and that is sufficient for adaptive evolution. Under certain circum-

stances, morphological characters may be subject to nonadaptive changes in evolution.

2. Natural selection is primarily a process to save beneficial mutations or to eliminate unfit genotypes.

3. New mutations spread through the population either by selection or by genetic drift, but a large proportion of them are eliminated by chance.

4. Populations do not necessarily have the genetic variability needed for new adaptation, though the variability at the molecular level is usually very large. When there is not enough genetic variability available, the population stays unchanged until new mutations occur or the population becomes extinct.

Lewontin (1974) coined the phrase *neoclassical theory* for this type of view, and I shall use this phrase for convenience, though his definition is somewhat different from what I visualize here and has a pessimistic connotation. The main difference between this neoclassical theory and Morgan and Muller's classical theory is that in the former theory a large number of neutral or nearly neutral mutations are assumed to exist in addition to the adaptive and deleterious mutations considered in the latter. [Muller was partly neo-Darwinian in the sense that he believed that all kinds of genetic variability are stored in the population; Muller (1949).] Thus, both the rate of gene substitution and the level of heterozygosity per locus are expected to be much higher in the former than in the latter. Genetic drift is also considered to be more important in the former. As noted by Lewontin (1974), however, both classical and neoclassical theories belong to the realm of Darwinism, where adaptive evolution occurs by selection of advantageous variations. They are different from the rigid form of neo-Darwinism defined earlier in that creative power is given to mutation and recombination rather than to natural selection. They also deemphasize the importance of overdominance in evolution.

In recent years, Eldredge and Gould (1972) and Gould (1982a) have argued that on a geological time scale evolution does not occur gradually, as is assumed in neo-Darwinism, but follows a pattern of "punctuated equilibrium." That is, the evolutionary change of a species is usually very rapid, but once the change is completed, the species remains unchanged for many millions of years, i.e., in stasis. The classical explanation of this type of evolutionary change is that stasis is caused by stabilizing selection in a constant environment, whereas rapid change is due to mass selection under a sudden environmental change (e.g., Simpson 1949; Charlesworth et al. 1982). Gould (1982a), however, questions this explanation, claiming that stasis usually persists through millions of years and through the extensive climatic changes that any species undergoes during such a long period of time. Nevertheless, neither Eldredge nor Gould present any convincing genetic explanation for their theory. While he is fully aware that Goldschmidt's systemic mutation is not acceptable with the current knowledge of genetics, Gould (1982b) tends to agree with Goldschmidt (1940) that macroevolution generating new species is different from microevolution that causes interracial or intersubspecific genetic differences. He seems to believe that macroevolution occurs mainly by species selection, whereas microevolution is caused by traditional Darwinian selection. As indicated by Stebbins and Ayala (1985), however, species selection cannot occur without interspecific genetic differences that have been generated by Darwinian selection within species, unless previously neutral changes later turn out to be adaptive because of environmental changes.

In my view, however, it is not difficult to explain the observations on punctuated equilibrium in terms of classical or neoclassical theory. The only assumption necessary is that the mutational change of highly adaptive or morphologically distinctive characters (Simpson's key mutation) occurs with a much lower frequency than that of less significant ones, mainly because the former would require a series of specific mutations (specific nucleotide sequence changes). For example, evolution of the mammary gland probably occurred through a series of mutational changes at the DNA level. The probability of occurrence of these mutational changes in one evolutionary lineage would be extremely small, so that the character could not evolve in many lineages independently. However, once the mammary gland evolved in a formerly reptilian line, it opened up a new way of life and consequently a new niche, and this led later to a rapid mammalian radiation. Similar but less conspicuous breakthroughs in adaptive evolution seem to have occurred even at a lower taxonomic level such as order, family, or genus. Thus, evolution of the human brain probably required a series of specific mutations, and these occurred (and became fixed) in the human line with a very low probability. At any rate, if certain adaptive or morphologically distinctive mutation occurs with a very low probability, evolution cannot occur gradually because the events of mutation and fixation are stochastic. If we consider this factor in addition to the classical explanation by environmental change and stability, there seems to be no problem in explaining the punctuated pattern of evolution.

The mechanism of species formation or speciation is an old but still unsolved problem. One of the most important aspects of speciation is the evolution of reproductive isolation. Individuals belonging to the same species are mutually fertile, but those belonging to different species usually do not mate or, even if they mate, they produce inviable or infertile offspring. How does this system of reproductive isolation evolve? In neo-Darwinism, it is customary to assume that reproductive isolation evolves as a by-product of adaptation of populations to different environments, i.e., as a result of pleiotropic effect of adaptive genes (Dobzhansky 1937; Muller 1942; Mayr 1963; Carson 1971). No detailed theoretical study has, however, been made on this hypothesis except for a few special cases (e.g., Sawyer and Hartl 1981; Sved 1981).

In my view, evolution of reproductive isolation can be explained more easily by neoclassical theory. Recent studies on isozyme expression in interspecific hybrids suggest that regulatory genes from one species are often incompatible with structural genes from the other and that this incompatibility of genes in developmental or gametogenetic process leads to hybrid inviability or sterility (e.g., Ohno 1969; Whitt et al. 1972; Frankel 1983; Parker et al. 1985). From this and other pieces of information, Nei (1976) and Nei et al. (1983) have proposed that hybrid inviability or sterility is caused by incompatibility of alleles at development-controlling or gametogenesis-controlling loci and that evolution of reproductive isolation occurs by random fixation of different mutations at these loci in different populations. In this view, there is no need of selection to intensify reproductive isolation; reproductive isolation is automatically developed as long as two populations are geographically or ecologically isolated and different incompatibility mutations are fixed in the two populations. In this case, the process of evolution of reproductive isolation is stochastic, so that it may occur relatively rapidly or extremely slowly, depending on chance. The speed of evolution also depends on the number of loci involved and the extent of selection due to incompatibility between mutant alleles. Nei (1976) and Nei et al. (1983b) developed a mathematical model of evolution of reproductive isolation based on this idea and showed that a variety of data on hybrid inviability or sterility, including the relationship between the extent of hybrid inviability and the time since divergence between two species (Wilson et al. 1974), can be explained by this model. Mayr's (1970) observation that speciation occurs more rapidly in small populations than in large populations is also explainable by this model. Furthermore, Nei et al. have shown that evolution of sexual or ethological isolation can be explained by essentially the same model of incompatibility genes.

Of course, the above theory is merely a hypothesis at the present time. There are many other possible ways of evolution of reproductive isolation, as discussed by Grant (1971), Bush (1975), White (1978), Templeton (1981), and Barton and Charlesworth (1984). However, the above theory can be tested experimentally. Particularly if we identify genes that control hybrid inviability or sterility and study the nature of their gene action at the molecular level, we will be able to gain deeper insight into the mechanism of speciation. Since gene cloning can now be done easily, it will not be long before speciation mechanisms are studied at the molecular level.

In this book, I have examined various aspects of molecular evolution and concluded that mutation is the driving force of evolution at the molecular level. I have also extended this view to the level of phenotypic evolution and speciation, though I do not deny the importance of natural selection in evolution. I have challenged the prevailing view that a population of organisms contains virtually all sorts of variation and that the only force necessary for a particular character to evolve is natural selection. I have also emphasized the unpredictability of the evolutionary fate of organisms caused by uncontrollable external factors such as rapid climatic changes, geological catastrophes, or even asteroid impacts.

In his book *Chance and Necessity,* Monod (1971) discussed the role of mutation in evolution. Recognizing that the occurrence of mutation has nothing to do with the teleonomy (purpose) of an organism, he concluded "chance alone is at the source of every innovation, of all creation in the biosphere. Pure chance, absolutely free, but blind, at the very root of the stupendous edifice of evolution." Chance plays an important role not only in the occurrence of mutation but also in the fixation of mutant genes, whether or not these genes are advantageous. Survival of a species also depends on various chance factors. Yet present-day organisms are not a random sample of mutant forms. They are products of screening by natural selection from an immense number of different kinds of organisms that have been produced by mutation. Innovation of organisms by single mutations may be minuscule, but their cumulative effect over a long evolutionary time can be colossal, as we can see in many higher organisms.

Bibliography

Abramowitz, M. and I. A. Stegun. 1964. *Handbook of Mathematical Functions with Formulas, Graphs, and Mathematical Tables.* Washington, D.C.: U.S. Dept. Commerce.

Adest, G. A. 1977. Genetic relationships in the genus *Uma* (Iguanidae). *Copeia* 1977:47–52.

Air, G. M. 1981. Sequence relationships among the hemagglutinin genes of 12 subtypes of influenza A virus. *Proc. Natl. Acad. Sci. USA* 78:7639–43.

Allen, G. E. 1978. *Thomas Hunt Morgan: The Man and His Science.* Princeton, N.J.: Princeton University Press.

Allendorf, F. W. 1979. Rapid loss of duplicate gene expression by natural selection. *Heredity* 43:247–58.

Allendorf, F. W., K. L. Knudsen, and R. F. Leary. 1983. Adaptive significance of differences in the tissue specific expression of a phosphoglucomutase gene in rainbow trout. *Proc. Natl. Acad. Sci. USA* 80:1397–1400.

Allendorf, F. W., F. M. Utter, and B. P. May. 1975. Gene duplication within the family *Salmonidae.* II. Detection and determination of the genetic control of duplicate loci through inheritance studies and the examination of populations. In C. L. Markert, ed., *Isozymes.* Vol. 4: *Genetics and Evolution,* pp. 415–32. New York: Academic Press.

Allison, A. C. 1955. Aspects of polymorphism in man. *Cold Spring Harbor Symp. Quant. Biol.* 20:239–55.

—— 1964. Polymorphism and natural selection in human populations. *Cold Spring Harbor Symp. Quant. Biol.* 29:137–49.

Alvarez, L. W., W. Alvarez, F. Asaro, and H. V. Michel. 1980. Extraterrestrial cause for the Cretaceous-Tertiary extinction. *Science* 208:1095–1108.

Ambros, V. and H. R. Horvitz. 1984. Heterochronic mutants of the nematode *Caenorhabditis elegans. Science* 226:409–16.

Aquadro, C. F. and J. C. Avise. 1981. Genetic divergence between rodent species assessed by using two-dimensional electrophoresis. *Proc. Natl. Acad. Sci. USA* 78:3784–88.

Aquadro, C. F., S. F. Deese, M. M. Bland, C. H. Langley, and C. C. Laurie-Ahlberg. 1986. Molecular population genetics of the alcohol dehydrogenase gene region of *Drosophila melanogaster. Genetics* 114:919–42.

Aquadro, C. F. and B. D. Greenberg. 1983. Human mitochondrial DNA variation and evolution: Analysis of nucleotide sequences from seven individuals. *Genetics* 103:287–312.

Aquadro, C. F., N. Kaplan, and K. J. Risko. 1984. An analysis of the dynamics of mammalian mitochondrial DNA sequence evolution. *Mol. Biol. Evol.* 1:423–34.

Archie, J. W. 1985. Statistical analysis of heterozygosity data: Independent sample comparisons. *Evolution* 39:623–37.

Avise, J. C. and C. F. Aquadro. 1982. A comparative summary of genetic distances in the vertebrates: Patterns and correlations. *Evol. Biol.* 15:151–85.

Avise, J. C., R. A. Lansman, and R. O. Shade. 1979. The use of restriction endonucleases to measure mitochondrial DNA sequence relatedness in natural populations. I. Population structure and evolution in the genus *Peromyscus. Genetics* 92:279–95.

Avise, J. C., J. C. Patton, and C. F. Aquadro. 1980a. Evolutionary genetics of birds. I. Relationships among North American thrushes and allies. *The Auk* 97:135–47.

—— 1980b. Evolutionary genetics of birds. III. Comparative molecular evolution in New World warblers and rodents. *Heredity* 71:303–10.

Avise, J. C. and R. K. Selander. 1972. Evolutionary genetics of cave-dwelling fishes of the genus *Astyanax. Evolution* 26:1–19.

Avise, J. C., J. F. Shapira, S. W. Daniel, C. F. Aquadro, and R. A. Lansman. 1983. Mitochondrial DNA differentiation during the speciation process in *Peromyscus. Mol. Biol. Evol.* 1:38–56.

Avise, J. C., J. J. Smith, and F. J. Ayala. 1975. Adaptive differentiation with little genic change between two native California minnows. *Evolution* 29:411–26.

Awramik, S. M., J. W. Schopf, and M. R. Walter. 1983. Filamentous fossil bacteria from the Archean of Western Australia. *Precambr. Res.* 20:357–74.

Ayala, F. J. 1975. Genetic differentiation during the speciation process. *Evol. Biol.* 8:1–78.

—— ed. 1976. *Molecular Evolution.* Sunderland, Mass.: Sinauer Associates.

Ayala, F. J. and M. L. Tracey. 1974. Genetic differentiation within and between species of the *Drosophila willistoni* group. *Proc. Natl. Acad. Sci. USA* 71:999–1003.

Ayala, F. J., M. L. Tracey, L. G. Barr, J. F. McDonald, and S. Perez-Salas. 1974. Genetic variation in natural populations of five *Drosophila* species and the hypothesis of the selective neutrality of protein polymorphisms. *Genetics* 77:343–84.

Baglioni, C. 1962. The fusion of two polypeptide chains in hemoglobin Lepore and its interpretation as a genetic deletion. *Proc. Natl. Acad. Sci. USA* 48:1880–86.

Bailey, G. S., R. T. M. Poulter, and P. A. Stockwell. 1978. Gene duplication in tetraploid fish: Model for gene silencing at unlinked duplicated loci. *Proc. Natl. Acad. Sci. USA* 75:5575–79.

Balakrishnan, V. and L. D. Sanghvi. 1968. Distance between populations on the basis of attribute data. *Biometrics* 24:859–65.

Barghoorn, E. S. and J. W. Schopf. 1966. Microorganisms three billion years old from the Precambrian of South Africa. *Science* 152: 758–63.

Barnard, E. A., M. S. Cohen, M. H. Gold, and J.-K. Kim. 1972. Evolution of ribonuclease in relation to polypeptide folding mechanisms. *Nature* 240:395–98.

Barr, T. C., Jr. 1968. Cave ecology and the evolution of troglobites. *Evol. Biol.* 2:35–102.

Barton, N. H. and B. Charlesworth. 1984. Genetic revolutions, founder effects, and speciation. *Ann. Rev. Ecol. Sys.* 15:133–64.

Bell, G. I., M. J. Selby, and W. J. Rutter. 1982. The highly polymorphic region near the human insulin gene is composed of simple tandemly repeating sequences. *Nature* 295:31–35.

Bender, W., M. Adam, F. Karch, P. A. Beachy, M. Peifer, P. Spierer, E. B. Lewis, and D. S. Hogness. 1983. Molecular genetics of the bithorax complex in *Drosophila melanogaster. Science* 221:23–29.

Benjamin, D. C. and 14 other authors. 1984. The antigenic structure of proteins: A reappraisal. *Ann. Rev. Immunol.* 2:67–101.

Bennett, J. H. 1954. On the theory of random mating. *Ann. Eugen.* 184:311–17.

—— 1965. Estimation of the frequencies of linked gene pairs in random mating populations. *Amer. J. Hum. Genet.* 17:51–53.

Benton, M. J. 1985. Interpretations of mass extinction. *Nature* 314: 496–97.

Benveniste, R. E. and G. J. Todaro. 1974. Evolution of C-type viral genes: Inheritance of exogenously acquired viral genes. *Nature* 252:456–59.

—— 1976. Evolution of type C viral genes: Evidence for an Asian origin of man. *Nature* 261:101–8.

Benyajati, C., A. R. Place, D. A. Powers, and W. Sofer. 1981. Alcohol dehydrogenase gene of *Drosophila melanogaster:* Relationship of intervening sequences to functional domains in the protein. *Proc. Natl. Acad. Sci. USA* 78:2717–21.

Berg, D. E., C. M. Berg, and C. Sasakawa. 1984. Bacterial transposon Tn5: Evolutionary inferences. *Mol. Biol. Evol.* 1:411–22.

Bermingham, E. and J. C. Avise. 1986. Molecular zoogeography of freshwater fishes in the southeastern United States. *Genetics* 113: 939–65.

Bernstein, S. C., L. H. Throckmorton, and J. L. Hubby. 1973. Still more genetic variability in natural populations. *Proc. Natl. Acad. Sci. USA* 70:3928–31.

Betz, J. L., P. R. Brown, M. J. Smyth, and P. H. Clarke. 1974. Evolution in action. *Nature* 247:261–64.

Beverley, S. M. and A. C. Wilson. 1984. Molecular evolution in *Drosophila* and the higher diptera. II. A time scale for fly evolution. *J. Mol. Evol.* 21:1–13.

—— 1985. Ancient origin for Hawaiian Drosophilinae inferred from protein comparisons. *Proc. Natl. Acad. Sci. USA* 82:4753–57.

Bhatia, K. K., N. M. Blake, and R. L. Kirk. 1979. The frequency of private electrophoretic variants in Australian aborigines and indirect estimates of mutation rate. *Amer. J. Hum. Genet.* 31:731–40.

Bhattacharyya, A. 1946. On a measure of divergence between two multinomial populations. *Sankhya* 7:401–6.

Bingham, P. M., M. G. Kidwell, and G. M. Rubin. 1982. The molecular basis of P-M hybrid dysgenesis: The role of the P element, a P-strain-specific transposon family. *Cell* 29:995–1004.

Bishop, J. M. 1981. Enemies within: The genesis of retrovirus oncogenes. *Cell* 23:5–6.

Blaisdell, B. E. 1985. A method for estimating from two aligned present-day DNA sequences, their ancestral composition and subsequent rates of substitution, possibly different in the two lineages, corrected for multiple and parallel substitutions at the same site. *J. Mol. Evol.* 22:69–81.

Blake, C. C. F. 1985. Exons and the evolution of proteins. *Intl. Rev. Cytol.* 93:149–85.

Blanken, R. L., L. C. Klotz, and A. G. Hinnebusch. 1982. Computer comparison of new and existing criteria for constructing evolutionary trees from sequence data. *J. Mol. Evol.* 19:9–19.

Blok, J. and G. M. Air. 1982. Sequence variation at the 3' end of the neuraminidase gene from 39 influenza type A viruses. *Virology* 121:211–29.

Bodmer, M. and M. Ashburner. 1984. Conservation and change in the DNA sequence coding for alcohol dehydrogenase in sibling species of *Drosophila*. *Nature* 309:425–30.

Bodmer, W. F. and J. Felsenstein. 1967. Linkage and selection: Theoretical analysis of the deterministic two locus random mating model. *Genetics* 57:237–65.

Botstein, D., R. L. White, M. Skolnick, and R. W. Davis. 1980. Construction of a genetic linkage map in man using restriction fragment length polymorphisms. *Amer. J. Hum. Genet.* 32:314–31.

Bowman, B. H. and A. Kurosky. 1982. Haptoglobin: The evolutionary product of duplication, unequal crossing over, and point mutation. *Adv. Hum. Genet.* 12:189–261.

Bowman, C. M., G. Bonnard, and T. A. Dyer. 1983. Chloroplast DNA variation between species of *Triticum* and *Aegilops*. Location of the variation of the chloroplast genome and its relevance to the inheritance and classification of the cytoplasm. *Theor. Appl. Genet.* 65:247–62.

Bridges, C. B. 1936. Genes and chromosomes. *Teaching Biol.* 11:17–23.

Britten, R. J. 1986. Rates of DNA sequence evolution differ between taxonomic groups. *Science* 231:1393–98.

Britten, R. J., D. E. Graham, and B. R. Neufeld. 1974. Analysis of repeating DNA sequences by reassociation. In L. Grossman and K. Moldave, eds., *Methods in Enzymology,* pp. 363–406. New York: Academic Press.

Britten, R. J. and D. E. Kohne. 1968. Repeated sequences in DNA. *Science* 161:529–40.

Brookfield, J. F. Y. 1982. Interspersed repetitive DNA sequences are unlikely to be parasitic. *J. Theor. Biol.* 94:281–99.

Brown, A. D. H. 1975. Sample sizes required to detect linkage disequilibrium between two or three loci. *Theor. Popul. Biol.* 8:184–201.

Brown, D. D. 1973. The isolation of genes. *Sci. Amer.* 229(2):20–29.

Brown, G. G. and M. V. Simpson. 1981. Intra- and interspecific variation of the mitochondrial genome in *Rattus norvegicus* and *Rattus rattus:* Restriction enzyme analysis of variant mitochondrial DNA molecules and their evolutionary relationships. *Genetics* 97:125–43.

Brown, W. M. 1983. Evolution of animal mitochondrial DNA. In M. Nei and R. Koehn, eds., *Evolution of Genes and Proteins,* pp. 62–88. Sunderland, Mass.: Sinauer Associates.

Brown, W. M., M. George, Jr., and A. C. Wilson. 1979. Rapid evolution of animal mitochondrial DNA. *Proc. Natl. Acad. Sci. USA* 76:1967–71.

Brown, W. M., E. M. Prager, A. Wang, and A. C. Wilson. 1982. Mitochondrial DNA sequences of primates: Tempo and mode of evolution. *J. Mol. Evol.* 18:225–39.

Bruce, E. J. and F. J. Ayala. 1979. Phylogenetic relationships between man and the apes: Electrophoretic evidence. *Evolution* 33:1040–56.

Bryant, E. H. 1974. On the adaptive significance of enzyme polymorphisms in relation to environmental variability. *Amer. Natur.* 108:1–19.

Burrows, P. M. and C. C. Cockerham. 1974. Distributions of time to fixation of neutral genes. *Theor. Popul. Biol.* 5:192–207.

Bush, G. L. 1975. Modes of animal speciation. *Ann. Rev. Ecol. Syst.* 6:339–64.

Bush, G. L., S. M. Case, A. C. Wilson, and J. L. Patton. 1977. Rapid speciation and chromosomal evolution in mammals. *Proc. Natl. Acad. Sci. USA* 74:3942–46.

Busslinger, M., S. Rusconi, and M. L. Birnstiel. 1982. An unusual evolutionary behaviour of a sea urchin histone gene cluster. *EMBO J.* 1:27–33.

Buth, D. G. 1979. Duplicate gene expression in tetraploid fishes of the tribe *Moxostomatini (Cypriniformes, Catastomidae). Comp. Biochem. Physiol.* 63B:7–12.

Buth, D. G. and C. B. Crabtree. 1982. Genetic variability and population structure of *Castomus santaanae* in the Santa Clara drainage. *Copeia* 1982:439–44.

Cabrera, V. M., A. M. Gonzalez, J. M. Larruga, and A. Gullon. 1983.

Genetic distance and evolutionary relationships in the *Drosophila obscura* group. Evolution 37:675–89.

Calder, N. 1983. *Timescale: An Atlas of the Fourth Dimension.* New York: Viking Press.

Campbell, A. 1983. Transposons and their evolutionary significance. In M. Nei and R. Koehn, eds., *Evolution of Genes and Proteins,* pp. 258–79. Sunderland, Mass.: Sinauer Associates.

Cann, R. L. 1982. The evolution of human mitochondrial DNA. Ph.D. thesis, University of California, Berkeley.

Cann, R. L., W. M. Brown, and A. C. Wilson. 1982. Evolution of human mitochondrial DNA: A preliminary report. In B. Bonné-Tamir, ed., *Human Genetics Part A: The Unfolding Genome,* pp. 157–65. New York: Alan R. Liss.

—— 1984. Polymorphic sites and the mechanism of evolution in human mitochondrial DNA. *Genetics* 106:479–99.

Cann, R. L. and A. C. Wilson. 1983. Length mutations in human mitochondrial DNA. *Genetics* 104:699–711.

Carson, H. L. 1971. Speciation and the founder principle. *Stadler Genet. Symp.* 3:51–70.

Cavalli-Sforza, L. L. 1969. Human diversity. In *Proc. 12th Intl. Cong. Genet.,* vol. 3, pp. 405–16. Tokyo.

Cavalli-Sforza, L. L. and W. F. Bodmer. 1971. *The Genetics of Human Populations.* San Francisco: Freeman.

Cavalli-Sforza, L. L. and A. W. F. Edwards. 1964. Analysis of human evolution. In *Proc. 11th Intl. Cong. Genet.,* pp. 923–33.

—— 1967. Phylogenetic analysis: Models and estimation procedures. *Amer. J. Hum. Genet.* 19:233–57.

Cavalli-Sforza, L. L. and A. Piazza. 1975. Analysis of evolution: Evolutionary rates, independence, and treeness. *Theor. Popul. Biol.* 8:127–65.

Cavender, J. A. 1981. Tests of phylogenetic hypotheses under generalized models. *Math. Biosci.* 54:217–29.

Cavener, D. R. and M. T. Clegg. 1981. Evidence for biochemical and physiological differences between enzyme genotypes in *Drosophila melanogaster. Proc. Natl. Acad. Sci. USA* 78:4444–47.

Chakraborty, R. 1977. Estimation of time of divergence from phylogenetic studies. *Can. J. Genet. Cytol.* 19:217–23.

—— 1980. Gene-diversity analysis in nested subdivided populations. *Genetics* 96:721–23.

—— 1985. Genetic distance and gene diversity: Some statistical considerations. In P. Krishnaiah, ed., *Multivariate Analysis,* pp. 77–96. Amsterdam: Elsevier.

Chakraborty, R., P. A. Fuerst, and M. Nei. 1977. A comparative study of genetic variation within and between populations under the neutral mutation hypothesis and the model of sequentially advantageous mutation. *Genetics* 86:s10–s11.

—— 1978. Statistical studies on protein polymorphism in natural populations. II. Gene differentiation between populations. *Genetics* 88:367–90.

—— 1980. Statistical studies on protein polymorphism in natural populations. III. Distribution of allele frequencies and the number of alleles per locus. *Genetics* 94:1039–63.

Chakraborty, R., M. Haag, N. Ryman, and G. Stahl. 1982. Hierarchical gene diversity analysis and its application to brown trout population data. *Heriditas* 97:17–21.

Chakraborty, R. and O. Leimar, 1987. Genetic variation within a subdivided population. In N. Ryman and F. Utter, eds., *Population Genetics and Fishery Management,* pp. 89–120. Seattle: University of Washington Press.

Chakraborty, R. and M. Nei. 1974. Dynamics of gene differentiation between incompletely isolated populations of unequal sizes. *Theor. Popul. Biol.* 5:460–69.

—— 1977. Bottleneck effects on average heterozygosity and genetic distance with the stepwise mutation model. *Evolution* 31:347–56.

Chakraborty, R. and N. Ryman. 1983. Relationship of mean and variance of genotypic values with heterozygosity per individual in a natural population. *Genetics* 103:149–52.

Chakravarti, A., K. H. Buetow, S. E. Antonarakis, P. G. Waber, C. D. Boehm, and H. H. Kazazian. 1984. Nonuniform recombination within the human β-globin gene cluster. *Amer. J. Hum. Genet.* 36:1239–58.

Champion, A. B., E. M. Prager, D. Wachter, and A. C. Wilson. 1974. Microcomplement fixation. In C. A. Wright, ed., *Biochemical and Immunological Taxonomy of Animals,* pp. 397–416. London: Academic Press.

Chao, L., C. Vargas, B. B. Spear, and E. C. Cox. 1983. Transposable elements as mutator genes in evolution. *Nature* 303:633–35.

Chapman, B. S., K. A. Vincent, and A. C. Wilson. 1986. Persistence or rapid generation of DNA length polymorphism at the zeta-locus of humans. *Genetics* 112:79–92.

Charlesworth, B. and D. Charlesworth. 1973. A study of linkage disequilibrium in populations of *Drosophila melanogaster. Genetics* 73:351–59.

—— 1983. The population dynamics of transposable elements. *Genet. Res.* 42:1–27.

Charlesworth, B., R. Lande, and M. Slatkin. 1982. A neo-Darwinian commentary on macroevolution. *Evolution* 36:474–98.

Christiansen, F. B. and M. W. Feldman. 1975. Subdivided populations: A review of the one- and two-locus deterministic theory. *Theor. Popul. Biol.* 7:13–38.

Clarke, B. 1970. Selective constraints on amino acid substitutions during the evolution of proteins. *Nature* 228:159–60.

—— 1975. The contribution of ecological genetics to evolutionary theory: Detecting the direct effects of natural selection on particular polymorphic loci. *Genetics* 79:101–13.

Clarke, P. H. 1984. Amidases of *Pseudomonas aeruginosa.* In R. P. Mortlock, ed., *Microorganisms as Model Systems for Studying Evolution,* pp. 187–231. New York: Plenum.

Clayton, G. A. and A. Robertson. 1955. Mutation and quantitative variation. *Amer. Natur.* 89:151–58.

Cleary, M. L., E. A. Schon, and J. B. Lingrel. 1981. Two related pseudogenes are the result of a gene duplication in the goat β-globin locus. *Cell* 26:181–90.

Cleland, R. E. 1923. Chromosome arrangements during meiosis in certain oenotheras. *Amer. Natur.* 57:562–66.

—— 1972. *Oenothera: Cytogenetics and Evolution.* New York: Academic Press.

Clifford, H. T. and W. Stephenson. 1975. *An Introduction to Numerical Classification.* New York: Academic Press.

Cockerham, C. C. 1969. Variance of gene frequencies. *Evolution* 23:72–84.

—— 1973. Analyses of gene frequencies. *Genetics* 74:679–700.

Coen, E. S., J. M. Thoday, and G. Dover. 1982. Rate of turnover of structural variants in the rDNA gene family of *Drosophila melanogaster. Nature* 295:564–68.

Cohn, V. H., M. A. Thompson, and G. P. Moore. 1984. Nucleotide sequence comparison of the *Adh* gene in three drosophilids. *J. Mol. Evol.* 20:31–37.

Collier, G. E. and S. J. O'Brien. 1985. A molecular phylogeny of the felidae: Immunological distance. *Evolution* 39:473–87.

Coon, C. S. 1965. *The Living Races of Man.* New York: Knopf.

Coyne, J. A. 1982. Gel electrophoresis and cryptic protein variation. *Isozymes* 5:1–32.

Crenshaw, J. W. 1965. Radiation-induced increases in fitness in the flour bettle *Tribolium confusum. Science* 149:426–27.

Crick, F. H. C. 1958. On protein synthesis. In *The Biological Replication of Macromolecules: 12th Symp. Soc. Expt. Biol.,* pp. 138–63. Cambridge: Cambridge University Press.

—— 1968. The origin of the genetic code. *J. Mol. Biol.* 38:367–79.

Crow, J. F. 1954. Breeding structure of populations. II. Effective population number. In T. A. Bancroft, J. W. Gowen, and J. L. Lush, eds., *Statistics and Mathematics in Biology,* pp. 543–56. Ames: Iowa State College Press.

Crow, J. F. and K. Aoki. 1984. Group selection for a polygenic behavioral trait: Estimating the degree of population subdivision. *Proc. Natl. Acad. Sci. USA* 81:6073–77.

Crow, J. F. and M. Kimura. 1970. *An Introduction to Population Genetics Theory.* New York: Harper and Row.

—— 1972. The effective number of a population with overlapping generations: A correction and further discussion. *Amer. J. Hum. Genet.* 24:1–10.

Crow, J. F. and N. E. Morton. 1955. Measurement of gene frequency drift in small populations. *Evolution* 9:202–14.

Curtis, S. E. and M. T. Clegg. 1984. Molecular evolution of chloroplast DNA sequences. *Mol. Biol. Evol.* 1:291–301.

Czekanowski, J. 1909. Zur Differentialdiagnose der Neandertalgruppe. *Korrespondenzblatt Deutsch. Ges. Anthropol. Ethnol. Urgesch.* 40:44–47.

Czelusniak, J., M. Goodman, D. Hewett-Emmett, M. L. Weiss, P. J. Venta, and R. E. Tashian. 1982. Phylogenetic origins and adaptive evolution of avian and mammalian haemoglobin genes. *Nature* 298:297–300.

Darnell, D. W. and K. M. Klotz. 1975. Subunit constitution of proteins: A table. *Arch. Biochem. and Biophys.* 166:651–82.

Darnell, J. E. 1978. Implications of RNA-RNA splicing in evolution of eukaryotic cells. *Science* 202:1257–60.

Darwin, C. 1859. *On the Origin of Species.* London: Murray.

Dayhoff, M. O., ed. 1969. *Atlas of Protein Sequence and Structure,* vol. 4. Silver Springs, Md.: Natl. Biomed. Res. Found.

—— 1972. *Atlas of Protein Sequence and Structure,* vol. 5. Silver Springs, Md.: Natl. Biomed. Res. Found.

Dayhoff, M. O. 1978. Survey of new data and computer methods of analysis. In M. O. Dayhoff, ed., *Atlas of Protein Sequence and Structure,* vol. 5, supp. 3, pp. 2–8. Silver Springs, Md.: Natl. Biomed. Res. Found.

Dayhoff, M. O. and W. C. Barker. 1972. Mechanisms in molecular evolution: Examples. In M. O. Dayhoff, ed., *Atlas of Protein Sequence and Structure,* vol. 5, pp. 41–45. Silver Springs, Md.: Natl. Biomed. Res. Found.

Dayhoff, M. O., R. E. Dayhoff, and L. T. Hunt. 1976. Composition of proteins. In M. O. Dayhoff, ed., *Atlas of Protein Sequence and Structure,* vol. 5, supp. 2, pp. 301–10. Silver Springs, Md.: Natl. Biomed. Res. Found.

Dayhoff, M. O. and C. M. Park. 1969. Cytochrome c: Building a phylogenetic tree. In M. O. Dayhoff, ed., *Atlas of Protein Sequence and Structure,* vol. 4, pp. 7–16. Silver Springs, Md.: Natl. Biomed. Res. Found.

Dayhoff, M. O., R. M. Schwartz, and B. C. Orcutt. 1978. A model of evolutionary change in proteins. In M. O. Dayhoff, ed., *Atlas of Protein Sequence and Structure,* vol. 5, supp. 3, pp. 345–52. Silver Springs, Md.: Natl. Biomed. Res. Found.

de Bruijn, M. H. L. 1983. *Drosophila melanogaster* mitochondrial DNA, a novel organization and genetic code. *Nature* 204:234–41.

Densmore, L. D., W. M. Brown, and J. W. Wright. 1985. Length variation and heteroplasmy are frequent in mitochondrial DNA from parthenogenetic and bisexual lizards (Genus *Cnemidophorus*). *Genetics* 110:689–707.

de Vries, H. 1901–3. *Die Mutationstheorie.* Leipzig: Von Veit. (1909–10 English translation, *The Mutation Theory,* trans. J. B. Farmer and A. D. Darbishire. Chicago: Open Court.)

Dickerson, R. E. 1971. The structure of cytochrome c and the rates of molecular evolution. *J. Mol. Evol.* 1:26–45.

Dickerson, R. E. and I. Geis. 1983. *Hemoglobin.* Menlo Park, Calif.: Benjamin/Cummings.

DiMichele, L. and D. A. Powers. 1982. Physiological basis for swimming endurance differences between LDH-B genotypes of *Fundulus heteroclitus. Science* 216:1014–16.

Dobzhansky, T. 1937. *Genetics and the Origin of Species.* New York: Columbia University Press.

—— 1951. *Genetics and the Origin of Species.* 3d ed. New York: Columbia University Press.

—— 1955. A review of some fundamental concepts and problems of population genetics. *Cold Spring Harbor Symp. Quant. Biol.* 20: 1–15.

—— 1970. *Genetics of the Evolutionary Process.* New York: Columbia University Press.

—— 1980. Morgan and his school in the 1930's. In E. Mayr and W. B. Provine, eds., *The Evolutionary Synthesis,* pp. 445–52. Cambridge, Mass.: Harvard University Press.

Dolan, R. and A. Robertson. 1975. The effect of conditioning the medium in *Drosophila* in relation to frequency-dependent selection. *Heredity* 35:311–16.

Doolittle, R. F. 1985. The genealogy of some recently evolved vertebrate proteins. *Trends in Biochem. Sci.* 10:233–37.

Doolittle, W. F. 1978. Genes in pieces: Were they ever together? *Nature* 272:581–82.

Doolittle, W. F. and C. Sapienza. 1980. Selfish genes, the phenotype paradigm, and genome evolution. *Nature* 284:601–3.

Dover, G. A. and R. B. Flavell, eds. 1982. *Genome Evolution.* New York: Academic Press.

Drake, J. W. 1966. Spontaneous mutations accumulating in bacteriophage T4 in the complete absence of DNA replication. *Proc. Natl. Acad. Sci. USA* 55:738–43.

Dykhuizen, D. and D. Hartl. 1980. Selective neutrality of 6PGD allozymes in *E. coli* and the effects of genetic background. *Genetics* 96:801–17.

Eanes, W. F. and R. K. Koehn. 1977. The correlation of rare alleles with heterozygosity: Determination of the correlation for the neutral models. *Genet. Res.* 29:223–30.

Eck, R. V. and M. O. Dayhoff. 1966. *Atlas of Protein Sequence and Structure.* Silver Springs, Md.: Natl. Biomed. Res. Found.

Easteal, S. 1985. Generation time and the rate of molecular evolution. *Mol. Biol. Evol.* 2:450–53.

Edgell, M. H. and 9 other authors. 1983. Evolution of the mouse β-globin complex locus. In M. Nei and R. Koehn, eds., *Evolution of Genes and Proteins,* pp. 1–13. Sunderland, Mass.: Sinauer Associates.

Efstratiadis, A. and 14 other authors. 1980. The structure and evolution of the human β-globin gene family. *Cell* 21:653–68.

Elandt-Johnson, R. C. 1970. *Probability Models and Statistical Methods in Genetics.* New York: Wiley.

Eldredge, N. and J. Cracraft. 1980. *Phylogenetic Patterns and the Evolutionary Process.* New York: Columbia University Press.

Eldredge, N. and S. J. Gould. 1972. Punctuated equilibria: An alternative to phyletic gradualism. In T. J. M. Schopf, ed., *Models in Paleobiology,* pp. 82–115. San Francisco: Freeman.

Endler, J. A. 1977. *Geographic Variation, Speciation, and Clines.* Princeton, N J : Princeton University Press.

Engels, W. R. 1981a. Estimating genetic divergence and genetic variability with restriction endonucleases. *Proc. Natl. Acad. Sci. USA* 78:6329–33.

—— 1981b. Hybrid dysgenesis in *Drosophila* and the stochastic loss hypothesis. *Cold Spring Harbor Symp. Quant. Biol.* 45:561–65.

Erickson, J. M., M. Rahire, P. Bennoun, P. Delepelaire, B. Diner, and J.-D. Rochaix. 1984. Herbicide resistance in *Chlamydomonas reinhardtii* results from a mutation in the chloroplast gene for the 32-kilodalton protein of photosystem II. *Proc. Natl. Acad. Sci. USA* 81:3617–21.

Estabrook, G. F., C. S. Johnson, and F. R. McMorris. 1975. An idealized concept of the true cladistic character. *Math. Biosci.* 23:263–72.

Ewens, W. J. 1963. The mean time for absorption in a process of genetic type. *J. Aust. Math. Soc.* 3:375–83.

—— 1964a. The pseudo-transient distribution and its uses in genetics. *J. Appl. Prob.* 1:141–56.

—— 1964b. The maintenance of alleles by mutation. *Genetics* 50:891–98.

——1972. The sampling theory of selectively neutral alleles. *Theor. Popul. Biol.* 3:87–112.

—— 1979. *Mathematical Population Genetics.* Berlin: Springer-Verlag.

Ewens, W. J. and W.-H. Li. 1980. Frequency spectra of neutral and deleterious alleles in a finite population. *J. Math. Biol.* 10:155–66.

Ewens, W. J., R. S. Spielman, and H. Harris. 1981. Estimation of genetic variation at the DNA level from restriction endonuclease data. *Proc. Natl. Acad. Sci. USA* 78:3748–50.

Faith, D. P. 1985. Distance methods and the approximation of most-parsimonious trees. *Syst. Zool.* 34:312–25.

Falconer, D. S. 1981. *Introduction to Quantitative Genetics.* 2d ed. London: Longman.

Farris, J. S. 1972. Estimating phylogenetic trees from distance matrices. *Amer. Natur.* 106:645–68.

—— 1977. On the phenetic approach to vertebrate classification. In M. K. Hecht, P. C. Goody, and B. M. Hecht, eds., *Major Patterns in Vertebrate Evolution,* pp. 823–50. New York: Plenum Press.

Farris, J. S., A. G. Kluge, and M. J. Eckardt. 1970. A numerical approach to phylogenetic systematics. *Syst. Zool.* 19:172–89.

Feldman, M. W. and F. B. Christiansen. 1975. The effect of population subdivision on two loci without selection. *Genet. Res.* 24:151–62.

Feldman, M. W., I. Franklin, and G. Thomson. 1974. Selection in complex genetic systems. I. The symmetric equilibria of the three-locus symmetric viability model. *Genetics* 76:135–62.

Felsenstein, J. 1965. The effect of linkage on directional selection. *Genetics* 52:349–63.

—— 1973a. Maximum likelihood estimation of evolutionary trees from continuous characters. *Amer. J. Hum. Genet.* 25:471–92.

——1973b. Maximum likelihood and minimum-steps methods for estimating evolutionary trees from data on discrete characters. *Syst. Zool.* 22:240–49.

—— 1976. The theoretical population genetics of variable selection and migration. *Ann. Rev. Genet.* 10:253–80.

——1978a. The number of evolutionary trees. *Syst. Zool.* 27:27–33.

——1978b. Cases in which parsimony or compatibility methods will be positively misleading. *Syst. Zool.* 27:401–10.

—— 1981. Evolutionary trees from DNA sequences: A maximum likelihood approach. *J. Mol. Evol.* 17:368–76.

—— 1982. Numerical methods for inferring evolutionary trees. *Quart. Rev. Biol.* 57:379–404.

—— 1985. Confidence limits on phylogenies with a molecular clock. *Syst. Zool.* 34:152–61.

Ferris, S. D., R. D. Sage, C.-M. Huang, J. T. Nielsen, U. Ritte, and A. C. Wilson. 1983. Flow of mitochondrial DNA across a species boundary. *Proc. Natl. Acad. Sci. USA* 80:2290–94.

Ferris, S. D., R. D. Sage, E. M. Prager, U. Ritte, and A. C. Wilson. 1983. Mitochondrial DNA evolution in mice. *Genetics* 105:681–721.

Ferris, S. D. and G. S. Whitt. 1977. Loss of duplicate gene expression after polyploidization. *Nature* 265:258–60.

—— 1979. Evolution of the differential regulation of duplicate genes after polyploidization. *J. Mol. Evol.* 12:267–317.

Ferris, S. D., A. C. Wilson, and W. M. Brown. 1981. Evolutionary tree for apes and humans based on cleavage maps of mitochondrial DNA. *Proc. Natl. Acad. Sci. USA* 78:2432–36.

Fiala, K. L. and R. R. Sokal. 1985. Factors determining the accuracy of cladogram estimation: Evaluation using computer simulation. *Evolution* 39:609–22.

Fisher, R. A. 1922. On the dominance ratio. *Proc. Royal Soc. Edinburgh* 42:321–341.

—— 1930. *The Genetical Theory of Natural Selection.* Oxford: Clarendon Press.

—— 1935. The sheltering of lethals. *Amer. Natur.* 69:446–55.

—— 1936. The use of multiple measurements in taxonomic problems. *Ann. Eugen.* 7:179–88.

Fitch, W. M. 1967. Evidence suggesting a non-random character to nucleotide replacements in naturally occurring mutations. *J. Mol. Biol.* 26:499–507.

—— 1971a. Toward defining the course of evolution: Minimum change for a specific tree topology. *Syst. Zool.* 20:406–16.

—— 1971b. Evolution of clupeine Z, a probable crossover product. *Nature New Biol.* 229:245–47

—— 1976. Molecular evolutionary clocks. In F. J. Ayala, ed., *Molecular Evolution,* pp. 160–78. Sunderland, Mass.: Sinauer Associates.

—— 1977. On the problem of discovering the most parsimonious tree. *Amer. Natur.* 3:223–57.

Fitch, W. M. and E. Margoliash. 1967a. Construction of phylogenetic trees. *Science* 155:279–84.

——— 1967b. A method for estimating the number of invariant amino acid coding positions in a gene using cytochrome c as a model case. *Biochem. Genet.* 1:65–71.

Flatz, G. and W. H. Rotthauwe. 1977. The human lactase polymorphism: Physiology and genetics of lactose absorption and malabsorption. *Prog. Med. Genet.* 2:205–49.

Flavell, A. 1985. Introns continue to amaze. *Nature.* 316:574–75.

Ford, E. B. 1964. *Ecological Genetics.* London: Metheun.

Fox, G. E., L. J. Magrum, W. E. Balch, R. S. Wolfe, and C. R. Woese. 1977. Classification of methogenic bacteria by 16S ribosomal RNA characterization. *Proc. Natl. Acad. Sci. USA* 74:4537–41.

Fox, G. E. and 18 other authors. 1980. The phylogeny of prokaryotes. *Science* 209:457–63.

Fox, T. D. 1985. Diverged genetic codes in protozoans and a bacterium. *Nature* 314:132–33.

Frankel, J. S. 1983. Allelic asynchrony during *Barbus* hybrid development. *J. Hered.* 74:311–12.

Franklin, I. and R. C. Lewontin. 1970. Is the gene the unit of selection? *Genetics* 65:707–34.

Fuerst, P. A., R. Chakraborty, and M. Nei. 1977. Statistical studies on protein polymorphism in natural populations. I. Distribution of single locus heterozygosity. *Genetics* 86:455–83.

Fuerst, P. A. and R. E. Ferrell. 1980. The stepwise mutation model: An experimental evaluation utilizing hemoglobin variants. *Genetics* 94:185–201.

Gall, J. G., E. H. Cohen, and D. D. Atherton. 1974. The satellite DNAs of *Drosophila virilis. Cold Spring Harbor Symp. Quant. Biol.* 38:417–21.

Gehring, W. J. 1985. The homeo box: A key to the understanding of development? *Cell* 40:3–5.

Gershowitz, H., P. C. Junqueira, F. M. Salzano, and J. V. Neel. 1967. Further studies on the Xavante Indians. III. Blood groups and the *ABH-Le^a* secretor types in the Simoes Lopes and Sao Marcos Xavantes. *Amer. J. Hum. Genet.* 19:502–13.

Gilbert, W. 1978. Why genes in pieces? *Nature* 271:501.

Gill, A. E. 1976. Genetic divergence of insular populations of deer mice. *Biochem. Genet.* 14:835–48.

Gillespie, J. H. 1978. A general model to account for enzyme variation in natural populations. VI. The SAS-CFF model. *Theor. Popul. Biol.* 14:1–45.

—— 1979. Molecular evolution and polymorphism in a random environment. *Genetics* 93:737–54.

—— 1983. Some properties of finite populations experiencing strong selection and weak mutation. *Amer. Natur.* 121:691–708.

—— 1984. The molecular clock may be an episodic clock. *Proc. Natl. Acad. Sci. USA* 81:8009–13.

—— 1985. The interaction of genetic drift and mutation with selection in a fluctuating environment. *Theor. Popul. Biol.* 27:222–37.

—— 1986. Natural selection and the molecular clock. *Mol. Biol. Evol.* 3:138–155.

Gillespie, J. H. and C. H. Langley. 1974. A general model to account for enzyme variation in natural populations. *Genetics* 76:837–48.

—— 1979. Are evolutionary rates really variable? *J. Mol. Evol.* 13:27–34.

Ginzburg, L. R., P. M. Bingham, and S. Yoo. 1984. On the theory of speciation induced by transposable elements. *Genetics* 107:331–41.

Gō, M. 1981. Correlation of DNA exonic regions with protein structural units in haemoglobin. *Nature* 291:90–92.

Gojobori, T. 1983. Codon substitution in evolution and the "saturation" of synonymous changes. *Genetics* 105:1011–27.

Gojobori, T., K. Ishii, and M. Nei. 1982a. Estimation of average number of nucleotide substitutions when the rate of substitution varies with nucleotide. *J. Mol. Evol.* 18:414–23.

Gojobori, T., W.-H. Li, and D. Graur. 1982b. Patterns of nucleotide substitution in pseudogenes and functional genes. *J. Mol. Evol.* 18:360–69.

Gojobori, T. and M. Nei. 1981. Inter-RNA homology and possible roles of small RNAs. *J. Mol. Evol.* 17:245–50.

—— 1984. Concerted evolution of the immunoglobulin V_H gene family. *Mol. Biol. Evol.* 1:195–212.

—— 1986. Relative contributions of germline gene variation and somatic mutation to immunoglobulin diversity in the mouse. *Mol. Biol. Evol.* 3:156–67.

Gojobori, T. and S. Yokoyama. 1985. Rates of evolution of the retroviral oncogene of Moloney murine sarcoma virus and of its cellular homologues. *Proc. Natl. Acad. Sci. USA* 82:4198–4201.

Golden, S. S. and R. Haselkorn. 1985. Mutation to herbicide resistance maps within the *psbA* gene of *Anacystis nidulans* R2. *Science* 229:1104–7.

Goldman, D., P. R. Giri, and S. J. O'Brien. 1987. A molecular phylogeny of the hominoid primates as indicated by two-dimensional protein electrophoresis. *Proc. Natl. Acad. Sci. USA* 84:3307–11.

Goldschmidt, R. 1940. *The Material Basis of Evolution.* New Haven: Yale University Press.

Goodman, M. 1962. Immunochemistry of the primates and primate evolution. *Ann. N.Y. Acad. Sci.* 102:219–34.

Goodman, M., B. F. Koop, J. Czelusniak, M. L. Weiss, and J. L. Slightom. 1984. The η-globin gene: Its long evolutionary history in the β-globin gene family of mammals. *J. Mol. Biol.* 180:803–23.

Goodman, M., G. W. Moore, and J. Barnabas. 1974. The phylogeny of human globin genes investigated by the maximum parsimony method. *J. Mol. Evol.* 3:1–48.

Goodman, M., G. W. Moore, and G. Matsuda. 1975. Darwinian evolution in the genealogy of hemoglobin. *Nature* 253:603–8.

Goodman, M., A. E. Romero-Herrera, H. Dene, J. Czelusniak, and R. E. Tashian. 1982. Amino acid sequence evidence on the phylogeny of primates and other eutherians. In M. Goodman, ed., *Macromolecular Sequences in Systematic and Evolutionary Biology,* pp. 115–91. New York: Plenum Press.

Goossens, M., A. M. Dozy, S. H. Embury, Z. Zachariades, M. G. Hadjiminas, G. Stamatoyanopoulos, and Y. W. Kan. 1980. Triplicated α-globin loci in humans. *Proc. Natl. Acad. Sci. USA* 77:518–21.

Gorman, G. C. and Y. J. Kim. 1977. Genotypic evolution in the face of phenotypic conservativeness: *Abudefduf* (Pomacentridae) from the Atlantic and Pacific sides of Panama. *Copeia* 1977:694–97.

Gorman, G. C. and J. Renzi Jr. 1979. Genetic distance and heterozygosity estimates in electrophoretic studies: Effects of sample size. *Copeia* 1979:242–49.

Gotoh, O. 1982. An improved algorithm for matching biological sequences. *J. Mol. Biol.* 162:705–8.

Gotoh, O., J.-I. Hayashi, H. Yonekawa, and Y. Tagashira. 1979. An improved method for estimating sequence divergence between related DNAs from changes in restriction endonuclease cleavage sites. *J. Mol. Evol.* 14:301–10.

Gould, S. J. 1977. *Ontogeny and Phylogeny*. Cambridge, Mass.: Harvard University Press.

—— 1982a. Darwinism and the expansion of evolutionary theory. *Science* 216:380–87.

—— 1982b. The uses of heresy: An introduction to R. Goldschmidt's *The Material Basis of Evolution*. New Haven: Yale University Press.

Gould, S. J. and R. C. Lewontin. 1979. The spandrels of San Marco and the Panglossian paradigm: A critique of the adaptationist programme. *Proc. Royal Soc. Lond. B* 205:581–98.

Gowen, J. W., ed. 1952. *Heterosis*. Ames: Iowa State University Press.

Grant, V. 1971. *Plant Speciation*. New York: Columbia University Press.

Grantham, R. 1974. Amino acid difference formula to help explain protein evolution. *Science* 185:862–64.

—— 1980. Workings of the genetic code. *Trends in Biochem. Sci.* 5:327–31.

Graur, D. 1985. Amino acid composition and the evolutionary rates of protein-coding genes. *J. Mol. Evol.* 22:53–62.

Greenwald, I. 1985. *lin-12*, a nematode homeotic gene, homologous to a set of mammalian proteins that include epidermal growth factor. *Cell* 43:583–90.

Griffiths, R. C. 1980a. Genetic identity between populations when mutation rates vary within and across loci. *J. Math. Biol.* 10:195–204.

—— 1980b. Lines of descent in the diffusion approximation of neutral Wright-Fisher models. *Theor. Popul. Biol.* 17:37–50.

Griffiths, R. C. and W.-H. Li. 1983. Simulating allele frequencies in a population and the genetic differentiation of populations under mutation pressure. *Theor. Popul. Biol.* 23:19–33.

Grunstein, M., P. Schedl, and L. Kedes. 1976. Sequence analysis and evolution of sea urchin (*Lytechinus pictus* and *Strongylocentrotus purpuratus*) histone H4 messenger RNAs. *J. Mol. Biol.* 104:351–69.

Gyllensten, U., N. Ryman, C. Reuterwall, and P. Dratch. 1983. Genetic differentiation in four European subspecies of red deer (*Cervus elephus L.*). *Heredity* 51:561–80.

Haldane, J. B. S. 1927. The mathematical theory of natural and artificial selection. Part V. *Proc. Cambridge Philos. Soc.* 23:838–44.

—— 1932. *The Causes of Evolution*. London: Longmans and Green.

—— 1939. The equilibrium between mutation and random extinction. *Ann. Eugen.* 9:400–5.

Haldane, J. B. S. and S. D. Jayakar. 1963. Polymorphism due to selection of varying direction. *J. Genet.* 58:237–42.

Hall, B. G. 1983. Evolution of new metabolic functions in laboratory organisms. In M. Nei and R. Koehn, eds., *Evolution of Genes and Proteins,* pp. 234–57. Sunderland, Mass.: Sinauer Associates.

Hall, B. G., S. Yokoyama, and D. H. Calhoun. 1983. Role of cryptic genes in microbial evolution. *Mol. Biol. Evol.* 1:109–24.

Hamrick, J. L., Y. B. Linhart, and J. B. Mitton. 1979. Relationships between life history characters and electrophoretically detectable genetic variation in plants. *Ann. Rev. Ecol. Syst.* 10:173–200.

Hardies, S. C., S. L. Martin, C. F. Voliva, C. A. Hutchison III, and M. H. Edgell. 1986. An analysis of replacement and synonymous changes in the rodent L1 repeat family. *Mol. Biol. Evol.* 3:109–25.

Hardison, R. C. 1984. Comparison of the β-globin gene families of rabbits and humans indicates that the gene cluster $5'$-ϵ-γ-δ-β-$3'$ predates the mammalian radiation. *Mol. Biol. Evol.* 1:390–410.

Hardy, G. H. 1908. Mendelian proportions in a mixed population. *Science* 28:49–50.

Harris, H. 1966. Enzyme polymorphisms in man. *Proc. Royal Soc. Lond. B* 164:298–310.

Harris, H., D. A. Hopkinson, and Y. H. Edwards. 1977. Polymorphism and the subunit structure of enzymes: A contribution to the neutralist-selectionist controversy. *Proc. Natl. Acad. Sci. USA* 74:698–701.

Harris, H., D. A. Hopkinson, and E. B. Robson. 1974. The incidence of rare alleles determining electrophoretic variants: Data on 43 enzyme loci in man. *Ann. Hum. Genet.* 37:237–53.

Hartl, D. L. 1980. *Principles of Population Genetics.* Sunderland, Mass.: Sinauer Associates.

Hartl, D. L. and D. E. Dykhuizen. 1985. The neutral theory and the molecular basis of preadaptation. In T. Ohta and K. Aoki, eds., *Population Genetics and Molecular Evolution,* pp. 107–24. Tokyo: Japan Scientific Societies Press.

Hartl, D. L., D. E. Dykhuizen, R. D. Miller, L. Green, and J. de-Framond. 1983. Transposable element IS50 improves growth rate of *E. coli* cell without transposition. *Cell* 35:503–10.

Hayashida, H., H. Toh, R. Kikuno, and T. Miyata. 1985. Evolution of influenza virus genes. *Mol. Biol. Evol.* 2:289–303.

Hays, J. D., J. Imbrie, and N. J. Shackleton. 1976. Variations in the earth's orbit: Pacemaker of the ice ages. *Science* 194:1121–32.

Hedge, P. J. and B. G. Spratt. 1985. Resistance to β-lactam antibiotics by remodeling the active site of an *E. coli* penicillin-binding protein. *Nature* 318:478–80.

Hedrick, P. W. 1974. Genetic variation in a heterogeneous environment. I. Temporal heterogeneity and the absolute dominance model. *Genetics* 78:757–70.

—— 1981. The establishment of chromosomal variants. *Evolution* 35:322–32.

—— 1983. *Genetics of Populations.* Portola Valley, Calif.: Science Books Intl.

Hedrick, P. W., M. E. Ginevan, and E. P. Ewing. 1976. Genetic polymorphism in heterogeneous environments. *Ann. Rev. Ecol. Syst.* 7:1–32.

Hedrick, P. W., S. Jain, and L. Holden. 1978. Multilocus systems in evolution. *Evol. Biol.* 11:101–84.

Hedrick, P. W. and G. Thomson. 1983. Evidence for balancing selection at HLA. *Genetics* 104:449–56.

Heindell, H. C., A. Liu, G. V. Paddock, G. M. Studnicka, and W. A. Salser. 1978. The primary sequence of rabbit α-globin mRNA. *Cell* 15:43–54.

Heuch, I. 1975. The relationship between separation time and genetic distance based on angular transformations of gene frequencies. *Biometrics* 31:685–700.

Highton, R. and A. Larson. 1979. The genetic relationships of the salamanders of the genus *Plethodon. Syst. Zool.* 28:579–99.

Hill, W. G. 1974a. Disequilibrium among several linked neutral genes in finite populations. I. Mean changes in disequilibrium. *Theor. Popul. Biol.* 5:366–92.

—— 1974b. Estimation of linkage disequilibrium in randomly mating populations. *Heredity* 33:229–39.

—— 1979. A note on effective population size with overlapping generations. *Genetics* 92:317–22.

Hill, W. G. and A. Robertson. 1968. Linkage disequilibrium in finite populations. *Theor. Appl. Genet.* 38:226–31.

Hinegardner, R. 1976. Evolution of genome size. In F. J. Ayala, ed., *Molecular Evolution,* pp. 179–99. Sunderland, Mass.: Sinauer Associates.

Hiraizumi, Y. and J. F. Crow. 1960. Heterozygous effects on viability, fertility, rate of development, and longevity of *Drosophila* chromosomes that are lethal when homozygous. *Genetics* 45:1071–83.

Hirschberg, J. and L. McIntosh. 1983. Molecular basis of herbicide resistance in *Amaranthus hybridus*. *Science* 222:1346–49.

Hixson, J. E. and W. M. Brown. 1986. A comparison of the small ribosomal RNA genes from the mitochondrial DNA of the great apes and humans: Sequence, structure, evolution and phylogenetic implications. *Mol. Biol. Evol.* 3:1–18.

Hochachka, P. W. and G. N. Somero. 1984. *Biochemical Adaptation*. Princeton, N.J.: Princeton University Press.

Hoffmann, H. J. and J. W. Schopf. 1983. Early proterozoic microfossils. In J. W. Schopf, ed., *Earth's Earliest Biosphere: Its Origin and Evolution,* pp. 329–60. Princeton, N.J.: Princeton University Press.

Holland, J., K. Spindler, F. Horodyski, E. Grabau, S. Nichol, and S. VandePol. 1982. Rapid evolution of RNA genomes. *Science* 215:1577–85.

Holt, S. 1961. *Dermatoglyphic patterns*. New York: Pergamon Press.

Hood, L., J. H. Campbell, and S. C. R. Elgin. 1975. The organization, expression, and evolution of antibody genes and other multigene families. *Ann. Rev. Genet.* 9:305–53.

Hood, L., M. Kronenberg, and T. Hunkapiller. 1985. T cell antigen receptors and the immunoglobulin supergene family. *Cell* 40:225–29.

Hopkinson, D. A. and H. Harris. 1969. Red cell acid phosphatase, phosphoglucomutase, and adenylate kinase. In G. Yunis, ed., *Biochemical Methods in Red Cell Genetics,* pp. 337–75. New York: Academic Press.

Hori, H. and S. Osawa. 1979. Evolutionary change in 5S RNA secondary structure and a phylogenetic tree of 54 5S RNA species. *Proc. Natl. Acad. Sci. USA* 76:381–85.

—— 1980. Recent studies on the evolution of 5S RNA. In S. Osawa, H. Ozeki, H. Uchida, and T. Yura, eds., *Genetics and Evolution of RNA Polymerase, tRNA, and Ribosomes,* pp. 539–51. Tokyo: University of Tokyo Press.

Hörz, W. and W. Altenburger. 1981. Nucleotide sequence of mouse satellite DNA. *Nucleic Acids Research* 9:683–97.

Howard, D. J., G. L. Bush, and J. A. Breznak. 1985. The evolutionary significance of bacteria associated with *Rhagoletis*. *Evolution* 39:405–17.

Hubby, J. L. and R. C. Lewontin. 1966. A molecular approach to the study of genic heterozygosity in natural populations. I. The number of alleles at different loci in *Drosophila pseudoobscura*. *Genetics* 54:577–94.

Hudson, R. R. 1982. Estimating genetic variability with restriction endonucleases. *Genetics* 100:711–19.

——— 1983. Testing the constant rate neutral allele model with protein sequence data. *Evolution* 37:203–17.

Hudson, R. R. and N. L. Kaplan. 1985. Statistical properties of the number of recombination events in the history of a sample of DNA sequences. *Genetics* 111:147–64.

Hunkapiller, T., H. Huang, L. Hood, and J. H. Campbell. 1982. The impact of modern genetics on evolutionary theory. In R. Milkman, ed., *Perspectives on Evolution*, pp. 164–89. Sunderland, Mass.: Sinauer Associates.

Hunt, J. A., T. J. Hall, and R. J. Britten. 1981. Evolutionary distances in Hawaiian *Drosophila* measured by DNA reassociation. *J. Mol. Evol.* 17:361–67.

Hunt, L. T., S. Hurst-Calderone, and M. O. Dayhoff. 1978. Globins. In M. O. Dayhoff, ed., *Atlas of Protein Sequence and Structure*, pp. 229–49. Silver Springs, Md.: Natl. Biomed. Res. Found.

Huxley, J. S. 1942. *Evolution: The Modern Synthesis*. London: Allen and Unwin.

Ikemura, T. 1985. Codon usage and tRNA content in unicellular and multicellular organisms. *Mol. Biol. Evol.* 2:13–34.

Ingram, V. M. 1963. *The Hemoglobins in Genetics and Evolution*. New York: Columbia University Press.

Ishii, K. and B. Charlesworth. 1977. Associations between allozyme loci and gene arrangements due to hitch-hiking effects of new inversions. *Genet. Res.* 30:93–106.

Jablonski, D. 1986. Background and mass extinctions: The alternation of macroevolutionary regimes. *Science* 231:129–33.

Jelinek, W. R. and 10 other authors. 1980. Ubiquitous, interspersed repeated sequences in mammalian genomes. *Proc. Natl. Acad. Sci. USA* 77:1398–1402.

Jensen, L. and E. Pollak. 1969. Random selective advantages of a gene in a finite population. *J. Appl. Probab.* 6:19–37.

Johnson, G. B. 1974. Enzyme polymorphism and metabolism. *Science* 184:28–37.

Johnson, M. J., D. C. Wallace, S. D. Ferris, M. C. Rattazzi, and L. L. Cavalli-Sforza. 1983. Radiation of human mitochondrial DNA types analyzed by restriction endonuclease cleavage patterns. *J. Mol. Evol.* 19:255–71.

Johnson, W. E. and R. K. Selander. 1971. Protein variation and systematics in kangaroo rats (genus *Dipodomys*). *Syst. Zool.* 20:377–405.

Jolles, P., F. Schoentgen, J. Jolles, D. E. Dobson, E. M. Prager, and A. C. Wilson. 1984. Stomach lysozymes of ruminants. II. Amino acid sequence of cow lysozyme 2 and immunological comparisons with other lysozymes. *J. Biol. Chem.* 259:11617–25.

Jones, D. F. 1917. Dominance of linked factors as a means of accounting for heterosis. *Genetics* 2:466–79.

Jones, J. S., S. H. Bryant, R. C. Lewontin, J. A. Moore, and T. Prout. 1981. Gene flow and the geographical distribution of a molecular polymorphism in *Drosophila pseudoobscura*. *Genetics* 98:157–78.

Jones, J. S. and T. Yamazaki. 1974. Genetic background and the fitness of allozymes. *Genetics* 78:1185–89.

Jorde, L. B. 1980. The genetic structure of subdivided human populations. In J. H. Mielke and M. H. Crawford, eds., *Current Developments in Anthropological Genetics,* pp. 135–208. New York: Plenum Press.

Joysey, K. A. 1981. Molecular evolution and vertebrate phylogeny in perspective. *Symp. Zool. Soc. London* 46:189–218.

Jukes, T. H. 1971. Comparisons of the polypeptide chains of globins. *J. Mol. Evol.* 1:46–62.

—— 1983. Evolution of the amino acid code. In M. Nei and R. Koehn, eds., *Evolution of Genes and Proteins,* pp. 191–207. Sunderland, Mass.: Sinauer Associates.

Jukes, T. H. and C. R. Cantor. 1969. Evolution of protein molecules. In H. N. Munro, ed., *Mammalian Protein Metabolism,* pp. 21–132. New York: Academic Press.

Kacser, H. and J. A. Burns. 1981. The molecular basis of dominance. *Genetics* 97:639–66.

Kafatos, F. C., A. Efstratiadis, B. G. Forget, and S. M. Weissman. 1977. Molecular evolution of human and rabbit β-globin mRNAs. *Proc. Natl. Acad. Sci. USA* 74:5618–22.

Kahler, A. L., R. W. Allard, and R. D. Miller. 1984. Mutation rates for enzyme and morphological loci in barley (*Hordeum vulgare* L.). *Genetics* 106:729–34.

Kan, Y. W. and A. M. Dozy. 1978. Polymorphism of DNA sequence adjacent to human β-globin structural gene: Relationship to sickle mutation. *Proc. Natl. Acad. Sci. USA* 75:5631–35.

—— 1980. Evolution of the hemoglobin S and C genes in world populations. *Science* 209:381–88.

Kaplan, N. 1983. Statistical analysis of restriction enzyme map data and nucleotide sequence data. In B. S. Weir, ed., *Statistical Analysis of DNA Sequence Data*, pp. 75–106. New York: Marcel Dekker.

Kaplan, N. and J. F. Y. Brookfield. 1983. Transposable elements in Mendelian populations. III. Statistical results. *Genetics* 104:485–95.

Kaplan, N. and C. H. Langley. 1979. A new estimate of sequence divergence of mitochondrial DNA using restriction endonuclease mapping. *J. Mol. Evol.* 13:295–304.

Kaplan, N. and K. Risko. 1981. An improved method for estimating sequence divergence of DNA using restriction endonuclease mappings. *J. Mol. Evol.* 17:156–62.

Karlin, S. 1969. *Equilibrium Behavior of Population Genetic Models with Nonrandom Mating.* New York: Gordon and Breach.

Karlin, S. and M. W. Feldman. 1969. Linkage and selection: New equilibrium properties of the two-locus symmetric viability model. *Proc. Natl. Acad. Sci. USA* 62:70–74.

Karlin, S., G. Ghandour, and D. E. Foulser. 1985. DNA sequence comparisons of the human, mouse, and rabbit immunoglobulin kappa gene. *Mol. Biol. Evol.* 2:35–52.

Karlin, S. and U. Lieberman. 1974. Temporal fluctuations in selection intensities: Case of large population size. *Theor. Popul. Biol.* 6:355–82.

Karlin, S. and J. McGregor. 1972. Application of method of small parameters to multiniche population genetic model. *Theor. Popul. Biol.* 3:186–209.

Karn, M. N. and L. S. Penrose. 1951. Birth weight and gestation time in relation to maternal age, parity, and infant survival. *Ann. Eugen.* 16:147–64.

Kato, I., W. J. Kohr, and M. J. Laskowski. 1978. Evolution of avian ovomucoids. In S. Magnuson, M. Ottesen, B. Taltmann, K. Dano, and H. Neurath, eds., *Proc. 11th Feder. Eur. Biol. Sci.,* pp. 197–206. New York: Pergamon Press.

Kawamoto, Y., O. Takenaka, and E. Brotoisworo. 1982. Preliminary report on genetic variations within and between species of Sulawesi macaques. *Kyoto University Res. Inst.* 2:23–37.

Kazazian, H. H., Jr., A. Chakravarti, S. H. Orkin, and S. E. Antonarakis. 1983. DNA polymorphisms in the human β-globin gene cluster. In M. Nei and R. Koehn, eds., *Evolution of Genes and Proteins,* pp. 137–46. Sunderland, Mass.: Sinauer Associates.

Keith, T. P., L. D. Brooks, R. C. Lewontin, J. C. Martinez-Cruzado, and D. L. Rigby. 1985. Nearly identical allelic distributions of xanthine dehydrogenase in two populations of *Drosophila pseudoobscura. Mol. Biol. Evol.* 2:206–16.

Kessler, L. G. and J. C. Avise. 1985. A comparative description of mitochondrial DNA differentiation in selected avian and other vertebrate genera. *Mol. Biol. Evol.* 2:109–25.

Kettlewell, H. B. D. 1955. Selection experiments on industrial melanism in the *Lepidoptera. Heredity* 9:323–42.

Kidwell, M. G. 1979. Hybrid dysgenesis in *Drosophila melanogaster:* The relationship between the P-M and I-R interaction systems. *Genet. Res.* 33:205–17.

—— 1983a. Intraspecific hybrid sterility. In M. Ashburner, H. L. Carson, and J. N. Thompson, eds., *The Genetics and Biology of Drosophila,* vol. 3c, pp. 125–53. New York: Academic Press.

—— 1983b. Evolution of hybrid dysgenesis determinants in *Drosophila melanogaster. Proc. Natl. Acad. Sci. USA* 80:1655–59.

Kidwell, M. G., J. F. Kidwell, and J. A. Sved. 1977. Hybrid dysgenesis in *Drosophila melanogaster:* A syndrome of aberrant traits including mutation, sterility, and male recombination. *Genetics* 86:813–33.

Kikuno, R. and T. Miyata. 1986. Slowly evolving *Drosophila* mitochondrial genes. *Mol. Biol. Evol.* (submitted).

Kim, S., M. Davis, E. Sinn, P. Patten, and L. Hood. 1981. Antibody diversity: Somatic hypermutation of rearranged V_H genes. *Cell* 27:573–81.

Kim, Y. J., G. C. Gorman, T. Papenfuss, and A. K. Roychoudhury. 1976. Genetic relationships and genetic variation in the amphisbaenian genus *Bipes. Copeia* 1976:120–24.

Kimura, M. 1954. Process leading to quasi-fixation of genes in natural populations due to random fluctuation of selection intensities. *Genetics* 39:280–95.

—— 1955a. Solution of a process of random genetic drift with a continuous model. *Proc. Natl. Acad. Sci. USA* 41:144–50.

—— 1955b. Stochastic processes and distribution of gene frequencies under natural selection. *Cold Spring Harbor Symp. Quant. Biol.* 20:33–53.

—— 1956. A model of a genetic system which leads to closer linkage under natural selection. *Evolution* 10:278–87.

—— 1957. Some problems of stochastic processes in genetics. *Ann. Math. Stat.* 28:882–901.

—— 1962. On the probability of fixation of mutant genes in a population. *Genetics* 47:713–19.

—— 1964. Diffusion models in population genetics. *J. Appl. Probab.* 1:177–232.

—— 1965a. Attainment of quasi linkage equilibrium when gene frequencies are changing by natural selection. *Genetics* 52:875–90.

—— 1965b. A stochastic model concerning the maintenance of genetic variability in quantitative characters. *Proc. Natl. Acad. Sci. USA* 54:731–36.

—— 1968a. Evolutionary rate at the molecular level. *Nature* 217:624–26.

—— 1968b. Genetic variability maintained in a finite population due to mutational production of neutral and nearly neutral isoalleles. *Genet. Res.* 11:247–69.

—— 1969. The number of heterozygous nucleotide sites maintained in a finite population due to steady flux of mutations. *Genetics* 61:893–903.

—— 1970. The length of time required for a selectively neutral mutant to reach fixation through random frequency drift in a finite population. *Genet. Res.* 15:131–33.

—— 1971. Theoretical foundation of population genetics at the molecular level. *Theor. Popul. Biol.* 2:174–208.

—— 1974. Gene pool of higher organisms as a product of evolution. *Cold Spring Harbor Symp. Quant. Biol.* 38:515–24.

—— 1977. Preponderance of synonymous changes as evidence for the neutral theory of molecular evolution. *Nature* 267:275–76.

—— 1979. Model of effectively neutral mutations in which selective constraint is incorporated. *Proc. Natl. Acad. Sci. USA* 76:3440–44.

—— 1980. A simple method for estimating evolutionary rate of base substitutions through comparative studies of nucleotide sequences. *J. Mol. Evol.* 16:111–20.

—— 1981a. Was globin evolution very rapid in its early stages? A dubious case against the rate-constancy hypothesis. *J. Mol. Evol.* 17:110–13.

—— 1981b. Estimation of evolutionary distances between homologous nucleotide sequences. *Proc. Natl. Acad. Sci. USA* 78:454–58.

—— 1981c. Possibility of extensive neutral evolution under stabilizing selection with special reference to non-random usage of synonymous codons. *Proc. Natl. Acad. Sci. USA* 78:5773–77.

—— 1983a. *The Neutral Theory of Molecular Evolution.* Cambridge: Cambridge University Press.

—— 1983b. The neutral theory of molecular evolution. In M. Nei and R. Koehn, eds., *Evolution of Genes and Proteins,* pp. 208–33, Sunderland, Mass.: Sinauer Associates.

—— 1983c. Rare variant alleles in the light of the neutral theory. *Mol. Biol. Evol.* 1:84–93.

Kimura, M. and J. F. Crow. 1964. The number of alleles that can be maintained in a finite population. *Genetics* 49:725–38.

Kimura, M., and J. L. King. 1979. Fixation of a deleterious allele at one of two "duplicate" loci by mutation pressure and random drift. *Proc. Natl. Acad. Sci. USA* 76:2858–61.

Kimura, M. and T. Ohta. 1969. The average number of generations until fixation of a mutant gene in a finite population. *Genetics* 61:763–71.

—— 1971a. *Theoretical Aspects of Population Genetics.* Princeton, N.J.: Princeton University Press.

—— 1971b. Protein polymorphism as a phase of molecular evolution. *Nature* 229:467–69.

—— 1972. On the stochastic model for estimation of mutational distance between homologous proteins. *J. Mol. Evol.* 2:87–90.

—— 1973a. Eukaryotes-prokaryotes divergence estimated by 5S ribosomal RNA sequences. *Nature New Biol.* 243:199–200.

—— 1973b. Mutation and evolution at the molecular level. *Genetics* 73:s19–s35.

—— 1974. On some principles governing molecular evolution. *Proc. Natl. Acad. Sci. USA* 71:2848–52.

—— 1978. Stepwise mutation model and distribution of allelic frequencies in a finite population. *Proc. Natl. Acad. Sci. USA* 75:2868–72.

King, J. L. 1972. The role of mutation in evolution. In *Proc. 6th Berkeley Symp. Math. Stat. and Probab.*, pp. 69–100. Berkeley: University of California Press.

King, J. L. and T. H. Jukes. 1969. Non-Darwinian evolution. *Science* 164:788–98.

King, M. C. and A. C. Wilson. 1975. Evolution at two levels in humans and chimpanzees. *Science* 188:107–16.

Kingman, J. F. C. 1982. On the genealogy of large populations. *J. Appl. Probab.* 19A:27–43.

Kirby, G. C. 1975. Heterozygote frequencies in small subpopulations. *Theor. Popul. Biol.* 8:31–48.

Klein, J. and F. Figueroa. 1986. Evolution of the major histocompatibility complex. *CRC Crit. Rev. Immunol.* 6:295–386.

Klotz, L. C. and R. L. Blanken. 1981. A practical method for calculating evolutionary trees from sequence data. *J. Theor. Biol.* 91:261–72.

Klotz, L. C., N. Komar, R. L. Blanken, and R. M. Mitchell. 1979. Calculation of evolutionary trees from sequence data. *Proc. Natl. Acad. Sci. USA* 76:4516–20.

Koehn, R. K. 1985. Adaptive aspects of biochemical and physiological variability. In *Proc. Eur. Mar. Biol. Symp. 1985,* pp. 425–41. Cambridge: Cambridge University Press.

Koehn, R. K. and W. F. Eanes. 1977. Subunit size and genetic variation of enzymes in natural populations of *Drosophila. Theor. Popul. Biol.* 11:330–41.

—— 1978. Molecular structure and protein variation within and among populations. *Evol. Biol.* 11:39–100.

Koehn, R. K. and S. R. Shumway. 1982. A genetic/physiological explanation for differential growth rate among individuals of the American oyster *Crassostrea virginica* (Gmelin). *Mar. Biol. Lett.* 3:35–42.

Koehn, R. K., A. J. Zera, and J. G. Hall. 1983. Enzyme polymorphism and natural selection. In M. Nei and R. Koehn, eds., *Evolution of Genes and Proteins,* pp. 115–36. Sunderland, Mass.: Sinauer Associates.

Kohne, D. E. 1970. Evolution of higher-organism DNA. *Quart. Rev. Biophys.* 3:327–75.

Kohne, D. E., J. A. Chiscon, and B. H. Hoyer. 1972. Evolution of primate DNA sequences. *J. Hum. Evol.* 1:627–44.

Kojima, K., J. Gillespie, and Y. N. Tobari. 1970. A profile of *Drosophila* species' enzymes assayed by electrophoresis. I. Number of alleles, heterozygosities, and linkage disequilibrium in glucose-metabolizing systems and some other enzymes. *Biochem. Genet.* 4:627–37.

Kondo, S. 1977. Evolutionary considerations on DNA repair and mutagenesis. In M. Kimura, ed., *Molecular Evolution and Polymorphism,* pp. 313–31. Mishima, Japan: National Institute of Genetics.

Konkel, D. A., S. M. Tilghman, and P. Leder. 1978. The sequence of the chromosomal mouse β-globin major gene: Homologies in capping, splicing and poly(A) sites. *Cell* 15:1125–32.

Koop, B. F., M. Goodman, P. Xu, K. Chan, and J. L. Slightom. 1986. Primate η-globin DNA sequences and man's place among the great apes. *Nature* 319:234–38.

Kreitman, M. 1983. Nucleotide polymorphism at the alcohol dehydrogenase locus of *Drosophila melanogaster. Nature* 304:412–17.

Krystal, M., D. Buonagurio, J. F. Young, and P. Palese. 1983. Sequential mutations in the NS genes of influenza virus field strains. *J. Virol.* 45:547–54.

Krzakowa, M. and J. Szweykowski. 1979. Isozyme polymorphism in natural populations of a liverwort, *Plagiochila asplenioides. Genetics* 93:711–19.

Kubitschek, H. E. 1970. *Introduction to Research with Continuous Cultures.* Englewood Cliffs, N.J.: Prentice-Hall.

Kurczynski, T. W. 1970. Generalized distance and discrete variables. *Biometrics* 26:525–34.

Kurnit, D. M. 1979. Evolution of sickle variant gene. *Lancet* 1:104.

Lake, J. A. 1985. Evolving ribosome structure: Domains in archaebacteria, eubacteria, eocytes, and eukaryotes. *Ann. Rev. Biochem.* 54:507–30.

Lakovaara, S., A. Saura, and C. T. Falk. 1972. Genetic distance and evolutionary relationships in the *Drosophila obscura* group. *Evolution* 16:177–84.

Lanave, C., G. Preparata, C. Saccone, and G. Serio. 1984. A new method for calculating evolutionary substitution rates. *J. Mol. Evol.* 20:86–93.

Lande, R. 1979. Effective deme sizes during long-term evolution estimated from rates of chromosomal rearrangement. *Evolution* 33:234–51.

Langley, C. H., J. F. Y. Brookfield, and N. Kaplan. 1983. Transposable elements in Mendelian populations. I. A theory. *Genetics* 104:457–71.

Langley, C. H. and W. M. Fitch. 1974. An examination of the constancy of the rate of molecular evolution. *J. Mol. Evol.* 3:161–77.

Langley, C. H., E. A. Montgomery, and W. F. Quattlebaum. 1982. Restriction map variation in the *Adh* region of *Drosophila*. *Proc. Natl. Acad. Sci. USA* 79:5631–35.

Langley, C. H., Y. N. Tobari, and K. Kojima. 1974. Linkage disequilibrium in natural populations of *Drosophila melanogaster:* Seasonal variation. *Genetics* 86:447–54.

Langridge, J. 1974. Mutation spectra and the neutrality of mutations. *Aust. J. Biol. Sci.* 27:309–19.

Lansman, R. A., S. N. Stacey, T. A. Grigliatti, and H. W. Brock. 1985. Sequences homologous to the P mobile element of *Drosophila melanogaster* are widely distributed in the subgenus Sophophora. *Nature* 318:561–63.

Larson, A., D. B. Wake, and K. P. Yanev. 1984. Measuring gene flow among populations having high levels of genetic fragmentation. *Genetics* 106:293–308.

Latter, B. D. H. 1972. Selection in finite populations with multiple alleles. III. Genetic divergence with centripetal selection and mutation. *Genetics* 70:475–90.

—— 1973a. Measures of genetic distance. In N. E. Morton, ed., *Genetic Structure of Populations,* pp. 27–37. Honolulu: University of Hawaii Press.

—— 1973b. The island model of population differentiation: A general solution. *Genetics* 73:147–57.

—— 1975. Influence of selection pressures on enzyme polymorphisms in *Drosophila*. *Nature* 257:590–92.

Laurie-Ahlberg, C. C., P. T. Barnes, J. W. Curtsinger, T. H. Emigh, B. Karlin, R. Morris, R. A. Norman, and A. N. Wilton. 1985. Genetic variability of flight metabolism in *Drosophila melanogaster*.

II. Relationship between power output and enzyme activity levels. *Genetics* 11:845–68.

Lawn, R. M., A. Efstratiadis, C. O'Connell, and T. Maniatis. 1980. The nucleotide sequence of the human β-globin gene. *Cell* 21:647–51.

Leary, R. F., F. W. Allendorf, and K. L. Knudsen. 1984. Superior developmental stability of heterozygotes at enzyme loci in Salmonid fishes. *Amer. Natur.* 124:540–51.

Leder, A., D. Swan, F. Ruddle, P. D'Eustachio, and P. Leder. 1981. Dispersion of α-like globin genes of the mouse to three different chromosomes. *Nature* 293:196–200.

Leder, P. 1982. The genetics of antibody diversity. *Sci. Amer.* 246(5):102–15.

Ledig, F. T., R. P. Furies, and B. A. Bonefield. 1983. The relation of growth to heterozygosity in pitch pine. *Evolution* 37:1227–38.

Lee, Y. M., D. J. Friedman, and F. J. Ayala. 1985. Superoxidase dimutase: An evolutionary puzzle. *Proc. Natl. Acad. Sci. USA* 82:824–28.

Leigh Brown, A. J. and C. H. Langley. 1979. Reevaluation of genic heterozygosity in natural populations of *Drosophila melanogaster* in two-dimensional electrophoresis. *Proc. Natl. Acad. Sci. USA* 76:2381–84.

LeQuesne, W. J. 1969. A method of selection of characters in numerical taxonomy. *Syst. Zool.* 18:201–5.

Lessios, H. A. 1979. Use of Panamian sea urchins to test the molecular clock. *Nature* 280:599–601.

Levene, H. 1949. On a matching problem arising in genetics. *Ann. Math. Stat.* 20:91–94.

—— 1953. Genetic equilibrium when more than one ecological niche is available. *Amer. Natur.* 87:331–33.

Levin, D. A. 1975. Genic heterozygosity and protein polymorphism among local populations of *Oenothera biennis*. *Genetics* 79:477–91.

Levin, D. A. and W. L. Crepet. 1973. Genetic variation in *Lycopodium lucidulum:* A phylogenetic relic. *Evolution* 27:622–32.

Lewin, B. 1985. *Genes.* 2d ed. New York: Wiley.

Lewin, R. 1981. Evolutionary history written in globin genes. *Science* 214:426–29.

Lewontin, R. C. 1964. The interaction of selection and linkage. I. General considerations: Heterotic models. *Genetics* 49:49–67.

—— 1972. The apportionment of human diversity. *Evol. Biol.* 6:381–98.

—— 1974. *The Genetic Basis of Evolutionary Change.* New York: Columbia University Press.

Lewontin, R. C. and C. C. Cockerham. 1959. The goodness-of-fit test for detecting natural selection in random mating populations. *Evolution* 13:561–64.

Lewontin, R. C. and J. L. Hubby. 1966. A molecular approach to the study of genic heterozygosity in natural populations. II. Amount of variation and degree of heterozygosity in natural populations of *Drosophila pseudoobscura. Genetics* 54:595–609.

Lewontin, R. C. and K. Kojima. 1960. The evolutionary dynamics of complex polymorphisms. *Evolution* 14:458–72.

Li, C. C. 1976. *First Course in Population Genetics.* Pacific Grove, Calif.: Boxwood Press.

Li, C. C. and D. G. Horvitz. 1953. Some methods of estimating the inbreeding coefficient. *Amer. J. Hum. Genet.* 5:107–17.

Li, W.-H. 1976a. Electrophoretic identity of proteins in a finite population and genetic distance between taxa. *Genet. Res.* 28:119–27.

—— 1976b. Effect of migration on genetic distance. *Amer. Natur.* 110:841–47.

—— 1976c. A mixed model of mutation for electrophoretic identity of proteins within and between populations. *Genetics* 83:423–32.

—— 1977a. Maintenance of genetic variability under mutation and selection pressures in a finite population. *Proc. Natl. Acad. Sci. USA* 74:2509–13.

—— 1977b. Distribution of nucleotide differences between two randomly chosen cistrons in a finite population. *Genetics* 85:331–37.

—— 1978. Maintenance of genetic variability under the joint effect of mutation, selection, and random drift. *Genetics* 90:349–82.

—— 1980. Rate of gene silencing at duplicate loci: A theoretical study and interpretation of data from tetraploid fishes. *Genetics* 95:237–58.

—— 1981a. A simulation study of Nei and Li's model for estimating DNA divergence from restriction enzyme maps. *J. Mol. Evol.* 17:251–55.

—— 1981b. A simple method for constructing phylogenetic trees from distance matrices. *Proc. Natl. Acad. Sci. USA* 78:1085–89.

—— 1982. Evolutionary change of duplicate genes. *Isozymes* 6:55–92.

—— 1983. Evolution of duplicate genes and pseudogenes. In M. Nei and R. Koehn, eds., *Evolution of Genes and Proteins,* pp. 14–37. Sunderland, Mass.: Sinauer Associates.

—— 1984. Retention of cryptic genes in microbial populations. *Mol. Biol. Evol.* 1:213–19.

—— 1986. Evolutionary change of restriction cleavage sites and phylogenetic inference. *Genetics* 113:187–213.

Li, W.-H. and T. Gojobori. 1983. Rapid evolution of goat and sheep globin genes following gene duplication. *Mol. Biol. Evol.* 1:94–108.

Li, W.-H., T. Gojobori, and M. Nei. 1981. Pseudogenes as a paradigm of neutral evolution. *Nature* 292:237–39.

Li, W.-H., C.-C. Luo, and C.-I. Wu. 1985a. Evolution of DNA sequences. In R. J. MacIntyre, ed., *Molecular Evolutionary Genetics.* New York: Plenum Press.

Li, W.-H. and M. Nei. 1974. Stable linkage disequilibrium without epistasis in subdivided populations. *Theor. Popul. Biol.* 6:173–83.

—— 1975. Drift variances of heterozygosity and genetic distance in transient states. *Genet. Res.* 25:229–48.

—— 1977. Persistence of common alleles in two related populations or species. *Genetics* 86:901–14.

Li, W.-H. C.-I. Wu, and C.-C. Luo. 1984. Nonrandomness of point mutation as reflected in nucleotide substitutions in pseudogenes and its evolutionary implications. *J. Mol. Evol.* 21:58–71.

Li, W.-H., C.-I. Wu, and C.-C. Luo. 1985b. A new method for estimating synonymous and nonsynonymous rates of nucleotide substitution considering the relative likelihood of nucleotide and codon changes. *Mol. Biol. Evol.* 2:150–74.

Litman, G. W., L. Berger, K. Murphy, R. Litman, K. Kinds, and B. W. Erickson. 1985. Immunoglobulin V_H structure and diversity in *Heterodontus,* a phylogenetically primitive shark. *Proc. Natl. Acad. Sci. USA* 82:2082–86.

Livingstone, F. B. 1962. Anthropological implications of sickle cell gene distribution in West Africa. In M. F. A. Montagu, ed., *Culture and the Evolution of Man,* pp. 271–99. Oxford: Oxford University Press.

—— 1967. *Abnormal Hemoglobins in Human Populations.* Chicago: Aldine.

Lomedico, P., N. Rosenthal, A. Efstratiadis, W. Gilbert, R. Kolodner, and R. Tizard. 1979. The structure and evolution of the two non-allelic rat preproinsulin genes. *Cell* 18:545–58.

Luciw, P. A., S. J. Potter, K. Steimer, D. Dina, and J. A. Levy. 1984. Molecular cloning of AIDS-associated retroviruses. *Nature* 312:760–63.

Lynch, M. 1984. The selective value of alleles underlying polygenic traits. *Genetics* 108:1021–33.

McCarthy, B. J. and M. N. Farquhar. 1972. The rate of change of DNA in evolution. *Brookhaven Symp. Biol.* 23:1–43.

McClintock, B. 1951. Chromosome organization and genic expression. *Cold Spring Harbor Symp. Quant. Biol.* 16:13–47.

McCommas, S. A. 1983. A taxonomic approach to evaluation of the charge state model using twelve species of sea anemone. *Genetics* 103:741–52.

McGinnis, W., R. L. Garber, J. Wirz, A. Kuroiwa, and W. J. Gehring. 1984 A homologous protein-coding sequence in *Drosophila* homeotic genes and its conservation in other metazoans. *Cell* 37:403–8.

MacIntyre, R. J. and T. R. F. Wright. 1966. Responses of esterase 6 alleles of *Drosophila melanogaster* and *D. simulans* to selection in experimental populations. *Genetics* 53:371–87.

McLaughlin, P. J. and M. O. Dayhoff. 1970. Eukaryotes versus prokaryotes: An estimate of evolutionary distance. *Science* 168:1469–70.

—— 1972. Evolution of species and proteins: A time scale. In M. O. Dayhoff, ed., *Atlas of Protein Sequence and Structure,* vol. 5, pp. 47–66. Silver Springs, Md.: Natl. Biomed. Res. Found.

McLellan, T. 1984. Molecular charge and electrophoretic mobility in cetacean myoglobins of known sequence. *Biochem. Genet.* 22:181–200.

McLellan, T., F.-L. Ames, and K. Nikaido. 1983. Genetic variation in proteins: Comparison of one-dimensional and two-dimensional gel electrophoresis. *Genetics* 104:381–90.

McLeod, M. J., S. I. Guttman, W. H. Eshbaugh, and R. E. Rayle. 1982. An electrophoretic study of evolution in *Capsicum (Solanaceae). Evolution* 37:562–74.

Maeda, N., F. Yang, D. R. Barnett, B. H. Bowman, and O. Smithies. 1984. Duplication within the haptoglobin Hp^2 gene. *Nature* 309:131–35.

Magni, G. E. 1969. Spontaneous mutations. In *Proc. 12th Intl. Cong. Genet.*, vol. 3, pp. 247–59. Tokyo.

Mahalanobis, P. C. 1936. On the generalized distance in statistics. *Proc. Natl. Inst. Sci. India* 2:49–55.

Malécot, G. 1948. *Les Mathématiques de l'hérédité.* Paris: Masson.

—— 1952. Les processus stochastiques et la methode des fonctions generatrices ou caracteristiques. *Extrait des Publications de l'ISUP* 3:1–16.

Margoliash, E. 1963. Primary structure and evolution of cytochrome c. *Proc. Natl. Acad. Sci. USA* 50:672–79.

Margoliash, E. and E. L. Smith. 1965. Structural and functional aspects of cytochrome c in relation to evolution. In V. Bryson and H. J. Vogel, eds., *Evolving Genes and Proteins,* pp. 221–42. New York: Academic Press.

Marshall, D. R. and A. H. D. Brown. 1975. The charge-state model of protein polymorphisms in natural populations. *J. Mol. Evol.* 6:149–63.

Maruyama, T. 1970. Effective number of alleles in a subdivided population. *Theor. Popul. Biol.* 1:273–306.

—— 1973. Isolation by distance, genetic variability, the time required for a gene substitution, and local differentiation in a finite, geographically structured population. In N. E. Morton, ed., *Genetic Structure of Populations,* pp. 80–81. Honolulu: University of Hawaii Press.

—— 1977. *Stochastic Problems in Population Genetics.* Berlin: Springer-Verlag.

Maruyama, T. and P. A. Fuerst. 1984. Population bottlenecks and non-equilibrium models in population genetics. I. Allele numbers when populations evolve from zero variability. *Genetics* 108:745–63.

Maruyama, T. and M. Kimura. 1974. Geographical uniformity of selectively neutral polymorphisms. *Nature* 249:30–32.

—— 1980. Genetic variability and effective population size when local extinction and recolonization of subpopulations are frequent. *Proc. Natl. Acad. Sci. USA* 77:6710–14.

Maruyama, T. and M. Nei. 1981. Genetic variability maintained by mutation and overdominant selection in finite populations. *Genetics* 98:441–59.

Maxson, L. R., 1984. Molecular probes of phylogeny and biogeography in toads of the widespread genus *Bufo. Mol. Biol. Evol.* 1:345–56.

Maxson, L. R., V. M. Sarich, and A. C. Wilson. 1975. Continental drift and the use of albumin as an evolutionary clock. *Nature* 255:397–400.

Maxson, L. R. and A. C. Wilson. 1974. Convergent morphological evolution detected by studying proteins of tree frogs in the *Hyla eximia* species group. *Science* 185:66–68.

—— 1975. Albumin evolution and organismal evolution in tree frogs (Hylidae). *Syst. Zool.* 24:1–15.

Maynard Smith, J. and J. Haigh. 1974. The hitch-hiking effect of a favourable gene. *Genet. Res.* 23:23–35.

Mayr, E. 1942. *Systematics and the Origin of Species.* New York: Columbia University Press.

—— 1963. *Animal Species and Evolution.* Cambridge, Mass.: Harvard University Press.

—— 1965. Discussion. In V. Bryson and H. J. Vogel, eds., *Evolving Genes and Proteins,* pp. 293–94. New York: Academic Press.

—— 1970. *Populations, Species, and Evolution.* Cambridge, Mass.: Harvard Universisty Press.

—— 1982. *The Growth of Biological Thought.* Cambridge, Mass.: Harvard University Press.

Mayr, E. and W. B. Provine. 1980. *The Evolutionary Synthesis.* Cambridge, Mass.: Harvard University Press.

Mellor, A. L., E. H. Weiss, K. Ramachandran, and R. A. Flavell. 1983. A potential donor gene for the *bm*1 gene conversion event in the C57BL mouse. *Nature* 306:792–95.

Milkman, R. 1976. Selection is the major determinant. *Trend Biochem. Sci.* 1:152–54.

—— 1982. Toward a unified selection theory. In R. Milkman, ed., *Perspectives on Evolution,* pp. 105–18. Sunderland, Mass.: Sinauer Associates.

Milkman, R., and I. P. Crawford. 1983. Clustered third-base substitutions among wild strains of *Escherichia coli*. *Science* 221:378–79.

Mitton, J. B. 1978. Relationship between heterozygosity for enzyme loci and variation of morphological characters in natural populations. *Nature* 273:661–62.

Mitton, J. B. and M. C. Grant. 1984. Associations among protein heterozygosity, growth rate, and developmental homeostasis. *Ann. Rev. Ecol. Syst.* 15:479–99.

Miyata, T. 1982. Evolutionary changes and functional constraints in

DNA sequences. In M. Kimura, ed., *Molecular Evolution, Protein Polymorphism, and the Neutral Theory,* pp. 233–66. Tokyo: Japan Scientific Societies Press.

Miyata, T., S. Miyazawa, and T. Yasunaga. 1979. Two types of amino acid substitutions in protein evolution. *J. Mol. Evol.* 12:219–36.

Miyata, T. and T. Yasunaga. 1980. Molecular evolution of mRNA: A method for estimating evolutionary rates of synonymous and amino acid substitutions from homologous nucleotide sequences and its application. *J. Mol. Evol.* 16:23–36.

—— 1981. Rapidly evolving mouse α-globin related pseudogene and its evolutionary history. *Proc. Natl. Acad. Sci. USA* 78:450–53.

Miyata, T., T. Yasunaga, and T. Nishida. 1980. Nucleotide sequence divergence and functional constraint in mRNA evolution. *Proc. Natl. Acad. Sci. USA* 77:7328–32.

Monod, J. 1971. *Chance and Necessity.* New York: Knopf.

Montanucci, R. R., R. W. Axtell, and H. C. Dessauer. 1975. Evolutionary divergence among collared lizards *(Crotaphytus),* with comments on the status of *Gambellia. Herpetologica* 31:336–47.

Montgomery, E. A. and C. H. Langley. 1983. Transposable elements in Mendelian populations. II. Distribution of three copia-like elements in a natural population of *Drosophila melanogaster. Genetics* 104:473–83.

Moran, P. A. P. 1975. Wandering distributions and the electrophoretic profile. *Theor. Popul. Biol.* 8:318–30.

—— 1976. Wandering distributions and the electrophoretic profile. II. *Theor. Popul. Biol.* 10:145–49.

Morgan, K. and C. Strobeck. 1979. Is intragenic recombination a factor in the maintenance of genetic variation in natural populations? *Nature* 277:383–84.

Morgan, T. H. 1925. *Evolution and Genetics.* Princeton, N.J.: Princeton University Press.

—— 1932. *The Scientific Basis of Evolution.* New York: Norton.

Morgan, T. H., H. J. Muller, A. H. Sturtevant, and C. B. Bridges. 1915. *The Mechanism of Mendelian Heredity.* New York: Holt.

Morizot, D. C. and M. J. Siciliano. 1982. Protein polymorphisms, segregation in genetic crosses, and genetic distances among fishes of the genus *Xiphorus (Poeciliidae). Genetics* 102:539–56.

Morton, N. E., J. F. Crow, and H. J. Muller. 1956. An estimate of the mutational damage in man from data on consanguineous marriages. *Proc. Natl. Acad. Sci. USA* 42:855–63.

Motulsky, A. G. 1964. Hereditary red cell traits and malaria. *Amer. J. Trop. Med.* 13:147–55.

Mourant, A. E., A. C. Kopec, and K. Domaniewska-Sobczak. 1976. *The Distribution of the Human Blood Groups and Other Polymorphisms.* Oxford: Oxford University Press.

Mueller, L. D. and F. J. Ayala. 1982. Estimation and interpretation of genetic distance on empirical studies. *Genet. Res.* 40:127–37.

Mukai, T. 1964. The genetic structure of natural populations of *Drosophila melanogaster.* I. Spontaneous mutation rate of polygenes controlling viability. *Genetics* 50:1–19.

Mukai, T. and C. C. Cockerham. 1977. Spontaneous mutation rates at enzyme loci in *Drosophila melanogaster. Proc. Natl. Acad. Sci. USA* 74:2514–17.

Mukai, T., L. E. Mettler, and S. I. Chigusa. 1971. Linkage disequilibrium in a local population of *Drosophila melanogaster. Proc. Natl. Acad. Sci. USA* 68:1056–69.

Mukai, T., H. Tachida, and M. Ichinose, 1980. Selection for viability at loci controlling protein polymorphisms in *Drosophila melanogaster* is very weak at most. *Proc. Natl. Acad. Sci. USA* 77:4857–60.

Mukai, T., T. K. Watanabe, and O. Yamaguchi. 1974. The genetic structure of natural populations of *Drosophila melanogaster.* XII. Linkage disequilibrium in a large local population. *Genetics* 77:771–93.

Mukai, T. and T. Yamazaki. 1980. Test for selection on polymorphic isozyme genes using the population cage method. *Genetics* 96:537–42.

Muller, H. J. 1929. The method of evolution. *Sci. Monthly* 29:481–505.

—— 1940. Bearings of the *Drosophila* work on systematics. In J. S. Huxley, ed., *The New Systematics,* pp. 185–268. Oxford: Clarendon Press.

—— 1942. Isolating mechanisms, evolution and temperature. In T. Dobzhansky, ed., *Biological Symposia* 6, pp. 71–125. Lancaster, Pa.: Jaques Cattell Press.

—— 1949. Reintegration of the symposium on genetics, paleontology, and evolution. In G. L. Jepsen, G. G. Simpson, and E. Mayr, eds., *Genetics, Paleontology, and Evolution,* pp. 421–45. Princeton, N.J.: Princeton University Press.

—— 1950. Our load of mutations. *Amer. J. Hum. Genet.* 2:111–76.

—— 1959. Advances in radiation mutagenesis through studies on *Dro-*

sophila. In *Progress in Nuclear Energy.* Ser. 6, vol. 2, pp. 146–60. New York: Pergamon Press.

Nagylaki, T. 1974. Quasilinkage equilibrium and the evolution of two-locus systems. *Proc. Natl. Acad. Sci. USA* 71:526–30.

—— 1983. The robustness of neutral models of geographical variation. *Theor. Popul. Biol.* 24:268–94.

—— 1984. The evolution of multigene families under intrachromosomal gene conversion. *Genetics* 106:529–48.

Na-Nakorn, S. and P. Wasi. 1970. Alpha-thalassemia in Northern Thailand. *Amer. J. Hum. Genet.* 22:645–51.

Narain, P. 1970. A note on the diffusion approximation for the variance of the number of generations until fixation of a neutral mutant gene. *Genet. Res.,* 15:251–55.

Needleman, S. B. and C. D. Wunsch. 1970. A general method applicable to the search for similarities in the amino acid sequence of two proteins. *J. Mol. Biol.* 48:443–53.

Neel, J. V. 1973. "Private" genetic variants and the frequency of mutation among South American Indians. *Proc. Natl. Acad. Sci. USA* 70:3311–15.

Neel, J. V., H. W. Mohrenweiser, and M. H. Meisler. 1980. Rate of spontaneous mutation at human loci encoding protein structure. *Proc. Natl. Acad. Sci. USA* 77:6037–41.

Neel, J. V. and E. D. Rothman. 1978. Indirect estimates of mutation rates in tribal Amerindians. *Proc. Natl. Acad. Sci. USA* 75:5585–88.

Neel, J. V. and R. H. Ward. 1972. The genetic structure of a tribal population, the Yanomama Indians. VI. Analysis by *F*-statistics (including a comparison with the Makiritare and Xavante). *Genetics* 72:639–66.

Nei, M. 1965. Variation and covariation of gene frequencies in subdivided populations. *Evolution* 19:256–58.

—— 1969. Gene duplication and nucleotide substitution in evolution. *Nature* 221:40–42.

—— 1971. Interspecific gene differences and evolutionary time estimated from electrophoretic data on protein identity. *Amer. Natur.* 105:385–98.

—— 1972. Genetic distance between populations. *Amer. Natur.* 106:283–92.

—— 1973a. Analysis of gene diversity in subdivided populations. *Proc. Natl. Acad. Sci. USA* 70:3321–23.

—— 1973b. The theory and estimation of genetic distance. In N. E. Morton, ed., *Genetic Structure of Populations,* pp. 45–54. Honolulu: University Press of Hawaii.

—— 1975. *Molecular Population Genetics and Evolution.* Amsterdam: North-Holland.

—— 1976. Mathematical models of speciation and genetic distance. In S. Karlin and E. Nevo, eds., *Population Genetics and Ecology.* pp. 723–65. New York: Academic Press.

—— 1977a. Estimation of mutation rate from rare protein variants. *Amer. J. Hum. Genet.* 29:225–32.

—— 1977b. Standard error of immunological dating of evolutionary time. *J. Mol. Evol.* 9:203–111.

—— 1977c. *F*-statistics and analysis of gene diversity in subdivided populations. *Ann. Hum. Genet.* 41:225–33.

—— 1978a. Estimation of average heterozygosity and genetic distance from a small number of individuals. *Genetics* 89:583 90.

—— 1978b. The theory of genetic distance and evolution of human races. *Japan. J. Hum. Genet.* 23:341–69.

—— 1980. Stochastic theory of population genetics and evolution. In C. Barigozzi, ed., *Vito Volterra Symposium on Mathematical Models in Biology,* pp. 17–47. Berlin: Springer-Verlag.

—— 1983. Genetic polymorphism and the role of mutation in evolution. In M. Nei and R. Koehn, eds., *Evolution of Genes and Proteins,* pp. 165–90. Sunderland, Mass.: Sinauer Associates.

—— 1985. Human evolution at the molecular level. In T. Ohta and K. Aoki, eds., *Population Genetics and Molecular Evolution,* pp. 41–64. Tokyo: Japan Scientific Societies Press.

Nei, M. and R. Chakraborty. 1973. Genetic distance and electrophoretic identity of proteins between taxa. *J. Mol. Evol.* 2:323–28.

Nei, M., R. Chakraborty, and P. A. Fuerst. 1976a. Infinite allele model with varying mutation rate. *Proc. Natl. Acad. Sci. USA* 73:4164–68.

Nei, M., A. Chakravarti, and Y. Tateno. 1977. Mean and variance of F_{ST} in a finite number of incompletely isolated populations. *Theor. Popul. Biol.* 11:291–306.

Nei, M. and R. K. Chesser. 1983. Estimation of fixation indices and gene diversities. *Ann. Hum. Genet.* 47:253–59.

Nei, M. and M. W. Feldman. 1972. Identity of genes by descent within and between populations under mutation and migration pressures. *Theor. Popul. Biol.* 3:460–65.

Nei, M., P. A. Fuerst, and R. Chakraborty. 1976b. Testing the neutral mutation hypothesis by distribution of single locus heterozygosity. *Nature* 262:491–93.

—— 1978. Subunit molecular weight and genetic variability of proteins in natural populations. *Proc. Natl. Acad. Sci. USA* 75:3359–62.

Nei, M. and T. Gojobori. 1986. Simple methods for estimating the numbers of synonymous and nonsynonymous nucleotide substitutions. *Mol. Biol. Evol.* 3:418–26.

Nei, M. and D. Graur. 1984. Extent of protein polymorphism and the neutral mutation theory. *Evol. Biol.* 17:73–118.

Nei, M. and Y. Imaizumi. 1966a. Genetic structure of human populations. I. Local differentiation of blood group gene frequencies in Japan. *Heredity* 21:9–25.

—— 1966b. Genetic structure of human populations. II. Differentiation of blood group gene frequencies among isolated populations. *Heredity* 21:183–90.

Nei, M. and R. K. Koehn, eds. 1983. *Evolution of Genes and Proteins.* Sunderland, Mass.: Sinauer Associates.

Nei, M. and W.-H. Li. 1973. Linkage disequilibrium in subdivided populations. *Genetics* 75:213–19.

—— 1975. Probability of identical monomorphism in related species. *Genet. Res.* 26:31–43.

—— 1979. Mathematical model for studying genetic variation in terms of restriction endonucleases. *Proc. Natl. Acad. Sci. USA* 76:5269–73.

—— 1980. Non-random association between electromorphs and inversion chromosomes in finite populations. *Genet. Res.* 35:65–83.

Nei, M., T. Maruyama, and R. Chakraborty. 1975. The bottleneck effect and genetic variability in populations. *Evolution* 29:1–10.

Nei, M., T. Maruyama, and C.-I. Wu. 1983b. Models of evolution of reproductive isolation. *Genetics* 103:557–79.

Nei, M. and A. K. Roychoudhury. 1973a. Probability of fixation and mean fixation time of an overdominant mutation. *Genetics* 74:371–80.

—— 1973b. Probability of fixation of nonfunctional genes at duplicate loci. *Amer. Natur.* 107:362–72.

—— 1974a. Sampling variance of heterozygosity and genetic distance. *Genetics* 76:379–90.

—— 1974b. Genetic variation within and between the three major races of man, Caucasoids, Negroids, and Mongoloids. *Amer. J. Hum. Genet.* 26:421–43.

—— 1982. Genetic relationship and evolution of human races. *Evol. Biol.* 14:1–59.

Nei, M. and N. Saitou. 1986. Genetic relationship of human populations and ethnic differences in reaction to drugs and food. In W. Kalow, H. W. Goedde, and D. P. Agarwal, eds., *Ethnic Differences in Reactions to Drugs and Xenobiotics,* pp. 21–37. New York: Alan R. Liss.

Nei, M., J. C. Stephens, and N. Saitou. 1985. Methods for computing the standard errors of branching points in an evolutionary tree and their application to molecular data from humans and apes. *Mol. Biol. Evol.* 2:66–85.

Nei, M. and F. Tajima. 1981. DNA polymorphism detectable by restriction endonucleases. *Genetics* 97:145–63.

     1983. Maximum likelihood estimation of the number of nucleotide substitutions from restriction sites data. *Genetics* 105:207–17.

—— 1985. Evolutionary change of restriction cleavage sites and phylogenetic inference for man and apes. *Mol. Biol. Evol.* 2:189–205.

Nei, M., F. Tajima, and Y. Tateno. 1983a. Accuracy of estimated phylogenetic trees from molecular data. II. Gene frequency data. *J. Mol. Evol.* 19:153–70.

Nei, M. and Y. Tateno. 1975. Interlocus variation of genetic distance and the neutral mutation theory. *Proc. Natl. Acad. Sci. USA* 72:2758–60.

Nei, M. and S. Yokoyama. 1976. Effects of random fluctuations of selection intensity on genetic variability in a finite population. *Japan. J. Genet.* 51:355–69.

Nelson, K. and D. Hedgecock. 1980. Enzyme polymorphism and adaptive strategy in the decapod *Crustacea. Amer. Natur.* 116:238–80.

Nevo, E. 1978. Genetic variation in natural populations: Patterns and theory. *Theor. Popul. Biol.* 13:121–77.

Nevo, E., A. Beiles, and R. Ben-Shlomo. 1984. The evolutionary significance of genetic diversity: Ecological, demographic and life history correlates. In G. S. Mani, ed., *Evolutionary Dynamics of Genetic Diversity,* pp. 13–213. Berlin: Springer-Verlag.

Nevo, E., Y. J. Kim, C. R. Shaw, and C. S. Thaeler, Jr. 1974. Genetic variation, selection, and speciation in *Thomomys talpoides* pocket gophers. *Evolution* 28:1–23.

Novacek, M. J. 1982. Information for molecular studies from anatomical and fossil evidence on higher eutherian phylogeny. In M. Goodman, ed., *Macromolecular Sequences in Systematic and Evolutionary Biology*, pp. 3–41. New York: Plenum Press.

Novick, A. and L. Szilard. 1950. Experiments with the chemostat on spontaneous mutations of bacteria. *Proc. Natl. Acad. Sci. USA* 36:708–19.

Nozawa, K., T. Shotake, Y. Kawamoto, and Y. Tanabe. 1982. Electrophoretically estimated genetic distance and divergence time between chimpanzee and man. *Primates* 23:432–43.

Nozawa, K., T. Shotake, Y. Ohkura, and Y. Tanabe. 1977. Genetic variations within and between species of Asian macaques. *Japan. J. Genet.* 52:15–30.

O'Brien, S. J., W. G. Nash, D. E. Wildt, M. E. Bush, and R. E. Benveniste. 1985a. A molecular solution to the riddle of the giant panda's phylogeny. *Nature* 317:140–44.

O'Brien, S. J. and 9 other authors. 1985b. Genetic basis for species vulnerability in the cheetah. *Science* 227:1428–34.

O'Farrell, P. H. 1975. High resolution two-dimensional electrophoresis of proteins. *J. Biol. Chem.* 250:4007–21.

O'Hare, K. and G. M. Rubin. 1983. Structure of P transposable elements and their sites of insertion and excision in the *Drosophila melanogaster* genome. *Cell* 34:25–35.

Ohnishi, S., M. Kawanishi, and T. K. Watanabe. 1983. Biochemical phylogenies of *Drosophila:* Protein differences detected by two-dimensional electrophoresis. *Genetica* 61:55–63.

Ohno, S. 1967. *Sex Chromosomes and Sex-linked Genes*. Berlin: Springer-Verlag.

—— 1969. The preferential activation of maternally derived alleles in development of interspecific hybrids. In V. Defendi, ed., *Heterospecific Genome Interaction*, pp. 137–50. Philadelphia, Pa.: Wister Institute Press.

—— 1970. *Evolution by Gene Duplication*. Berlin: Springer-Verlag.

—— 1972. So much "junk" DNA in our genome. *Brookhaven Symp. Biol.* 23:366–70.

—— 1981. Original domain for the serum albumin family arose from repeated sequences. *Proc. Natl. Acad. Sci. USA* 78:7657–61.

—— 1984. Repeats of base oligomers as the primordial coding sequences of the primeval earth and their vestiges in modern genes. *J. Mol. Evol.* 20:313–21.

Ohta, T. 1971. Associative overdominance caused by linked detrimental mutations. *Genet. Res.* 18:277–86.

—— 1973a. Effect of linkage on behavior of mutant genes in finite populations. *Theor. Popul. Biol.* 4:145–62.

—— 1973b. Slightly deleterious mutant substitutions in evolution. *Nature* 246:96–98.

—— 1974. Mutational pressure as the main cause of molecular evolution and polymorphism. *Nature* 252:351–54.

—— 1976. Role of very slightly deleterious mutations in molecular evolution and polymorphism. *Theor. Popul. Biol.* 10:254–75.

—— 1977. Extension to the neutral mutation random drift hypothesis. In M. Kimura, ed., *Molecular Evolution and Polymorphism,* pp. 148–67. Mishima, Japan: National Institute of Genetics.

—— 1980. *Evolution and Variation of Multigene Families.* Berlin: Springer-Verlag.

—— 1981. Genetic variation in small multigene families. *Genet. Res.* 37:133–49.

—— 1982a. Linkage disequilibrium due to random genetic drift in finite subdivided populations. *Proc. Natl. Acad. Sci. USA* 79:1940–44.

—— 1982b. Linkage disequilibrium with the island model. *Genetics* 101:139–55.

—— 1983a. On the evolution of multigene families. *Theor. Popul. Biol.* 23:216–40.

—— 1983b. Time until fixation of mutant belonging to a gene family. *Genet. Res.* 41:47–55.

—— 1984. Population genetics theory of concerted evolution and its application to the immunoglobulin V gene tree. *J. Mol. Evol.* 20:274–80.

Ohta, T. and G. A. Dover. 1984. The cohesive population genetics of molecular drive. *Genetics* 108:501–21.

Ohta, T. and M. Kimura. 1969. Linkage disequilibrium at steady state determined by random genetic drift and recurrent mutation. *Genetics* 63:229–38.

—— 1971a. On the constancy of the evolutionary rate of cistrons. *J. Mol. Evol.* 1:18–25.

—— 1971b. Linkage disequilibrium between two segregating nucleotide sites under the steady flux of mutations in a finite population. *Genetics* 68:571–80.

—— 1973. A model of mutation appropriate to estimate the number of electrophoretically detectable alleles in a finite population. *Genet. Res.* 22:201–4.

—— 1981. Some calculations on the amount of selfish DNA. *Proc. Natl. Acad. Sci. USA* 78:1129–32.

Orgel, L. E. and F. H. C. Crick. 1980. Selfish DNA: The ultimate parasite. *Nature* 284:604–7.

Page, D. C., M. E. Harper, J. Love, and D. Botstein. 1984. Occurrence of a transposition from the X-chromosome long arm to the Y-chromosome short arm during human evolution. *Nature* 311:119–23.

Palmer, J. D. and D. Zamir. 1982. Chloroplast DNA evolution and phylogenetic relationships in *Lycopersicon*. *Proc. Natl. Acad. Sci. USA* 79:5006–10.

Parker, H. R., D. P. Philipp, and G. S. Whitt. 1985. Gene regulatory divergence among species estimated by altered developmental patterns in interspecific hybrids. *Mol. Biol. Evol.* 2:217–50.

Patton, J. L., R. K. Selander, and M. H. Smith. 1972. Genic variation in hybridizing populations of gophers (genus *Thomomys*). *Syst. Zool.* 21:263–70.

Peacock, D. and D. Boulter. 1975. Use of amino acid sequence data in phylogeny and evaluation of methods using computer simulation. *J. Mol. Biol.* 95:513–27.

Pearson, K. 1926. On the coefficient of racial likeness. *Biometrika* 18:337–43.

Peetz, E. W., G. Thomson, and P. W. Hedrick. 1986. Charge changes in protein evolution. *Mol. Biol. Evol.* 3:84–94.

Perler, F., A. Efstratiadis, P. Lomedico, W. Gilbert, R. Kolodner, and J. Dodgeson. 1980. The evolution of genes: The chicken preproinsulin gene. *Cell* 20:555–66.

Perutz, M. F. 1983. Species adaptation in a protein molecule. *Mol. Biol. Evol.* 1:1–28.

Perutz, M. F. and 8 other authors. 1981. Allosteric regulation of crocodilian haemoglobin. *Nature* 291:682–84.

Perutz, M. F. and H. Lehman. 1968. Molecular pathology of human haemoglobin. *Nature* 219:902–9.

Pilbeam, D. 1984. The descent of hominoids and hominids. *Sci. Amer.* 250(3):84–96.

Pinsker, W. and D. Sperlich. 1981. Geographic pattern of allozyme and inversion polymorphism on chromosome O of *Drosophila subobscura* and its evolutionary origin. *Genetica* 57:51–64.

Place, A. R. and D. A. Powers. 1979. Genetic variation and relative catalytic efficiencies: Lactate dehydrogenase B allozymes of *Fundulus heteroclitus. Proc. Natl. Acad. Sci. USA* 76:2354–58.

Poncz, M., E. Schwartz, M. Ballantine, and S. Surrey. 1983. Nucleotide sequence analysis of the $\delta\beta$-globin gene region in humans. *J. Biol. Chem.* 258:11599–609.

Post, T. J. and T. Uzzell. 1981. The relationship of *Rana sylvatica* and the monophyly of the *Rana boylii* group. *Syst. Zool.* 30:170–80.

Potter, J., M.-W. Ho, H. Bolton, A. J. Furth, D. M. Swallow, and B. Griffiths. 1985. Human lactase and the molecular basis of lactase persistence. *Biochem. Genet.* 23:423–39.

Powell, J. R. 1975. Protein variation in natural populations. *Evol. Biol.* 8:79–119.

——— 1983. Interspecific cytoplasmic gene flow in the absence of nuclear gene flow: Evidence from *Drosophila. Proc. Natl. Acad. Sci. USA* 80:492–95.

Prager, E. M., A. H. Brush, R. A. Nolan, M. Nakanishi, and A. C. Wilson. 1974. Slow evolution of transferrin and albumin in birds according to micro-complement fixation analysis. *J. Mol. Evol.* 3:243–62.

Prager, E. M. and A. C. Wilson. 1971. The dependence of immunological cross-reactivity upon sequence resemblance among lysozymes. *J. Biol. Chem.* 246:5978–89.

Prakash, S. 1969. Genic variation in a natural population of *Drosophila persimilis. Proc. Natl. Acad. Sci. USA* 62:778–84.

Prakash, S. and R. C. Lewontin. 1968. A molecular approach to the study of genic heterozygosity in natural populations. III. Direct evidence of coadaptation in gene arrangements of *Drosophila. Proc. Natl. Acad. Sci. USA* 59:398–405.

Prevosti, A., M. P. Garcia, L. Serra, M. Aquade, G. Ribo, and E. Sagarra. 1983. Association between allelic isozyme alleles and chromosomal arrangements in European populations and Chilean colonizers of *Drosophila subobscura. Isozymes* 10:171–91.

Prevosti, A., J. Ocana, and G. Alonso. 1975. Distances between populations of *Drosophila subobscura* based on chromosome arrangement frequencies. *Theor. Appl. Genet.* 45:231–41.

Proudfoot, N. 1984. The end of the message and beyond. *Nature* 307:412–13.

Prout, T. 1973. Appendix to J. B. Mitton and R. K. Koehn, 1973, Population genetics of marine *pelecypods*. III. Epistasis between functionally related isoenzymes of *Mytilus edulis. Genetics* 73:487–96.

Provine, W. B. 1986. *Sewall Wright: Geneticist and Evolutionist.* Chicago: University of Chicago Press.

Quax-Jeuken, Y., S. Bruisten, E. Nevo, H. Bloemendal, and W. W. de Jong. 1985. Evolution of crystallins: Expression of lens-specific proteins in the blind mammals mole *(Talpa europaea)* and mole rat *(Spalax ehrenbergi). Mol. Biol. Evol.* 2:279–88.

Rabson, A. B. and M. A. Martin. 1985. Molecular organization of the AIDS retrovirus. *Cell* 40:477–80.

Radding, C. M. 1982. Strand transfer in homologous genetic recombination. *Ann. Rev. Genet.* 16:405–37.

Raff, R. A. and T. C. Kaufman. 1983. *Embryos, Genes, and Evolution.* New York: Macmillan.

Ramshaw, J. A. M., J. A. Coyne, and R. C. Lewontin. 1979. The sensitivity of gel electrophoresis as a detector of genetic variation. *Genetics* 93:1019–37.

Rao, C. R. 1952. *Advanced Statistical Methods in Biometric Research.* New York: Wiley.

Raup, D. M. and J. J. Sepkoski, Jr. 1984. Periodicity of extinctions in the geologic past. *Proc. Natl. Acad. Sci. USA* 81:801–5.

Reed, T. E. 1969. Caucasian genes in American Negroes. *Science* 165:762–68.

Renner, O. 1914. Befruchtung und Embryobilding bei *Oenothera lamarckiana* und einigen verwandten Arten. *Flora* 107:115–50.

Reynolds, J., B. S. Weir, and C. C. Cockerham. 1983. Estimation of the coancestry coefficient: Basis for a short-term genetic distance. *Genetics* 105:767–79.

Riggs, S. R. 1984. Paleoceanographic model of Neogene phosphorite deposition, U.S. Atlantic continental margin. *Science* 223:123–31.

Robertson, A. 1956. The effect of selection against extreme deviants based on deviation or on homozygosis. *J. Genet.* 54:236–48.

—— 1967. The nature of quantitative genetic variation. In R. A. Brink, ed., *Heritage from Mendel,* pp. 265–80. Madison, Wis.: University of Wisconsin Press.

——— 1968. The spectrum of genetic variation. In R. C. Lewontin, ed., *Population Biology and Evolution,* pp. 5–16. Syracuse, N.Y: Syracuse University Press.

Robinson, D. F. and L. R. Foulds. 1981. Comparison of phylogenetic trees. *Math. Biosci.* 53:131–47.

Rogers, J. 1985. Origins of repeated DNA. *Nature* 317:765–66.

Rogers, J. S. 1972. Measures of genetic similarity and genetic distance. In *Studies in Genetics VII,* pp. 145–53. University of Texas Publication 7213. Austin, Tex.: University of Texas.

Romero-Herrera, A. E., H. Lehmann, K. A. Joysey, and A. E. Friday. 1973. Molecular evolution of myoglobin and the fossil record: A phylogenetic synthesis. *Nature* 246:389–95.

——— 1978. On the evolution of myoglobin. *Phil. Trans. Roy. Soc. London (B)*283.61–163.

Rubin, G. M., W. J. Brorein, Jr., P. Dunsmuir, A. J. Flavell, R. Levis, E. Strobel, J. J. Toole, and E. Young. 1981. *Copia*-like transposable elements in the *Drosophila* genome. *Cold Spring Harbor Symp. Quant. Biol.* 45:619–28.

Rubin, G. M. and A. C. Spradling. 1982. Genetic transformation of *Drosophila* with transposable element vectors. *Science* 218:348–53.

Ryman, N. 1983. Patterns of distribution of biochemical genetic variation in salmonids: Differences between species. *Aquaculture* 33:1–21.

Sage, R. D. 1981. Wild mice. In H. L. Foster, J. D. Small, and J. G. Fox, eds., *The Mouse in Biomedical Research,* vol. 1, pp. 39–90. New York: Academic Press.

Saigo, K., W. Kugimiya, Y. Matsuo, S. Inouye, K. Yoshioka, and S. Yuki. 1984. Identification of the coding sequence for a reverse transcriptase-like enzyme in a transposable genetic element in *Drosophila melanogaster. Nature* 312:659–61.

Saitou, N. and M. Nei. 1986a. Polymorphism and evolution of influenza A virus genes. *Mol. Biol. Evol.* 3:57–74.

Saitou, N. and M. Nei. 1986b. The number of nucleotides required to determine the branching order of three species with special reference to the human-chimpanzee-gorilla divergence. *J. Mol. Evol.* (in press).

Sanger, F. and 8 other authors. 1977. Nucleotide sequence of bacteriophage ϕX174 DNA. *Nature* 265:687–95.

Sanghvi, L. D. 1953. Comparison of genetical and morphological methods for a study of biological differences. *Amer. J. Phys. Anthrop.* 11:385–404.

Sankoff, D. and R. J. Cedergren. 1983. Simultaneous comparison of three or more sequences related by a tree. D. Sankoff and J. B. Kruskal, eds., *Time Warps, String Edits, and Macromolecules: The Theory and Practice of Sequence Comparison,* pp. 253–63. Reading, Mass.: Addison-Wesley.

Sarich, V. M. 1973. The giant panda is a bear. *Nature* 245:218–20.

—— 1977. Rates, sample sizes, and the neutrality hypothesis for electrophoresis in evolutionary studies. *Nature* 265:24–28.

Sarich, V. M. and A. C. Wilson. 1966. Quantitative immunochemistry and the evolution of primate albumins: Micro-complement fixation. *Science* 154:1563–66.

—— 1967. Immunological time scale for hominid evolution. *Science* 158:1200–3.

—— 1973. Generation time and genomic evolution in primates. *Science* 179:1144–47.

Saunders, N. C., L. G. Kessler, and J. C. Avise. 1986. Genetic variation and geographic differentiation in mitochondrial DNA of the horseshoe crab, *Limulus polyphemus. Genetics* 112:613–27.

Sawyer, S. and D. Hartl. 1981. On the evolution of behavioral reproductive isolation: The Wallace effect. *Theor. Popul. Biol.* 19:261–73.

Schaal, B. A. and D. A. Levin. 1976. The demographic genetics of *Liatris cylindracea* Michx. (Compositae). *Amer. Natur.* 110:191–206.

Schnell, G. D. and R. K. Selander. 1981. Environmental and morphological correlates of genetic variation in mammals. In M. H. Smith and T. Toule, eds., *Mammalian Population Genetics,* pp. 60–99. Athens: University of Georgia Press.

Schopf, J. W., J. M. Hayes, and M. R. Walter. 1983. Evolution of Earth's earliest ecosystems: Recent progress and unsolved problems. In J. W. Schopf, ed., *Earth's Earliest Biosphere: Its Origin and Evolution,* pp. 361–84. Princeton, N.J.: Princeton University Press.

Schopf, J. M. and M. R. Walter. 1983. Archean microfossils: New evidence of ancient microbes. In J. W. Schopf, ed., *Earth's Earliest Biosphere: Its Origin and Evolution,* pp. 214–39. Princeton, N.J.: Princeton University Press.

Schreier, P. H., A. L. M. Bothwell, B. Mueller-Hill, and D. Baltimore. 1981. Multiple differences between the nucleic acid sequences of the $IgG2a^a$ and $IgG2a^b$ alleles of the mouse. *Proc. Natl.* *Acad. Sci. USA* 78:4495–99.

Selander, R. K. 1975. Stochastic factors in the genetic structure of populations. In G. F. Estabrook, ed., *Proc. of 8th Intl. Cong. Numerical Taxonomy,* pp. 284–332. San Francisco: Freeman.

—— 1976. Genic variation in natural populations. In F. J. Ayala, ed., *Molecular Evolution,* pp. 21–45. Sunderland, Mass: Sinauer Associates.

Selander, R. K., W. G. Hung, and S. Y. Yang. 1969. Protein polymorphism and genic heterozygosity in two European subspecies of the house mouse. *Evolution* 23:379–90.

Selander, R. K. and D. W. Kaufman. 1975. Genetic structure of populations of the brown snail *(Helix aspersa).* I. Microgeographic variation. *Evolution* 29:385–401.

Selander, R. K. and B. R. Levin. 1980. Genetic diversity and structure in *Escherichia coli* populations. *Science* 210:545–47.

Selander, R. K., M. H. Smith, S. Y. Yang, W. E. Johnson, and J. B. Gentry. 1971. Biochemical polymorphism and systematics in the genus *peromyscus.* I. Variation in the old-field mouse *(Peromyscus polionotus) Studies in Genetics VI,* pp. 49–90. University of Texas Publication 7103. Austin, Tex.: University of Texas.

Selander, R. K. and T. S. Whittam. 1983. Protein polymorphism and the genetic structure of populations. In M. Nei and R. Koehn, eds., *Evolution of Genes and Proteins,* pp. 89–114. Sunderland, Mass: Sinauer Associates.

Selander, R. K., S. Y. Yang, R. C. Lewontin, and W. E. Johnson. 1970. Genetic variations in the horseshoe crab *(Limulus polyphemus),* a phylogenetic "relic." *Evolution* 24:402–14.

Sellers, P. H. 1974. On the theory and computation of evolutionary distances. *SIAM J. Appl. Math.* 26:787–93.

Shah, D. M. and C. H. Langley. 1979. Inter- and intraspecific variation in restriction maps of *Drosophila* mitochondrial DNAs. *Nature* 281:696–99.

Shapiro, J. A., ed. 1983. *Movable Genetic Elements.* New York: Academic Press.

Shaw, C. R. 1970. How many genes evolve? *Biochem. Genet.* 4:275–83.

Shaw, G. M., B. H. Hahn, S. K. Arya, J. E. Groopman, R. C. Gallo, and F. Wong-Staal. 1984. Molecular characterization of human T-cell leukemia (lymphotropic) virus type III in the acquired immune deficiency syndrome. *Science* 226:1165–71.

Sheppard, H. W. and G. A. Gutman. 1981. Allelic forms of rat κ chain genes: Evidence for strong selection at the level of nucleotide sequence. *Proc. Natl. Acad. Sci. USA* 78:7064–68.

Shinozaki, K., C. Yamada, N. Takahata, and M. Sugiura. 1983. Molecular cloning and sequence analysis of the cyanobacterial gene for the large subunit of ribulose-1,5-bisphosphate carboxylase/oxygenase. *Proc. Natl. Acad. Sci. USA* 80:4050–54.

Sibley, C. G. and J. E. Ahlquist. 1981. The phylogeny and relationships of the ratite birds as indicated by DNA-DNA hybridization. In G. G. E. Scudder and J. L. Reveal, eds., *Evolution Today: Proc. 2d Intl. Cong. Syst. and Evol. Biol.*, pp. 301–35. Pittsburgh, Pa.: Carnegie-Mellon University.

—— 1984. The phylogeny of the hominoid primates, as indicated by DNA-DNA hybridization. *J. Mol. Evol.* 20:2–15.

Simmons, M. J. and J. F. Crow. 1977. Mutations affecting fitness in *Drosophila* populations. *Ann. Rev. Genet.* 11:49–78.

Simpson, G. G. 1944. *Tempo and Mode in Evolution.* New York: Columbia University Press.

—— 1949. *The Meaning of Evolution.* New Haven: Yale University Press.

—— 1953. *The Major Features of Evolution.* New York: Columbia University Press.

—— 1964. Organisms and molecules in evolution. *Science* 146:1535–38.

Singh, R. S., R. C. Lewontin, and A. A. Felton. 1976. Genic heterogeneity within electrophoretic "alleles" of xanthine dehydrogenase in *Drosophila pseudoobscura*. *Genetics* 84:609–29.

Singh, S. M. and E. Zouros. 1978. Genetic variation associated with growth rate in the American oyster *(Crassostrea virginica)*. *Evolution* 32:342–53.

Skibinski, D. O. F. and R. D. Ward. 1981. Relationship between allozyme heterozygosity and rates of divergence. *Genet. Res.* 38:71–92.

—— 1982. Correlations between heterozygosity and evolutionary rate of proteins. *Nature* 298:490–92.

Slatkin, M. 1972. On treating the chromosome as the unit of selection. *Genetics* 72:157–68.

Slatkin, M. and T. Maruyama. 1975. The influence of gene flow on genetic distance. *Amer. Natur.* 109:597–601.

Slightom, J. L., A. E. Blechl, and O. Smithies. 1980. Human fetal

$^{G}\gamma$- and $^{A}\gamma$-globin genes: Complete nucleotide sequences suggest that DNA can be exchanged between these duplicated genes. *Cell* 21:627–38.

Smith, D. G. and R. G. Coss. 1984. Calibrating the molecular clock: Estimates of ground squirrel divergence made using fossil and geological time markers. *Mol. Biol. Evol.* 1:249–59.

Smith, G. P. 1974. Unequal crossover and the evolution of multigene families. *Cold Spring Harbor Symp. Quant. Biol.* 38:507–13.

Smith, G. P., L. Hood, and W. M. Fitch. 1971. Antibody diversity. *Ann. Rev. Biochem.* 49:969–1012.

Smith, T. F., M. S. Waterman, and W. M. Fitch. 1981. Comparative biosequence metrics. *J. Mol. Evol.* 18:38–46.

Smithies, O., G. E. Connell, and G. H. Dixon. 1962. Chromosomal rearrangements and the evolution of haptoglobin genes. *Nature* 196:232–36.

Smouse, P. E. 1974. Likelihood analysis of recombinational disequilibrium in multiple-locus gametic frequencies. *Genetics* 76:557–65.

Sneath, P. H. A. and R. R. Sokal. 1973. *Numerical Taxonomy.* San Francisco: Freeman.

Sober, E. 1985. A likelihood justification of parsimony. *Cladistics* 1:209–33.

Sokal, R. R. and C. D. Michener. 1958. A statistical method for evaluating systematic relationships. *University of Kansas Sci. Bull.* 28:1409–38.

Soulé, M. 1976. Allozyme variation: Its determinants in space and time. In F. J. Ayala, ed., *Molecular Evolution,* pp. 61–77. Sunderland, Mass.: Sinauer Associates.

Sourdis, J. and C. Krimbas. 1987. Accuracy of phylogenetic trees estimated from DNA sequence data. *Mol. Biol. Evol.* 4:159–66. (submitted).

Southern, E. M. 1975. Long range periodicities in mouse satellite DNA. *J. Mol. Biol.* 94:51–69.

Sparrow, A. H., H. J. Price, and A. G. Underbrink. 1972. A survey of DNA content per cell and per chromosome of prokaryotic and eukaryotic organisms: Some evolutionary considerations. *Brookhaven Symp. Biol.* 23:451–94.

Spiess, E. B. 1977. *Genes in Populations.* New York: Wiley.

Spieth, P. T. 1975. Population genetics of allozyme variation in *Neurospora intermedia. Genetics* 80:785–805.

Spolsky, C. and T. Uzzell. 1984. Natural interspecies transfer of mitochondrial DNA in amphibians. *Proc. Natl. Acad. Sci. USA* 81:5802–5.

Stanley, S. M. 1984. Mass extinctions in the ocean. *Sci. Amer.* 250(6):64–72.

Stebbins, G. L. 1950. *Variation and Evolution in Plants.* New York: Columbia University Press.

Stebbins, G. L. and F. J. Ayala. 1985. The Evolution of Darwinism. *Sci. Amer.* 253(1):72–82.

Stein, J. P., J. F. Catterall, P. Kristo, A. R. Means, and B. W. O'Malley. 1980. Ovomucoid intervening sequences specify functional domains and generate protein polymorphism. *Cell* 21:681–87.

Steinberg, A. G., H. K. Bleibtreu, T. W. Kurczynski, A. O. Martin, and E. M. Kurczynski. 1967. Genetic studies on an inbred human isolate. In J. F. Crow and J. V. Neel, eds., *Proc. 3rd Intl. Cong. Hum. Genet.,* pp. 267–89. Baltimore, Md.: Johns Hopkins Press.

Stephens, J. C. 1985. Statistical methods of DNA sequence analysis: Detection of intragenic recombination or gene conversion. *Mol. Biol. Evol.* 2:539–56.

Stephens, J. C. and M. Nei. 1985. Phylogenetic analysis of polymorphic DNA sequences at the *Adh* locus in *Drosophila melanogaster* and its sibling species. *J. Mol. Evol.* 22:289–300.

Stern, C. 1973. *Principles of Human Genetics.* 3d. ed. San Francisco: W. H. Freeman.

Stewart, F. M. 1976. Variability in the amount of heterozygosity maintained by neutral mutations. *Theor. Popul. Biol.* 9:188–201.

Stout, D. L. and C. R. Shaw. 1974. Genetic distance among certain species of *Mucor. Mycologia* 66:969–77.

Strachan, T., D. Webb, and G. A. Dover. 1985. Transition stages of molecular drive in multiple-copy DNA families in *Drosophila. EMBO J.* 4:1701–8.

Suzuki, H. and 9 other authors. 1986. Evolutionary implication of heterogeneity of the non-transcribed spacer region of ribosomal DNA repeating units in various subspecies of *Mus musculus. Mol. Biol. Evol.* 3:126–37.

Sved, J. A. 1968. The stability of linked systems of loci with a small population size. *Genetics* 59:543–63.

—— 1972. Heterosis at the level of the chromosome and at the level of the gene. *Theor. Popul. Biol.* 3:491–506.

—— 1981. A two-sex polygenic model for the evolution of premating isolation. I. Deterministic theory for natural populations. *Genetics* 97:197–215.

Syvanen, M. 1984. The evolutionary implications of mobile genetic elements. *Ann. Rev. Genet.* 18:271–93.

Tajima, F. 1983a. Evolutionary relationship of DNA sequences in finite populations. *Genetics* 105:437–60.

—— 1983b. Mathematical studies on the evolutionary change of DNA sequences. Ph.D. thesis, University of Texas at Houston, Houston, Tex.

Tajima, F. and M. Nei. 1982. Biases of the estimates of DNA divergence obtained by the restriction enzyme technique. *J. Mol. Evol.* 18:115–20.

—— 1984. Estimation of evolutionary distance between nucleotide sequences. *Mol. Biol. Evol.* 1:269–85.

Takahata, N. 1981. Genetic variability and rate of gene substitution in a finite population under mutation and fluctuating selection. *Genetics* 98:427–40.

—— 1985. Introgression of extranuclear genomes in finite populations: Nucleo-cytoplasmic incompatibility. *Genet. Res.* 45:179–94.

Takahata, N. and M. Kimura. 1979. Genetic variability maintained in a finite population under mutation and autocorrelated random fluctuation of selection intensity. *Proc. Natl. Acad. Sci. USA* 76:5813–17.

—— 1981. A model of evolutionary base substitutions and its application with special reference to rapid change of pseudogenes. *Genetics* 98:641–57.

Takahata, N. and T. Maruyama. 1979. Polymorphism and loss of duplicate gene expression: A theoretical study with application to tetraploid fish. *Proc. Natl. Acad. Sci. USA* 76:4521–25.

Takahata, N. and M. Nei. 1984. F_{ST} and G_{ST} statistics in the finite island model. *Genetics* 107:501–4.

—— 1985. Gene genealogy and variance of interpopulational nucleotide differences. *Genetics* 110:325–44.

Tartof, K. D. 1975. Redundant genes. *Ann. Rev. Genet.* 9:355–85.

Tateno, Y., M. Nei, and F. Tajima. 1982. Accuracy of estimated phylogenetic trees from molecular data. I. Distantly related species. *J. Mol. Evol.* 18:387–404.

Tavaré, S. 1984. Line-of-descent and genealogical processes, and their applications in population genetics models. *Theor. Popul. Biol.* 26:119–64.

Temin, H. M. 1985. Reverse transcription in the eukaryotic genome: Retroviruses, pararetroviruses, retrotransposons, and retrotranscripts. *Mol. Biol. Evol.* 2:455–68.

Templeton, A. R. 1981. Mechanisms of speciation—a population genetic approach. *Ann. Rev. Ecol. Syst.* 12:23–48.

—— 1983. Phylogenetic inference from restriction endonuclease cleavage site maps with particular reference to the evolution of humans and apes. *Evolution* 37:221–44.

Terachi, T., Y. Ogihara, and K. Tsunewaki. 1984. The molecular basis of genetic diversity among cytoplasms of *Triticum* and *Aegilops*. III. Chloroplast genomes of the M and modified M genome-carrying species. *Genetics* 108:681–95.

Thompson, E. A. 1975. *Human Evolutionary Trees.* Cambridge: Cambridge University Press.

Thomson, G. 1977. The effect of a selected locus on linked neutral loci. *Genetics* 85:753–88.

Thorpe, J. P. 1979. Enzyme variation and taxonomy: The estimation of sampling errors in measurements of interspecific genetic similarity. *Biol. J. Linn. Soc.* 11:369–86.

—— 1982. The molecular clock hypothesis: Biochemical evolution, genetic differentiation, and systematics. *Ann. Rev. Ecol. Syst.* 13:139–68.

Tonegawa, S. 1983. Somatic generation of antibody diversity. *Nature* 302:575–81.

Topal, M. D. and J. R. Fresco. 1976. Complementary base pairing and the origin of substitution mutations. *Nature* 263:285–89.

Tracey, M. L., N. F. Bellet, and C. D. Gravem. 1975. Excess allozyme homozygosity and breeding population structure in the mussel *Mytilus californianus*. *Marine Biol.* 32:303–11.

Tsunewaki, K. and Y. Ogihara. 1983. The molecular basis of genetic diversity among cytoplasms of *Triticum* and *Aegilops* species. II. On the origin of polyploid wheat cytoplasms as suggested by chloroplast DNA restriction fragment patterns. *Genetics* 104:155–71.

Turner, J. R. G., M. S. Johnson, and W. F. Eanes. 1979. Contrasted modes of evolution in the same genome: Allozymes and adaptive change in *Heliconius*. *Proc. Natl. Acad. Sci. USA* 76:1924–28.

Ueda, S., O. Takenaka, and T. Honjo. 1985. A truncated immunoglobulin ε pseudogene is found in gorilla and man but not in chimpanzee. *Proc. Natl. Acad. Sci. USA* 82:3712–15.

Ullu, E. and C. Tschudi. 1984. *Alu* sequences are processed 7SL RNA genes. *Nature* 312:171–72.

Upholt, W. B. 1977. Estimation of DNA sequence divergence from comparison of restriction endonuclease digests. *Nucleic Acids Res.* 4:1257–65.

Uyenoyama, M. K. 1985. Quantitative models of hybrid dysgenesis: Rapid evolution under transposition, extrachromosomal inheritance, and fertility selection. *Theor. Popul. Biol.* 27:176–201.

Uzzell, T. and K. W. Corbin. 1971. Fitting discrete probability distributions to evolutionary events. *Science* 172:1089–96.

Valentine, J. W. 1976. Genetic strategies of adaptation. In F. J. Ayala, ed., *Molecular Evolution*, pp. 78–94. Sunderland, Mass.: Sinauer Associates.

Vawter, A. T., R. Rosenblatt, and G. C. Gorman. 1980. Genetic divergence among fishes of the Eastern Pacific and the Caribbean: Support for the molecular clock. *Evolution* 34:705–11.

Voelker, R. A., H. E. Schaffer, and T. Mukai. 1980. Spontaneous allozyme mutations in *Drosophila melanogaster:* Rate of occurrence and nature of the mutants. *Genetics* 94:961–68.

Vogel, F. 1972. Non-randomness of base replacement in point mutation. *J. Mol. Evol.* 1:334–67.

Waddington, C. H. 1957. *The Strategy of the Genes*. London: Allen and Unwin.

Wahlund, S. 1928. Zusammensetzung von Populationen und Korrelationserscheinungen vom Standpunkt der Vererbungslehre aus betrachtet. *Hereditas* 11:65–106.

Wainscoat, J. S. and 10 other authors. 1986. Evolutionary relationships of human populations from an analysis of nuclear DNA polymorphisms. *Nature* 319:491–93.

Walker, P. M. B. 1968. How different are the DNAs from related animals? *Nature* 219:228–32.

Wallace, B. 1959. The role of heterozygosity in *Drosophila* populations. *Proc. 10th Intl. Cong. Genet.* 1:408–19.

—— 1968. *Topics in Population Genetics*. New York: Norton.

—— 1981. *Basic Population Genetics.* New York: Columbia University Press.

Walsh, M. M. and D. R. Lowe. 1985. Filamentous microfossils from the 3,500 MY old onverwacht group, Barberton Mountain Land, South Africa. *Nature* 314:530–32.

Ward, R. D. 1977. Relationships between enzyme heterozygosity and quarternary structure. *Biochem. Genet.* 15:123.

—— 1978. Subunit size of enzymes and genetic heterozygosity in vertebrates. *Biochem. Genet.* 16:799–810.

Ward, R. D. and R. A. Galleguillos. 1978. Protein variation in the plaice, dab, and flounder and their genetic relationships. In B. Battaglia and J. A. Beard, eds., *Marine Organisms: Genetics, Ecology, and Evolution,* pp. 71–93. New York: Plenum Press.

Waterman, M. S. 1984. General methods of sequence comparison. *Bull. Math. Biol.* 46:473–500.

Waterman, M. S., T. F. Smith, and W. A. Beyer. 1976. Some biological sequence metrics. *Adv. Math.* 20:367–87.

Watt, W. B. 1985. Allelic isozymes and the mechanistic study of evolution. *Isozymes* 12:89–132.

Watt, W. B., R. C. Cassin, and M. S. Swan. 1983. Adaptation at specific loci. III. Field behavior and survivorship differences among colias PGI genotypes are predictable from in vitro biochemistry. *Genetics* 103:725–39.

Watterson, G. A. 1974. Models for the logarithmic species abundance distributions. *Theor. Popul. Biol.* 6:217–50.

—— 1975. On the number of segregating sites in genetical models without recombination. *Theor. Popul. Biol.* 7:256–76.

—— 1977. Heterosis or neutrality? *Genetics* 85:789–814.

—— 1978. The homozygosity test of neutrality. *Genetics* 88:405–17.

—— 1983. On the time for gene silencing at duplicate loci. *Genetics* 105:745–66.

—— 1984a. Allele frequencies after a bottleneck. *Theor. Popul. Biol.* 26:387–407.

—— 1984b. Lines of descent and the coalescent. *Theor. Popul. Biol.* 26:77–92.

—— 1986. The homozygosity test after a change in population size. *Genetics* 112:899–907.

Webster, R. G., W. G. Laver, G. M. Air, and G. C. Schild. 1982.

Molecular mechanisms of variation in influenza viruses. *Nature* 296:115–21.

Webster, T. P., R. K. Selander, and S. Y. Yang. 1972. Genetic variability and similarity in the *Anolis* lizards of Bimini. *Evolution* 26:523–35.

Weinberg, W. 1908. Über den Nachweis der Vererbung beim Menschen. *Jahresh. Verein f. vaterl. Naturk. Württem.* 64:368–82.

Weir, B. S., A. H. D. Brown, and D. R. Marshall. 1976. Testing for selective neutrality of electrophoretically detectable protein polymorphisms. *Genetics* 84:639–59.

Weir, B. S. and C. C. Cockerham. 1984. Estimating *F*-statistics for the analysis of population structure. *Evolution* 38:1358–70.

Weir, B. S. and W. G. Hill. 1986. Nonrandom recombination within the human β-globin gene cluster. *Amer. J. Hum. Genet.* (in press).

Weitkamp, L. R., T. Arends, M. L. Gallango, J. V. Neel, J. Schultz, and D. C. Shreffler. 1972. The genetic structure of a tribal population, the Yanomama Indians. III. Seven serum protein systems. *Ann. Hum. Genet.* 35:271–79.

Weitkamp, L. R. and J. V. Neel. 1972. The genetic structure of a tribal population, the Yanomama Indians. IV. Eleven erythrocyte enzymes and summary of protein variants. *Ann. Hum. Genet.* 35:433–44.

Wharton, K. A., K. M. Johansen, T. Xu, and S. A. Tsakonas. 1985. Nucleotide sequence from the neurogenic locus *Notch* implies a gene product that shares homology with proteins containing *EGF*-like repeats. *Cell* 43:567–81.

White, M. J. D. 1978. *Modes of Speciation.* San Francisco: Freeman.

Whitney, J. B., R. R. Cobb, R. A. Popp, and T. W. O'Rourke. 1985. Detection of neutral amino acid substitutions in proteins. *Proc. Natl. Acad. Sci. USA* 82:7646–50.

Whitt, G. S., P. L. Cho, and W. F. Childers. 1972. Preferential inhibition of allelic isozyme synthesis in an interspecific sunfish hybrid. *J. Exp. Zool.* 179:271–82.

Whittaker, R. H. 1969. New concepts of kingdoms or organisms. *Science* 163:150–60.

Whittam, T. S., H. Ochman, and R. K. Selander. 1983. Geographic components of linkage disequilibrium in natural populations of *Escherichia coli. Mol. Biol. Evol.* 1:67–83.

Wiesenfeld, S. L. 1967. Sickle-cell trait in human biological and cultural evolution. *Science* 157:1134–40.

Williams, S. M., R. DeSalle, and C. Strobeck. 1985. Homogenization of geographical variants at the nontranscribed spacer of rDNA in *Drosophila mercatorum. Mol. Biol. Evol.* 2:338–46.

Wills, C. 1981. *Genetic Variability.* Oxford: Oxford University Press.

Wills, C. and D. R. Londo. 1981. Is the doubly deleted α-thalassemia gene a "fugitive" allele? *Amer. J. Hum. Genet.* 33:217–26.

Wilson, A. C. 1975. Evolutionary importance of gene regulation. *Stadler Symp.* 7:117–34. Columbia, Mo.: University of Missouri.

—— 1985. Molecular evolution. *Sci. Amer.* 253(4):164–73.

Wilson, A. C., S. S. Carlson, and T. J. White. 1977a. Biochemical evolution. *Ann. Rev. Biochem.* 46:573–639.

Wilson, A. C., L. R. Maxson, and W. M. Sarich. 1974. Two types of molecular evolution: Evidence from studies of interspecific hybridization. *Proc. Natl. Acad. Sci. USA* 71:2843–47.

Wilson, A. C., T. J. White, S. S. Carlson, and L. M. Cherry. 1977b. Molecular evolution and cytogenetic evolution. In R. S. Sparks and D. E. Comings, eds., *Molecular Human Cytogenetics,* pp. 375–93. New York: Academic Press.

Woese, C. R. 1981. Archaebacteria. *Sci. Amer.* 244(6):98–122.

Wolstenholme, D. R. and D. O. Clary. 1985. Sequence evolution of *Drosophila* mitochondrial DNA. *Genetics* 109:725–44.

Workman, P. L. and J. D. Niswander. 1970. Population studies on southwestern Indian tribes. II. Local genetic differentiation in the Papago. *Amer. J. Hum. Genet.* 22:24–49.

Wozney, J., D. Hanahan, V. Tate, H. Boedtker, and P. Doty. 1981. Structure of the pro α2(I) collagen gene. *Nature* 294:129–35.

Wright, S. 1931. Evolution in Mendelian populations. *Genetics* 16:97–159.

—— 1932. The roles of mutation, inbreeding, crossbreeding, and selection in evolution. *Proc. 6th Intl. Cong. Genet.* 1:356–66.

—— 1935. The analysis of variance and the correlations between relatives with respect to deviations from an optimum. *J. Genet.* 30:243–56.

—— 1937. The distribution of gene frequencies in populations. *Proc. Natl. Acad. Sci. USA* 23:307–20.

—— 1938a. Size of population and breeding structure in relation to evolution. *Science* 87:430–31.

—— 1938b. The distribution of gene frequencies under irreversible mutations. *Proc. Natl. Acad. Sci. USA* 24:253–59.

—— 1942. Statistical genetics and evolution. *Bull. Amer. Math. Soc.* 48:223–46.

—— 1943. Isolation by distance. *Genetics* 28:114–38.

—— 1948. On the roles of directed and random changes in gene frequency in the genetics of populations. *Evolution* 2:279–94.

—— 1949. Adaptation and selection. In G. L. Jepsen, G. G. Simpson, and E. Mayr, eds., *Genetics, Paleontology, and Evolution,* pp. 365–89. Princeton, N.J.: Princeton University Press.

—— 1951. The genetical structure of populations. *Ann. Eugen.* 15:323–54.

—— 1952. The genetics of quantitative variability. In E. C. R. Reeve and C. H. Waddington, eds., *Quantitative Inheritance,* pp. 5–41. London: Her Majesty's Stationery Office.

—— 1964. Pleiotropy in the evolution of structural reduction and of dominance. *Amer. Natur.* 98:65–69.

—— 1965. The interpretation of population structure by *F*-statistics with special regard to systems of mating. *Evolution* 19:395–420.

—— 1969. *Evolution and the Genetics of Populations.* Vol. 2: *The Theory of Gene Frequencies.* Chicago: Ill.: University of Chicago Press.

—— 1970. Random drift and the shifting balance theory of evolution. In K. Kojima, ed., *Mathematical Topics in Population Genetics,* pp. 1–31. Berlin: Springer-Verlag.

—— 1977. *Evolution and the Genetics of Populations.* Vol. 3: *Experimental Results and Evolutionary Deductions.* Chicago, Ill.: University of Chicago Press.

—— 1978. *Evolution and the Genetics of Populations.* Vol. 4: *Variability Within and Among Natural Populations.* Chicago, Ill.: University of Chicago Press.

—— 1982. Character change, speciation, and the higher taxa. *Evolution* 36:427–43.

Wright, S. and T. Dobzhansky. 1946. Genetics of natural populations. XII. Experimental reproduction of some of the changes caused by natural selection in certain populations of *Drosophila pseudoobscura*. *Genetics* 31:125–56.

Wu, C.-I. and W.-H. Li. 1985. Evidence for higher rates of nucleotide substitution in rodents than in man. *Proc. Natl. Acad. Sci. USA.* 82:1741–45.

Wyles, J. S., J. G. Kunkel, and A. C. Wilson. 1983. Birds, behavior, and anatomical evolution. *Proc. Natl. Acad. Sci. USA* 80:4394–97.

Yager, L. N., J. F. Kaumeyer, and E. S. Weinberg. 1984. Evolving sea urchin histone genes—nucleotide polymorphisms in the H4 gene and spacers of *Strongylocentrotus purpuratus*. *J. Mol. Evol.* 20:215–26.

Yamazaki, T. 1977. Enzyme polymorphism and functional difference: Mean, variance, and distribution of heterozygosity. In M. Kimura, ed., *Molecular Evolution and Polymorphism*, pp. 127–47. Mishima, Japan: National Institute of Genetics.

—— 1981. Genetic variabilities in natural populations of haploid plant, *Coenophalum conicum*. I. The amount of heterozygosity. *Japan. J. Genet.* 56:373–383.

Yamazaki, T. and T. Maruyama. 1972. Evidence for the neutral hypothesis of protein polymorphism. *Science* 178:56–58.

—— 1974. Evidence that enzyme polymorphisms are selectively neutral, but blood group polymorphisms are not. *Science* 183:1091–92.

Yang, S. Y. and J. L. Patton. 1981. Genic variability and differentiation in the Galapagos finches. *The Auk* 98:230–42.

Yokoyama, S. 1983. Selection for the α-thalassemia genes. *Genetics* 103:143–48.

Yokoyama, S. and M. Nei. 1979. Population dynamics of sex-determining alleles in honey bees and self-incompatibility alleles in plants. *Genetics* 91:609–26.

Yoshimaru, H. and T. Mukai. 1979. Lack of experimental evidence for frequency-dependent selection at the alcohol dehydrogenase locus in *Drosophila melanogaster*. *Proc. Natl. Acad. Sci. USA* 76:876–78.

Yunis, J. J. and O. Prakash. 1982. The origin of man: A chromosomal pictorial legacy. *Science* 215:1525–30.

Zera, A. J., R. K. Koehn, and J. G. Hall. 1985. Allozymes and biochemical adaptation. In G. A. Kerkut and L. I. Gilbert, eds., *Comprehensive Insect Physiology, Biochemistry, and Pharmacology*, pp. 633–74. New York: Pergamon Press.

Zimmer, E. A., S. L. Martin, S. M. Beverley, Y. W. Kan, and A. C. Wilson. 1980. Rapid duplication and loss of genes coding for the α chains of hemoglobin. *Proc. Natl. Acad. Sci. USA* 77:2158–62.

Zimmerman, E. G. and M. E. Nejtek. 1977. Genetics and speciation of three semispecies of *Neotoma*. *J. Mammalogy* 58:391–402.

Zouros, E. 1976. Hybrid molecules and the superiority of the hetero-zygote. *Nature* 262:227–29.

—— 1979. Mutation rates, population sizes, and amounts of electro-phoretic variation of enzyme loci in natural populations. *Genetics* 92:623–46.

Zouros, E. and D. W. Foltz. 1984. Possible explanations of heterozy-gote deficiency in bivalve molluscs. *Malacologia* 25:583–91.

—— 1987. The use of allelic isozyme variation for the study of heter-osis. *Isozymes* 13:1–59.

Zouros, E., S. M. Singh, and H. E. Miles. 1980. Growth rate in oys-ters: An overdominant phenotype and its possible explanations. *Evolution* 34:856–67.

Zuckerkandl, E. and L. Pauling. 1962. Molecular disease, evolution, and genetic heterogeneity. In M. Kasha and B. Pullman, eds., *Horizons in Biochemistry,* pp. 189–225. New York: Academic Press.

—— 1965. Evolutionary divergence and convergence in proteins. In V. Bryson and H. J. Vogel, eds., *Evolving Genes and Proteins,* pp. 97–166. New York: Academic Press.

Zurawski, G., M. T. Clegg, and A. H. D. Brown. 1984. The nature of nucleotide sequence divergence between barley and maize chlo-roplast DNA. *Genetics* 106:735–49.

Author Index

Subject Index

Pseudogenes, 28, 32, 143–48, 403; processed, 135, 144; reactivation of, 147; relative mutation rates of, 28
Pseudomonas aeruginosa, 115, 413
Punctuated equilibrium, 428
Purifying selection, 32, 52

Rainbow trout, 419
Rana lessonae, 285
Rana ridibunda, 285
Random fluctuation of selection intensity, 357
Rat, 105
Recessive gene, 336, 382
Recognition sequence, 97
Recombination hot spot, 274
Red-cell acid phosphatase, 151
Reference OTU, 303
Reptiles, 10, 59
Restriction endonuclease, 96
Restriction fragment, 330
Restriction fragment data, 106
Restriction fragment length polymorphism (RFLP), 259
Restriction site, evolutionary change of, 97
Restriction site differences, 260, 279
Restriction site polymorphism, 259, 273
Retroposon, 144
Retrovirus, 136, 141, 273
Reverse transcription, 135
Ribonuclease, 51, 56
Ribosomes, 20
RNA: 18S rRNA, 22; 5S RNA, 12, 124; messenger (mRNA), 19; precursor messenger, 20; processed, 135; ribosomal (rRNA), 19, 113, 124; 7SL RNA, 135; small nuclear (snRNA), 19, 22; transfer (tRNA), 19, 113
RNA viruses, 34, 141
Rodents, 83
Roger's distance, 211, 311
Ruminants, 53

Salmonella, 147
Salmonids, 390

Saltatory replication, 129
Sanghvi's distance, 213
Scaptomyza, 63
Sea urchin, 83, 120, 249
Selection, 155, 406; advantageous, 53; balancing, 198, 420; centripetal (stabilizing), 414; diversity-enhancing, 199; diversity-reducing, 199; epistatic, 172; fluctuating, 201, 373; frequency-dependent, 350; genic, 381, 388; natural, 269, 335, 413; overdominant, 198, 346, 373; positive, 53, 388; slightly disadvantageous, 373; stabilizing, 422
Selection coefficient, 201, 336
Selection intensity, 84
Semidominant gene, 336
Separate sexes, 362
Sequence distance method, 89
Shark, 133
Sheep, 59
Shifting-balance theory, 408, 419
Sickle cell anemia, 206, 330, 347
Singular nucleotide, 315
Slime molds, 22
Small population, 422
Speciation, 430
Species population size, 194
Species tree, 288, 313, 401
Spermophilus, 181–84, 228
Spinach, 271
Stepwise mutation model, 236, 378
Stochastic change, 352
Stochastic theory, 327
Stomach lysozyme, 53
Subdivided population, 154, 187, 420
Superoxide dismutase, 42
Synonymous mutation (substitution), 29, 33, 73, 79
Synonymous site, 73
Synthetic theory of evolution, 1

Tandem duplication, 111
Taricha rivularis, 369
TATA box, 20
T cell receptor, 125